Atmospheric Health, Nature, and Well-being

Reimagining the garden as a vital metaphysical framework for understanding the intricate relationships between health, well-being, and the environment, the book proposes a holistic, ecologically sensitive model that integrates mind, body, nature, and community.

The text offers the reader a transformative perspective that, transcending traditional dualisms, embraces the atmospheric sense of deep interconnectedness between individuals, communities, and the natural world. Through a blend of philosophical analysis, autoethnographic narratives, and cultural reflections, the authors, proposing a transgressive metaphysical model for personal and communal engagement with the processes of life, revise the idea of health as a "lack of illness," redefining it as "flourishing." Such garden-inspired metaphysics promotes a dynamic, inclusive understanding of health, balancing modern medical approaches with a profound attunement to the rhythms of nature.

Atmospheric Health, Nature, and Well-being is an ideal resource for scholars, students, and other professionals in philosophy, environmental studies, health sciences, and cultural studies, as well as policymakers, urban planners, and anyone interested in innovative approaches to health, well-being, and the environment.

Paweł Jędrzejko, PhD, DLitt., is an associate professor at the University of Silesia in Katowice, Poland, and faculty member at "Sapienza" University of Rome, Italy. Author of over 200 publications, he is also co-editor in chief of *Review of International American Studies*. His research areas include literary theory and philosophy, the intersections of cognitive experience and creative activity, and 19th-century American literature. Jędrzejko, the past president of the International American Studies Association (2021–2023), is also an ocean-going yacht master and active musician.

Marcin Fabjański, PhD, philosopher and therapist, specializes in the intersections of Buddhism and Stoicism. A Harvard Divinity School Buddhist Ministry Fellow, he merges Eastern and Western philosophy, incorporating them into his meditative and therapeutic practice. He has authored 12 books, including

Embodied Nature and Health. Fabjański directs the Apennine School of Philosophy in Italy and teaches at the University of Silesia, exploring the application of philosophy to everyday life.

Dariusz Kubok, DLitt., PhD, is a full professor and Director of the Institute of Philosophy at the University of Silesia in Katowice, Poland. His research focuses on ancient philosophy, especially Greek epistemology and ontology, with particular emphasis on such problems as *apeiron* and *peras* in pre-Socratic philosophy, the Parmenidean thought, or forms of criticism in the classical tradition. Author of *Criticism, Skepticism, and Zeteticism in Early Greek Philosophy*, he also edits the journal *Folia Philosophica*.

Geographies of Health

Edited by Allison Williams
Associate Professor, School of Geography and Earth Sciences, McMaster University, Canada

Susan Elliott
Professor, Department of Geography and Environmental Management and School of Public Health and Health Systems, University of Waterloo, Canada

There is growing interest in the geographies of health and a continued interest in what has more traditionally been labeled medical geography. The traditional focus of 'medical geography' on areas such as disease ecology, health service provision and disease mapping (all of which continue to reflect a mainly quantitative approach to inquiry) has evolved to a focus on a broader, theoretically informed epistemology of health geographies in an expanded international reach. As a result, we now find this subdiscipline characterized by a strongly theoretically-informed research agenda, embracing a range of methods (quantitative; qualitative and the integration of the two) of inquiry concerned with questions of: risk; representation and meaning; inequality and power; culture and difference, among others. Health mapping and modeling has simultaneously been strengthened by the technical advances made in multilevel modeling, advanced spatial analytic methods and GIS, while further engaging in questions related to health inequalities, population health and environmental degradation.

This series publishes superior quality research monographs and edited collections representing contemporary applications in the field; this encompasses original research as well as advances in methods, techniques and theories. The *Geographies of Health* series will capture the interest of a broad body of scholars, within the social sciences, the health sciences and beyond.

Equity in Global Health Research
Edited by Elijah Bisung and Katrina Plamondon

Anxious Geographies
Worlds of Social Anxiety
Louise Boyle

Atmospheric Health, Nature, and Well-being
Towards a Philosophy of the Garden
Paweł Jędrzejko, Marcin Fabjański, and Dariusz Kubok

For a full list of titles in this series, please visit https://www.routledge.com/Geographies-of-Health-Series/book-series/GHS

Atmospheric Health, Nature, and Well-being

Towards a Philosophy of the Garden

Paweł Jędrzejko, Marcin Fabjański, and Dariusz Kubok

Routledge
Taylor & Francis Group
LONDON AND NEW YORK

First published 2025
by Routledge
4 Park Square, Milton Park, Abingdon, Oxon OX14 4RN

and by Routledge
605 Third Avenue, New York, NY 10158

Routledge is an imprint of the Taylor & Francis Group, an informa business

British Library Cataloguing-in-Publication Data
A catalogue record for this book is available from the British Library

Library of Congress Cataloging-in-Publication Data
Names: Jędrzejko, Paweł, 1970- author. | Fabjański, Marcin, 1970- author. | Kubok, Dariusz, author.
Title: Atmospheric health, nature, and well-being : towards a philosophy of the garden / Paweł Jędrzejko, Marcin Fabjański and Dariusz Kubok.
Description: Abingdon, Oxon ; New York, NY : Routledge, 2025. | Series: Geographies of health series | Includes bibliographical references and index.
Identifiers: LCCN 2024054007 (print) | LCCN 2024054008 (ebook) | ISBN 9781032357676 (hardback) | ISBN 9781032357683 (paperback) | ISBN 9781003328469 (ebook)
Subjects: LCSH: Environmental psychology. | Nature--Psychological aspects. | Human beings--Effect of environment on. | Well-being. | Mind and body. | Nature, Healing power of.
Classification: LCC BF353 .J43 2025 (print) | LCC BF353 (ebook) | DDC 155.9/1--dc23/eng/20250212
LC record available at https://lccn.loc.gov/2024054007
LC ebook record available at https://lccn.loc.gov/2024054008

ISBN: 978-1-032-35767-6 (hbk)
ISBN: 978-1-032-35768-3 (pbk)
ISBN: 978-1-003-32846-9 (ebk)

DOI: 10.4324/9781003328469

Typeset in Sabon
by SPi Technologies India Pvt Ltd (Straive)

To Anna, Julia, Karolina, and Monika,
in whose love we flourish.

Contents

Figures

Summary

The monograph *Atmospheric Health, Nature, and Well-being: Towards a Philosophy of the Garden* offers an in-depth exploration of how the garden, both as a metaphor and a lived reality, can serve as a foundational framework for understanding the complex relationships between health, well-being, and the environment. This book invites readers to engage critically with the philosophy of the garden, proposing it as a holistic, metaphysical model that transcends traditional dualisms and promotes a deep interconnectedness between individuals, communities, and the natural world. The book is organized into three parts. The first, "Back to the Garden (Toward Theory)," offers the readers the fundamentals of the theoretical language used throughout the book, the cultural background, against which the central premises of the presented argumentation are best visible, and the main precepts of the proposed revision of the attitudes to health still dominant in the Western world. Part Two, "Atmospheric Health Practices (Autoethnologies)," roots the readers in the lived experience of the attunement to the rhythms of nature, both personal (the individual experiences of the authors) and communal—the experiences of the communities, in whose lives the authors immersed themselves. The final part, sublimating the findings presented in the first two, combines theory, cultural practice, and individual experience in the formulation of a new, transgressive, metaphysics of the garden. Through a blend of philosophical analysis, autoethnography, and reflections on cultural practices, the authors posit a new paradigm for health that integrates ecological sensitivity and personal growth, aiming to address the pressing challenges of modern life.

The "Introduction" sets the stage for exploring the philosophy of the garden as a framework for understanding health, well-being, and the intricate interdependencies between humans and their environment. Addressing contemporary audiences, the authors invite them to engage in critical reflection, fostering a deep, personal connection with the themes presented. The garden is posited as a space of interconnectedness, where individual, communal, and environmental health are tended to as one. It is presented not merely as a metaphor but as an extension of the human being, an atmospheric phenomenon whose cycles align with the rhythms of life. In the introduction, the authors stress the garden's liminality, which serves as the basis for broader philosophical inquiries into the human-nature relationship, reflecting cultural and historical shifts. The section outlines the book's structure and methodologies, including philosophical analysis, multisensory autoethnography, and critical reflection on historical and contemporary perspectives. It advocates for a holistic understanding of health that transcends mind-body dualism, incorporating psychological, social, and environmental dimensions. Challenging prevailing paradigms of health, particularly in light of modern crises, the authors propose a shift toward an integrated, ecologically sensitive model of atmospheric health, rooted in the philosophy of the garden.

Exploring the origins of philosophical inquiry, Chapter 1, titled "Outcasts of the Garden: Uncertainty as Inspiration," opening Part One of this book, offers the reader essential vocabulary to navigate the central problems of this book, focusing in particular on the roles that wonder, doubt, and liminality play in philosophical reflection. The contemporary world amplifies these drivers, particularly through global crises like the recent pandemic, which exacerbated our sense of uncertainty and isolation. This experience has catalyzed critical reflection on our life models and values, emphasizing the need for a holistic reassessment of health. The pandemic's pervasive impact underscores the need to redefine health not merely as the absence of disease but as a dynamic process of growth, interconnectedness, and community. In this context, *crisis*, etymologically linked to *criticism*, necessitates intellectual vigilance that involves our readiness to re-evaluate core questions and recognize the limitations of narrowly focused disciplines. Such critical stance leads to the recognition of the necessity to adopt a holistic perspective, integrating multiple dimensions of human existence, to adequately address complex global issues. The text advocates for rethinking health through a broader, integrative lens that transcends traditional dichotomies, fostering a balanced and comprehensive understanding of human well-being in harmony with nature. This reexamination draws on the insights of Hans-Georg Gadamer emphasizing the importance of equilibrium and the interconnectedness of all life processes, and thereby challenges the isolationist tendencies of modern scientific approaches.

To substantiate the observations from the previous section, in Chapter 2, "From the Garden of Eden to *Project House*," the authors present them against the backdrop of the overview of the cultural and philosophical evolution of visual representations of the garden. In the light of selected canonical paintings, the narrative traces the development of Western culture through-philosophical oscillations between paradigms based on dichotomies (natural versus unnatural, natural versus artificial/artful, soul versus body, self versus other) and non-dualist paradigms, from Early Modern Era to the Age of Reason and to Romanticism's emphasis on unmediated experience, the blurring of the limits of the self, and the sublime's resistance to categorization. The chapter posits that the trajectory of shifts between dualist and non-dualist philosophies is that of an ascending spiral, with positions fostering metaphysics "transgressive" with respect to those dominant *always-already* present in culture. Concluding with contemporary perspectives, the chapter, emphasizing the evolution of the philosophical fundaments of present-day ecological consciousness and of a holistic understanding of health, demonstrates that the subcutaneous yearning to be one with the rhythms of nature, inherent in human beings, colors their actions, including art.

Drawing on the general tenets of the proposed philosophy of the garden presented in Chapter 1 and against the background of the cultural history of the Western conceptualizations of nature in canonical paintings, Chapter 3, "If Health Were a Cloud," zooms in to the issues of health and healing. Exploring the intersection of classical physics, Cartesian metaphysics, and atmospheric phenomena, the authors sketch out the essentials of a novel perspective on health and well-being. Adopting the poetic language of Goethe's *Faust* as its point of departure, the text illustrates how unsteady forms and atmospheric experiences influence human cognition and self-perception. The chapter discusses how Goethe's interaction with these forms leads to a transformation that transcends the Newtonian-Descartian worldview, suggesting that health, like atmospheric phenomena, is dynamic and elusive. Introducing Hermann Schmitz's phenomenology, the authors emphasize the significance of atmospheres and their impact on human emotions and cognition, arguing that health should be viewed as a rhythmic process, continuously shaped by interactions with the environment. To illustrate the validity of their claim, the authors examine the concept of "pathocentric epistemic injustice," highlighting how patients' subjective experiences are often marginalized in the biomedical model adopted in the West. By advocating for a holistic approach to health, the chapter suggests that understanding health requires a metaphysical humility based on the recognition of the interconnectedness between humans and their environment. In the light of the above, the authors posit the idea of the garden as a metaphysical framework for conceptualizing health, a framework emphasizing flourishing, adaptation, and the integration of various life processes, tended to by a compassionate gardener-doctor. Challenging traditional biomedical models, the perspective proposed by the authors promotes an inclusive, dynamic understanding of health and well-being.

In Chapter 4, "Sharing the (Common)Wealth of the Garden," opening the autoethnographic Part Two of this book, in which the authors offer narratives of their first-hand experiences of the atmospheric attunement to the processes of life, Dariusz Kubok reflects on his evolving connection to Greece, developed over two decades of immersive travel. He contrasts the *tourist*'s search for predictable experiences with the *traveler*'s embrace of uncertainty and cultural engagement. Kubok's travels emphasize the importance of profound presence, the deep engagement with local environment and people transforming each visit into a continuous inside journey. He draws on Lawrence Durrell's sentiment that Greece offers travelers self-discovery rather than mere observations, distinguishing between tourists seeking comfort and travelers seeking authentic experiences. The chapter discusses Blue Zones, particularly Ikaria, Greece, where longevity is linked to a lifestyle of physical activity, social connection, and natural diet, reflecting the deeply rooted philosophy of well-being at work. Against the background of the philosophies of Antiquity, Kubok resorts to the poetry of Cavafy and Herbert to explore the journey of life as a means of self-exploration and flourishing. The chapter concludes with the garden as the epitome of Greek life. Drawing on Epicurus, Hesiod, and Xenophon to advocate for a life of simplicity, self-sufficiency, and communal harmony, Kubok argues that the Greek lifestyle echoes Epicurean philosophy, where happiness stems from modesty and contentment, offering lessons for modern life.

Complementing Dariusz Kubok's journey of self-discovery, Paweł Jędrzejko's autoethnographic account in Chapter 5, "In Search of the Garden of the Isles," emphasizes the multisensory experience in processes leading to *oikeiôsis*. Linking Dariusz Kubok's insights on Greece with Marcin Fabjański's approach to philosophy as *therapeia*, the chapter begins with reflections drawn from a philosophical workshop organized aboard a sailing yacht in Greece, focusing on the five classical elements—earth, water, fire, air, and ether. The chapter explores how sensory exercises attune participants to life's rhythms, leading to the revision of their self-knowledge. These controlled sensory experiences are contrasted with the harrowing ordeal of the yacht *Oskard*, which sank in the Bay of Biscay. While the official Maritime Chamber report provides a detached account, Paweł Jędrzejko's narrative offers a poignant introspection into the psychological and existential impact of the experience. Against such background, the chapter further examines the difference between *oikeiôsis* and self-annihilation, integrating Herman Melville's critiques of Transcendentalism and Andrzej Chojecki's concept of the "Ontic Empire" into its discourse. A mental experiment conducted in the sauna highlights the philosophical and therapeutic benefits of the Scandinavian *bastu*, underscoring the importance of the philosophy of *lagom*—a concept embodying moderation and balance—as a key to personal and philosophical growth. By connecting theoretical and experiential elements, the chapter illustrates how immersion in nature fosters deep attunement and interconnectedness, challenging the notions of self-effacement and erasure.

In contrast with the spontaneity of some of the processes catalytic to attunement described in Paweł Jędrzejko's autoethnographic section, in Chapter 6, "*Bhāvanā*: The Thai Forest Tradition Revisited," Marcin Fabjański offers an autophenomenological account of his exploration of *meditative* multisensory experience and of its potential to lead to *oikeiôsis*. The author describes his experience of a three-month meditation retreat at the Sāsanārakkha Buddhist Sanctuary in Malaysia, whose rule follows the Thai Forest Tradition. This tradition, emphasizing living in harmony with nature, views health as a process depending on the attunement to its rhythms, and thereby—as an atmospheric phenomenon. The author details his daily routine of the practice of *bhāvanā*: the cultivation of mind to a state where it ceases to generate self-induced suffering, resulting from unhealthy engagements with mental constructs. Focusing on mindfulness and meditation on the four elements—earth, water, air, and fire—Fabjański resorts to Ayya Khema's teachings to demonstrate how such practices help to dissolve perceived barriers between the self and the environment. The narrative contrasts Fabjański's Malaysian experiences with those central to his previous retreat in Myanmar, highlighting the impact of expectations on meditative insights. Consequently, the author discusses the pedagogy of the Buddhist garden, paying special attention to the use of nature-related similes to facilitate the meditators' mental development. The chapter concludes with reflections on the Buddhist metaphysics and health. Juxtaposing early Buddhist teachings and the WHO's definition, Fabjański advocates for the atmospheric understanding of health as a natural process in the contemporary Western medical practice. Integrating theoretical insights, personal narratives, and cultural practices, the chapter illustrates how nature and meditation in nature, fostering attunement, interconnectedness, and personal growth, affect our sense of well-being and health.

Opening the third part of the book, which aims at a synthetic revision of theory in light of lived experience of immersion in the rhythms of nature and local communities, is Chapter 7, "*Friluftsliv*: The Cultural Practice of Attunement," which offers a case study providing an example of the application of the proposed philosophy of the garden in cultural practice. The key concept of the chapter is *friluftsliv*, or the philosophy of "open-air living." Deeply ingrained in Scandinavian culture, *friluftsliv* reflects the values of simplicity, sustainability, and a strong connection to nature, shared by individuals and communities alike. This philosophy is celebrated through seasonal festivals like Midsummer and Christmas, which incorporate outdoor rites that reinforce the bond between people and the natural world. At the same time, traditional practices—such as foraging, fishing, and hunting—continue to foster the Scandinavians' profound sense of interconnectedness with their environment. These practices, passed down through generations, preserving cultural heritage also solidify *friluftsliv* as a way of life. Scandinavian literature, art, and music often celebrate nature's inscrutability, conveying reverence for wilderness and the sense of the importance of the individual experience of nature. At the same time, the global influence of *friluftsliv* is evident in the rise of outdoor recreation and nature tourism, where international visitors and Scandinavians alike engage in activities like hiking and skiing, promoting a shared commitment to environmental stewardship. Crucially, *friluftsliv* in Scandinavian countries has become institutionalized, which offers a practical model for integrating the philosophy of the garden into legislation that promotes atmospheric health and well-being.

Although the Greek, Thai, and Scandinavian examples of the implementation of practices based on attunement to the rhythms of nature offer both inspiration and hope, in Chapter 8, "Walking through Atmospheres," the authors return to the dominant perceptions of thinking about health and well-being in the Western world. The section begins with a critique of the modern obsession with health, drawing on Byung-Chul Han's argument that Western society's focus on physical health and efficiency has replaced ancient concepts of flourishing and spiritual salvation. This shift has given rise to the "palliative society," where the "sterile self" prioritizes productivity over genuine well-being. The chapter contrasts this attitude with both ancient and contemporary views that recognize the importance of pain and discomfort for spiritual and psychological growth. Nietzsche's critique of "comfortable survival" and his own practice of seeking vitality through immersion in specific environmental contexts align with Han's position, which emphasizes an integrated approach to health. The chapter contrasts the vitality of health as an atmospheric phenomenon with the lifelessness, pertaining to digital technologies. The applicative dimension of the Białowieża Forest experiment, described in the chapter, underscores the importance of the philosophical reflections inspired by meditative practices from the Thai Forest Tradition, further illustrating how one's attunement to the rhythms of nature helps transcend egocentric perspectives, promoting a holistic understanding of health. In conclusion of the chapter, the authors propose a concretized health model that integrates modern medicine with a deeper, atmospheric approach, advocating for a balance that maintains sensitivity to natural rhythms and compassionate interaction in the process of healing, which jointly foster flourishing.

Chapter 9, "Toward the Metaphysics of the Garden," sublimating the reflections from the previous sections, sets the new model of health proposed in Chapter 8 upon the fundament of a garden-inspired metaphysics, transgressing the traditional view of humans as separate from the world. Health is articulated as a communal good within a holistic framework, emphasizing atmospheric ethics and the interconnectedness of all beings. The chapter explores the dissolution of rigid boundaries, advocating for a transgressive metaphysics that challenges models based on Cartesian-Newtonian dualisms (self versus non-self, soul versus body, or nature versus culture). This approach aligns with deep ecological principles, recognizing the intrinsic value of all life forms and their relational nature. Contrasting with modernist views critiqued by Zygmunt Bauman, where the garden is seen as a controlled, artificial space, the proposed metaphysics emphasizes trust, cooperation, and alignment with nature's rhythms. It recognizes the inherent value of wildness and asserts nature's unconsumability, advocating for a respectful, restrained interaction with nature, acknowledging the complexity and unpredictability of natural processes. This respect for wildness is seen as essential for true flourishing, echoing thinkers like Henry David Thoreau. The chapter concludes in our return to aesthetics. Framing the garden as poetry of nature, the authors elevate the language of poetry to a position of a cognitive tool, proposing that relationships and dependencies of the garden are best visible through the lens of poetic figurations, providing a language that sensitizes one to the transgressive character of the garden metaphysics, and thereby allows one to organically embrace the new concept of health.

The final section of the book, "Conclusions: On Health and Its (Not So) Obvious Connections," distills the central tenets of the transgressive metaphysics of the garden, presenting it as a form of first philosophy rooted in attentive participation in everyday experience. The authors argue that elements of this philosophy are evident in global practices, though it remains marginalized in the Western world. Reiterating key ideas from previous chapters, they emphasize that recognizing our deep connection with nature allows us to "live deliberately," fulfilling personal and social responsibilities without compromising eudaimonic well-being. Flourishing is positioned as a core component of health and healing, which should be viewed as a dynamic process of attunement to nature's rhythms. Healing, therefore, involves creating conditions that enable individuals to reconnect with the fabric of life and its processes. The section frames the garden as an operational concept transcending traditional metaphysical binaries, advocating for an understanding of health as inherently linked to environmental well-being. This approach, proposing a revision of ethics into "atmospheric ethics," fosters communal, individual, and environmental health as an element of well-being rooted in attentive attunement the natural world. Such metaphysics, the authors conclude, could underpin the functioning of new public institutions, and inform education and legislation, which, based on a radical rethinking of our relationship with health, nature, and community, should emerge in response to contemporary global challenges.

"Just a Little." An Introduction

This book was highly challenging to write: from the moment the idea of it was born, we understood that we could not expect our readers—used to commercial platitudes, and cautious of omnipresent clichés—to accept our philosophy of the garden uncritically. We knew that we needed to run a risk: to stand nude in front of our audiences, in hope that rather than derision, we will awaken a compassionate connection, allowing those who read our words to tune in with us, and thereby to discover their own tactile relationship with the surrounding world. At the same time, we could not leave them lost in the space of ineffability: we needed to offer our readers vocabulary to help them articulate their own insights, and to draw their own, very personal, very intimate, conclusions, for themselves. In a sense, this book *resists* language, uses it *against itself*, in order to touch upon what we wish to communicate. The sense of fundamental wholeness, the recognition of which is the starting point for our explorations of health and well-being, must inevitably remain beyond—or beside—language.

The space of the garden is replete with ambivalence.[1] On the one hand, etymologically speaking, the garden is a closed space, often demarcated by fences or walls discouraging "illegal" access: one may either enter a thus conceived garden by invitation, or as a trespasser. On the other hand, when the word "garden" is used in compounds referring to coastal archipelagos in Scandinavia, the notion manifests itself as both inclusive and non-binary: *Skärgården* (Swedish), *Skjærgården* (Norwegian), or *Eyjagarðurinn* (Icelandic), literally translate as "the Garden of the Isles," whose only limit is the limit of the open sea, while the legally sanctioned "All Men's Right" (*Allemansrätten*[2]), or "The Right to Roam," renders the idea of "trespassing" void. Such a garden encompasses both animate and inanimate components (whether man-made or coming into existence—and changing—independently of human intentions), including its human inhabitants. Yet, irrespective of how we construe it, the garden is a body of interconnected and co-dependent entities, whose collective rhythms are a function of biology attuned to seasons, but also one to which one can choose to consciously tend.

DOI: 10.4324/9781003328469-1

Naïvely, one could reduce the ambivalence (whose surface we have barely scratched), to "mere semantics." Yet, whether the space of the garden is inclusive or exclusive, whether its limits are physically marked or defined by natural divides, the idea of tending the garden remains indubitably pertinent to all of the forms of the latter. Importantly, however, the perception of what it means to "tend the garden" appears to very heavily depend on these, allegedly "mere," semantics. Quite comprehensibly, the latter have been evolving hand in hand with the formation of dominant stances concerning the position of the human in the universe, which is why the idea of the garden is never free from its multifaceted entanglements in the central metanarratives of culture.

Happily, the recent decades, embracing more and more of the tenets of post-humanist/transhumanist reflection, have generated stances conducive to de-centralization of the still dominant concepts of nature in the West. Despite the unchecked hegemony of late capitalist narratives, the evident outcomes of global warming and the recent world-wide experience of the COVID-19 pandemic have dealt a serious blow to human *hubris*: an increasingly large number of the representatives of the younger generation turn to activism oriented toward the revision of the destructive axiologies, replacing the understanding of the human in relation to other *entities* with the understanding of the human in relation to other *subjectivities*, whether human or non-human, whether representing animate or inanimate nature, or man-made artificial intelligences.[3] It is a commonplace first-hand experience that when one finds it hard to catch breath in the center of a city of concrete and glass, when one cannot leave one's apartment because of a mandatory lockdown, when one experiences a mild panic attack as a result of the density of smog, one quickly realizes the obvious fact: our lungs evolved in order to process oxygen produced as a result of photosynthesis. This simple "discovery" is likely to make one authentically embrace the futility of all projects that treat the environment as simply *complementing* the human. Understanding that the environment is also *us*, one begins to understand the importance of the commonwealth of nature and community, and thereby also of one's own personal responsibility to first understand the rhythms of the garden and then to serve it as a caring, compassionate gardener. Attuned to the cycles and rhythms of nature, our attention shifts from "ego" to "eco," allowing us to profoundly comprehend that locked in the concrete maze of the urban space, (dis)connected as a part of "the community" sharing the virtual space, we fail to prepare our bodies and minds to fend off pathogens that in densely populated areas spread like wildfire. Conversely, when we consciously choose to tend the garden of nature and allow it to thrive, we cultivate a physical space in which we do not only confront pathogens in concentrations that our antibodies can handle, but also a space, which our own actions protect. Simultaneously, such a garden invites many like-minded gardeners: it gives rise to the "new Epicurean Garden," a green, breathing, pulsing agora, epitomizing the commonwealth of nature, supporting growth and, ultimately, flourishing.

Such a garden, like the Scandinavian gardens of the isles, indeed, transcends inherited or imposed categorial limits. Furthermore, the cultivation of the mind—and of ethical conduct—typical of the Hellenistic orientation, is where the classical Western thought intersects with Buddhism, whose practitioners in the East and in the West alike see the garden as akin to the mind—or heart—that must be tended and nurtured through meditation, mindfulness, and adherence to the principles of "right livelihood" (Pali: *sammā ājīva*). Such a "garden," like any other garden, requires constant attention and care to maintain its health and vitality, extending beyond the physical realm into the domain of personal and spiritual development. In fact, collapsing binaries, the kind of garden that our book describes transcends categories embracing an individual, who, in turn, embracing such a garden, transcends their own self. And this is not a naïve claim.

In fact—on the contrary. We are well aware that a 21st-century book whose title includes keywords such as "health," "nature," "well-being," and the "garden" might, if judged only by the cover, stir some prejudice. Yet, proposing a philosophical insight into "atmospheric health," the fundamental concept of our philosophy of the garden, we could not be further from the platitudes of "pop-mindfulness," simple didacticism, or nostalgia for the bygone idyll—even though, as humanists, we are not entirely immune to some of the "insecurities" that Leszek Kołakowski laid painfully bare already in 1989:

> [...] The fact that people construct doctrines and ideologies to compensate for their insecurities is a natural, and widespread, phenomenon. Humanists tend to hide their complexes under the masks of anti-technological mythologies, while technologists resort to technocratic utopias, purportedly free of ideological distortions, to do the same. Referring to these insecurities as "complexes" is somewhat metaphorical, as they are, in fact, social phenomena that cannot be adequately interpreted using purely psychological terminology. [...] Humanists, who commonly feel out of place in the world and are horrified by how mainstream societal currents proceed without their involvement—leaving their small concerns, antiquarian preferences, problems that would better fit a museum, and their old-fashioned passions on the sidelines—often experience a distressing sense of redundancy. Observing how intellectual interests turn elsewhere, they tend to seek blame outside their own environment. Consequently, they are eager to attribute responsibility to the supposed overemphasis on technology in modern life, denouncing artificial needs created by technological advancements, and condemning the alleged barbarization of developed societies stemming from an excessive focus on technology. They weave Arcadian tales of a world with moderate needs and, therefore, a moderate demand for technological innovations, threatening with visions of a monstrous future dominated by brainless robots. They lament societies where

> everyone can drive but no longer reads Aeschylus. In short, they advocate a return to nature, in more or less veiled forms, while fully aware of the hopelessness of their program. They find solace in the melancholic sigh that they bear on their frail shoulders the values of a dying world. Their ideology consists of a warm nostalgia for the charm of the bygone—and unnecessary—and of spasms of horror triggered by a vision of the world of machines and automation.[4]

All true. But the three of us are entirely aware of the fact that falling into a "nostalgic" paradigm would not help us open anyone's eyes, and that fashioning the progress of technological civilization as the "arch-enemy" of the human would be nothing short of hypocrisy: all three of us drive cars, use computers, and welcome hip replacement surgeries. We depend on GPS, use smartphones, and leave carbon footprint with every travel upon which we embark, professional or otherwise. We have long come to terms with the fact that we live the Anthropocene[5] and that the Deleuzoguattarian schizophrenia of the late capitalist world is our reality. We realize, without a shadow of a doubt, that humankind, albeit occasionally romanticizing the "return to the age of innocence,"[6] is most unlikely to voluntarily choose to revert to a life without electricity, air travel, the Internet, or antibiotics (unless it is only for a brief vacation). But, at the same time, we second Eric Baratay's understanding of animal subjectivity,[7] and fully acknowledge the explanatory power of the numerous Indigenous methodologies of the world: Umeek's *Tsawalk* postulates "the nature of existence as an integrated and orderly whole, and thereby recognizes the intrinsic relationship between the physical and spiritual";[8] Leslie Marmon Silko's insight at the core of *Ceremony*[9] reveals that in the context of culturally induced trauma the success of any healing process depends on the inclusion of the allegedly "intrusive" elements of the present-day reality into a much older poem—a never-ending song of cosmic order, whose melody becomes diluted in the white noise of everyday bustle. Floyd Favel[10] and Tomson Highway[11] teach compassion through performance, storytelling, and the organization of space—and Jiddu Krishnamurti, having found a common language with David Bohm, has demonstrated that "the spiritual" does not have to clash with "the scientific."[12] And, like all of the thinkers and writers listed above (and many others could be added), we acknowledge the currency of the discourses born in the past for the sake of a well-grounded reflection on the healthy present. We do so without dismissing the value of what both industrial revolutions—and, more recently, the digital revolution—introduced to the world, yet also without downplaying the gravity of the pathological consequences of some of these awe-inspiring, and simultaneously terrifying, transformations.

Therefore, eschewing the naïvete of the "Arcadian tales," we propose an "atmospheric" philosophy of the garden, and—steering clear of the reefs and shallows of "pop-mindfulness"—we strive to root our experiences (both those unmediated and those contemplative) in the philosophical substratum of major metanarratives with the view to sharing with the Readers a *path of thought* that departs from the ones already in existence *just a little*. It is, of course, a challenge, because

> [i]t is very difficult to describe paths of thought where there are already many lines of thought laid down,—your own or other people's—and not to get into one of the grooves. It is difficult to deviate from an old line of thought just a little.[13]

And yet, the "just a little" from Wittgenstein's notebook entry makes a *radical* difference. Apples have been falling to our heads ever since both apples and our heads evolved, but it took the bloodshed of the 17th-century Wars of Religion to give rise to (critical) Deism, which effectively allowed Sir Isaac Newton to ask *the question* that propels our space probes today. Nominally—nothing changed: apples still fall and hit our heads. But, with Newton, deviating from the old line of thought "just a little"—*everything* changed. Similarly, Columbus's voyage has not affected the *shape* of the world, but it did transform the essential concepts of shipbuilding: knowing that there is a continent across the Atlantic, shipwroughts would accommodate for the requirements of a transoceanic voyage. Again—nominally, *nothing changed*, but square-rigging and extra space for live animal pens became a norm in the ship design.

By analogy, Wittgensteinian "just a little" proves to be "a lot" in the context of the lingering communal experience of the COVID-19 pandemic, when our homes—the safe spaces of love and peace—became our prisons, while the parks outside—green spaces that we used to take for granted because they had "always" been available—transmogrified into desperately desired Edens; the sheer ability to enjoy the aroma of the trees in the forest became the object of our dreams. Again: nominally, nothing changed. But, in fact, all of us, remembering the lockdowns, have come to understand that our well-being, our physical and mental health, depends fully upon our ability to interact with what is *outside* of the limits of our highly penetrable body-walls. An altogether fresh—albeit, essentially, very old—idea of humans as *extended beings* has become most tangibly evident. To some of us, it has transpired at the level of the discursive comprehension of our limits; to *all of us*—at the level of the sense, or sensation, of *well-being*. And even though the seasons still come and go, sun rises and sets, people are born, grow old, and die, the experience of the pandemic changed our line of thought "just a little"—enough for us to realize that we can no longer separate our sense of "being healthy" from our essential unity with what exists around us (again: we have lungs, because plants have chlorophyll). This is "old knowledge," no doubt. But its "revival" in our perceptions—changing things "just a little"—radically affects the ways in which we experience the flows and ebbs of the universe. Whether this perspectival change will last is impossible to say, but at least for as long as the "pestilence" lives in our memory our sensation that "every atom belonging to me as good belongs to you"[14] may be stronger than ever.

Adopting this shared sensation as our point of origin, we need to make one more provision: we are aware that the existing models of atmospheric health to which we refer throughout this book in order to illustrate the very essence of the concept are largely culture-specific. It would be naïve to believe that one can simply "transplant" the Greek, Scandinavian, or Thai models onto foreign soils. This is the reason why, in the following sections, we seek to sublimate a culture-independent, compassion-based, philosophical model: a model with which the willing reader—irrespective of his or her cultural rootedness—may have a chance to identify.

***

Therefore, we begin our common garden stroll with sketching out the territory of our interest. However, to contextualize the essential terms we employ throughout this book, we start off with the intellectual background of our observations in Part One, which consists of three chapters. Chapter 1, titled "Outcasts of the Garden: Uncertainty as Inspiration," opens with a philosophical reflection on the stimuli of philosophical inquiry, emphasizing the role of wonder, doubt, and liminality as primary catalysts for critical reflection in the context of the more than etymological affinity binding the concept of "criticism" with that of "crisis." Against such a backdrop, we present the crisis of the recent pandemic as the factor prompting the critical reassessment of existing norms and values—and as the driving force for rethinking health—to demonstrate how the crisis has enforced a revised understanding of health, one beyond its traditional concept as a binary opposition to "illness." Moving away from seeing health as merely the "absence of disease" to viewing it as a dynamic process of engagement with the world, we stress the inextricable connections between social, mental, and physical well-being, and health comprehended holistically, i.e., as a function of both communal and environmental interactions. Highlighting the need for a broader, more integrated view of health that extends beyond physical symptoms to include psychological and communal well-being, we argue for the processual understanding of health, which involves the need to cultivate individual awareness of one's interconnectedness with the wider community and the natural world. These reflections lead to the final section of the chapter—"Garden-Agora-Paradise"—which explores the relationship between human health and environment through the lens of philosophical traditions, particularly emphasizing the integrative models evolved in ancient Greece. This view is then contrasted with the reality based on isolation and digital disconnection, observed in modern times, suggesting the need—and possibility—of rebuilding these fundamental links by constructing a "new agora": a critically reconsidered, 21st-century communal space. It is on such premises that we further explore the concept of the "philosophical garden" as a model for living. We argue that gardens, physical and metaphorical, provide the space for critical reflection and as such are conducive to the cultivation of mind. Linking gardening with philosophical practices that address core human concerns—such as well-being, wisdom, and the nature of existence—we pass on to philosophy as a way of life. Drawing on Pierre Hadot's interpretation of ancient philosophy, we re-envisage philosophy not just as theoretical inquiry but as a lived practice ("spiritual exercises" or "practices of the self"), which positions philosophy as a transformative force, one involving self-care and an alignment with universal "reason" that transcends individualistic concerns, where the cultivation of the self and environment leads to a harmonious life. Calling for a philosophy that integrates internal transformation with external engagement, and rejecting the separation between self and world, we advocate for a philosophical practice that recognizes the interconnectedness of all life, promoting a holistic—and engaged—approach to living: an approach attentive to both personal and communal health. This leads us to the recognition of the grassroots need to revive the ancient "garden-agora" paradigm as a model for individuals to reconnect with nature and community. We posit these spaces as essential for fostering personal growth and communal health, advocating for a reinvention of the classical ideals in the context of modern disconnections and ensuing health crises.

Chapter 2, "From the Garden of Eden to *Project House*," proposing an insight into the evolving understanding of "nature" as a function of a "comprehensible" order, offers a brief overview of the evolution of the cultural imaginary of the garden in the context of the historical transformations of the dominant trends in the philosophical and cultural interpretations of the concepts of "the natural" and "the unnatural." Analyzing major works of Western art, we demonstrate how the cultural representations of the garden, beginning with works typifying early modern and finishing with those characteristic for postmodern aesthetic paradigms, implicate philosophical dilemmas concerning good (natural) and evil (unnatural), thereby challenging the simplicity of the dichotomy as subject to cultural, historical, and philosophical redefinitions. We argue that the concepts of "the natural" and "the unnatural," deeply entrenched in cultural narratives, serving as vehicles for philosophical inquiry and moral reflection, dynamically relate to the evolution of the often-contested concepts of what is "healthy," and what is "unhealthy." The chapter delves into the cultural and philosophical evolution of garden representations, tracing the development of Western culture in the light of philosophical oscillations between paradigms based on dichotomies (natural versus unnatural, natural versus artificial/artful, soul versus body, self versus other), from Cartesian-Newtonian dualism of the Age of Reason to Romanticism's emphasis on unmediated experience, the blurring of the limits of the self, and the sublime's resistance to categorization. We argue that the dynamics between binary and non-binary perceptions of the human in the universe—a persistent theme in cultural practice and artistic expression—have left an indelible mark on Western art, moving from anthropocentric and feudal representations of nature to decentralized visions of the world. The chapter posits that the trajectory of shifts between dualist and non-dualist philosophies is that of an ascending spiral, with positions fostering metaphysics "transgressive" with respect to those dominant, always present, albeit—for a time—occupying a marginal position in the cultural polysystem. We close the chapter with the discussion of the trauma of the world wars, presenting it as a turning point that intensified existential, non-dualist, reflections on human nature and the environment, and emphasizing how post-war literature, art, and philosophy—influenced by existentialist thinkers—grappled with the absurdity of existence and the need for ethical frameworks beyond mere survival, as artists and philosophers sought to reconcile humanity's place in a fractured world. Concluding with contemporary perspectives, we point to the evolution of the philosophical fundaments of present-day ecological consciousness and of a holistic understanding of health, demonstrating how the transformations of gardens—miniature cosmoses—reflected broader cultural and philosophical transformations in humanity's evolving relationship with nature.

On these grounds, Chapter 3, "If Health Were a Cloud," explores the metaphysical and phenomenological dimensions of health, drawing on literary examples and philosophical theories to propose a nuanced understanding of health as an "atmospheric" and dynamic condition, rather than a static "state of being." We argue that health is a fluid "atmospheric" phenomenon, akin to weather patterns that change and evolve, thus challenging traditional views. On the basis of the insights into Goethe's *Faust*—whose interactions with formless, misty figures may be interpreted in terms of the engagement with elusive health states dependent on both external and internal atmospheres—and drawing upon reflections offered by contemporary phenomenologist, Hermann Schmitz, the chapter illustrates how human cognition and emotional responses to the (more or less immediate) environments shape understanding—and experience—of health. The chapter further explores ways in which our internal landscapes are influenced by the atmospheres we inhabit, affecting our emotional and cognitive states, providing a deeper understanding of one's health in relation to the environment. Such a position offers a point of departure for a revised critique of the sharp separation of mind and body—typical of Descartian thinking—to demonstrate why such a view is insufficient for the understanding of the complexities of health. Promoting an alternative, in which the physical and mental are seen as interconnected and influencing each other, the chapter also emphasizes the consequences of the "epistemic injustice," born out of traditionally professed dualism. We argue that the "pathocentric" paradigm, in which patients' experiences—revealed in their reports—are undervalued by the medical establishment, leads not only to the lack of understanding, but also to the misrepresentation of the former's true health conditions, the recognition of which opens space for an even broader critical reflection on how health systems often fail to acknowledge the subjective dimensions of patient experiences expressed in the form of personal narratives. This discussion leads to the postulate of the reimagining of health practices as more akin to gardening, i.e., attentive to the rhythms and needs of the body and mind within their individually conditioned, particular environments. Advocating for holistic, attentive, and compassionate healthcare practice that recognizes the subjective experiences of patients, the chapter metaphysically extends the concept of a garden to health: a garden, requiring careful, responsive, interaction with its environment, is an analogon of the human, whose health requires a responsive and attentive care, attuned to, and integrating individual experiences and conditionings. Thus, the concept of "atmospheric health" is shown as potentially motivating a shift in how health should be conceptualized, and justifies the urge for a move away from rigid—objectifying—medical models toward a more fluid, responsive, and phenomenologically informed understanding of the subjects of medical care. Such a model necessitates compassionate understanding that respects and integrates in its procedures the complex ways in which particular individuals interact with their environments.

Part One, providing the vocabulary indispensable for the attunement to what the subsequent sections of this book have to offer, paves the way toward establishing a more intimate, compassionate rapport with a reader. In Part Two, we embrace the methodology of multisensory autoethnography, as detailed in Przemysław Rura's 2017 article on sensory experience in everyday life, titled "Doświadczenia zmysłowe w życiu codziennym: Autoetnografia wielozmysłowa" [Sensory Experience in Everyday Life: Multisensory Autoethnography],[15] and further expounded upon by many other authors representing a variety of disciplines, including, among others, such scholars as Jacquelyn Allen-Collinson,[16] Tony. E. Adams and Stacy Holman Jones,[17] Drick Boyd,[18] or Cali Prince.[19] As the very semantics of the concept of multisensory autoethnography suggests, the second part of the book is fashioned to emulate *experience through reading*, rather than to simply be *read*. It utilizes our autobiographical materials as the primary data, combining elements of analytic and evocative autoethnography, and emphasizing our own—physical and sensory—presence within the environment. Upholding methodological rigor, this section involves reflexivity as both a research orientation and a writing practice, and recognizes the processual character and cognitive importance of the fluid, "becoming," subjectivity. In Part Two, we embrace multiple forms of knowledge representation, viewing such type of research as a possible agent of change. Through the application of sensory autoethnography, *engaged* as we are, we also *engage* our readers in a critical dialogue concerning the limits of knowledge and certainty. To better represent the multifaceted nature of the "atmospheric" experience, we offer personal accounts, acknowledging sensory impressions as core components of cultural, social, and effectively—philosophical analysis. Importantly, the Whitmanian adage: "What I assume you shall assume"[20]—is not a requirement. It is a guideline pointing the reader toward the preferred—compassionate—reading of our experiences of the living "atmospheres" that the gardens we have blended with incarnate. Unsurprisingly, the methodology is analogous to that employed, for centuries, by North American Indigenous Nations. Communicating the existential experience, storytelling, with its emotive language and care for continuity, becomes the point of departure for philosophical revisions of one's own weltanschauung.

And thus Chapter 4, opening the second part of this book, is a narrative of Dariusz Kubok's personal transformation. In his story of the experience of Greece, the author, an expert in ancient Greek philosophy, departs from his academic prejudices, asserting that travel is not merely about visiting different locations: instead, it is an avenue for profound self-discovery. The process involves engaging deeply with the unique conditions of each visited place—its culture, people, and environment—which helps the traveler uncover a variety of aspects of their own identity and life philosophy. Such an engagement, in turn, requires embracing uncertainty: unlike *tourists*, who prefer predictability and comfort, the author champions the idea of the *traveler* who welcomes the unknown and unexpected. Such openness is emphasized as a *sine qua non* condition for experiencing genuine growth and learning, as it challenges one's preconceptions and pushes one toward new insights. Immersing oneself in the local culture as a way to truly understand and experience a place involves staying with locals, exploring less touristy areas, and engaging with the everyday life of the place rather than merely its tourist attractions. Yet, challenging as it might be, such an attitude warrants that the journey is both continuous and holistic. Kubok views each trip not as an isolated event, but as part of an ongoing journey of exploration that contributes to a lifelong process of learning and transformation. This perspective shifts the focus from individual trips to an ongoing narrative of personal and philosophical development, which raises the issue of the formative role of the philosophical engagement with travel. Kubok's personal narrative, by invoking writers and philosophers, situates personal travel experiences within broader philosophical and existential contexts. In this sense, the text draws a clear distinction between "the tourist" and "the traveler"—the latter seeking experiences that challenge and expand their understanding of the world. This distinction is crucial in highlighting the different attitudes and outcomes associated with each approach to travel. Kubok's personal and philosophical exploration serves as a testimony to the "atmospheric" experience of the "agora-garden," which we theorize in the first part of the book.

Similarly, Paweł Jędrzejko's autoethnographic Chapter 5, "In Search of the Garden of the Isles," emphasizes the significance of multisensory experiences in the processes that may lead to *oikeiôsis*. It begins with an account of a unique philosophical workshop held aboard a sailing yacht in Aegean Greece, which connects Dariusz Kubok's autoethnological insights and Marcin Fabjański's therapeutic philosophy. Central to the chapter is the exploration of the classical elements—earth, water, fire, air, and ether—whose lived experience facilitates one's reconnection with the natural world. Through sensory exercises designed to foster attunement to the rhythms of life, participants learn to "get their sea-legs," thereby revising their self-knowledge and understanding of the world. The sensory experiences of the elements, to which the workshop participants were exposed in a pedagogically controlled context, are then juxtaposed with the uncontrollability of the dramatic situation of the crew of the sailing yacht *Oskard*, which sustained damage and eventually sank in the Bay of Biscay in 2002. Unlike the impersonal and technical Maritime Chamber report, which focuses on procedural failures and factual details, the captain's (Paweł Jędrzejko's) narrative provides a deeply emotional and introspective account of the crisis, highlighting its psychological and existential dimensions. This contrast underscores the fundamental difference between *oikeiôsis* and self-annihilation, which is further explored in the perspective of Herman Melville's reflections on Emerson's pantheism, linking philosophical insights to the experiences aboard the sinking *Oskard*. Melville's critique of uncritical adherence to Transcendentalist propositions is expanded upon through the lens of Andrzej Chojecki's concept of the "Ontic Empire." This theoretical framework is put to the test in a mental experiment conducted in the sauna, a space that both epitomizes and enforces profound attunement to the rhythms of life. The sauna experience, integral to Scandinavian culture, enables the reduction of hypertelia to the non-verbal "truths" confirmed by lived experience. The discussion of the sauna's philosophical and therapeutic benefits opens avenues for broader reflection on life in the Gardens of the Isles, especially in the context of the Scandinavian practice of *lagom*, manifest in every area of everyday life of Swedes, Finns, and Norwegians. *Lagom* is "an inner state," "a ritual of immersion in life," a "mantra," it involves "the purification of mind," the recognition of the fact that objects with which we live convey "emotions," it entails compassion in one's relation to family, friends, and community members, and—last but not least—is a principle of "proud opposition to excess and of a rejection of originality at all costs." Underscoring the significance of direct engagement with nature in fostering personal and philosophical growth, *lagom* offers a philosophical underpinning to the Scandinavian practice of *friluftsliv*. This examination highlights the significance of direct engagement with nature in promoting both personal and philosophical growth. Through the integration of theoretical and experiential elements, the chapter illustrates how immersion in nature fosters a deep understanding of attunement, contrasting it with self-effacement and emphasizing "interconnectedness" over "self-erasure."

The same interconnectedness is key to the personal narrative of Marcin Fabjański's. In Chapter 5, "*Bhāvanā*: The Thai Forest Tradition Revisited," Marcin—whose work bridges Asia, Europe, and America—attests to the present-day integration of Buddhist practices with the teachings (and ethos) of ancient philosophies, rooted in the profound connection with nature. The monastery, as the author experiences it, embodies principles reminiscent of both Epicurean and Stoic philosophies, which focus on the elimination of fear (particularly the fear of death) and the correction of cognitive errors that lead to human suffering. This integration, in Fabjański's perception, incarnates the universal quest for wisdom and tranquility that transcends cultural and historical boundaries. Unlike Western philosophies that often remain confined to textual studies, in his chapter Buddhism is presented as a living tradition, its teachings—preserved and transmitted through an unbroken lineage of practitioners—manifesting themselves as *experiential* knowledge and the continuous *practice* of philosophy as a way of life. Fabjański's experience highlights the Thai Forest Tradition's profound rootedness in natural settings, emphasizing the Buddhist conviction that enlightenment, and by definition, wisdom, are best pursued in close, meditative communion with nature. This principle is a vital element of the philosophy itself, as natural settings are conducive to spiritual awakening through the coalescence of the bodily experience and contemplative insight. In this respect, Fabjański emphasizes the fact that the Thai Forest Tradition cannot be reduced to a mere set of monastic rules: it is a philosophical stance that prioritizes simplicity, asceticism, and the meditative life in a response to the corruption and materialism surfacing in the more "urban," conventional, Buddhist practices. As such, the Thai Forest Tradition advocates for basic, nature-integrated living as a form of ethical and philosophical purification—yet, it is far from the naïve seclusion from the present-day material and technological reality: it is important to remember that the core Buddhist tenets reflected in the narrative are the doctrines of interdependence and *Anatta* (non-self). These concepts are philosophically explored through the interaction with nature, where meditative practices help dissolve the illusion of a separate self, highlighting the interconnectedness of all life forms, which, in turn, leads to the idea of meditative realization as philosophical search. The *practice* of meditation is framed not only as a religious or spiritual exercise but as a form of philosophical inquiry. Through meditation, Marcin asserts, one engages deeply with fundamental questions about the nature of self, the universe, and the causes of suffering—and such an insight leads to profound personal transformations, including the incentive to embrace the central tenets of the philosophy of environmental ethics, in the practice of which living in harmony with nature is seen as essential for personal and communal well-being. This approach respects the natural world not merely as a backdrop for human activity, but as an integral part of philosophical practice and existential understanding.

Presented against the background of the cultural imaginary of nature and in the context of the continuity and change in philosophical reflection on the garden as the space of health and well-being, the three personal narratives allow us to propose a revision of the perspectives that, in the post-pandemic Western world, continue to dominate. Consequently, in Part Three, whose narrative reverts to the second-person plural, we collect our findings to propose a revised, philosophically grounded, culture-independent understanding of well-being, which provides the fundament for the systematically described "transgressive" model of atmospheric health.

And so, Chapter 7, "*Friluftsliv*: The Cultural Practice of Attunement," is dedicated to the institutionalization of the grassroots Scandinavian philosophy of *lagom* as manifested in the practice of "open-air living"—a practice deeply embedded in the cultural fabric of Nordic societies, whose cultural significance is evident in various traditions (seasonal festivals, such as Midsummer or Christmas, often incorporate outdoor elements), communal norms, but also in jurisprudence that renders access to "open air" a human right, thus institutionally reinforcing the bond between people and nature. Although it is rooted in traditional activities such as foraging, fishing, and hunting, contemporary "open-air living"—*friluftsliv*—is not reducible to "outdoor recreation" only. On the contrary: *friluftsliv* today is an inextricable component of the state legislature in countries such as Norway, Sweden, or Finland, whose governments, central and local alike, acknowledging the cyclical rhythms of the natural world, see it as a model for a holistic understanding of the human-in-the-universe, recognizing that its institutional implementation significantly contributes to well-being across various dimensions, enhancing physical, mental, social, and environmental health.

Both the communities and authorities, in whose decision-making the former actively participate, understand that the essence of *friluftsliv* is not merely about "being in nature" but about experiencing it deeply and personally, and that its practice fosters a state of mindful engagement that connects individuals to the environment and to others. As a result, Scandinavian policies embody a paradigm shift—away from neoclassical models founded on duality toward a more integrated, holistic view of the world. The profound understanding of the all-encompassing interconnectedness creates a spiritual sense of belonging and encourages individual and communal commitment to preserving the natural balance. The recognition of its value is evident in governmental initiatives, like the Norwegian "*Friluftsliv* i skolen," which integrates outdoor life into the education system to promote health, learning, and social interaction—but it is also visible in Scandinavian urban planning, in legislature related to sustainable energy production, laws regulating biodiversity offsetting, waste management, and other aspects of being-in-the-world. Reflecting the philosophy of *lagom*, *friluftsliv* offers a blueprint for a harmonious lifestyle that balances modern life with the timeless rhythms of nature, fostering a sustainable future grounded in respect, equilibrium, and interconnectedness—and thereby promoting atmospheric health. Yet, bearing in mind that the Scandinavian philosophy of well-being is culture-specific, and therefore, like the practices of Greece and Thailand, resistant to easy transplantation, in subsequent sections we return to the reflection aimed at the sublimation of a philosophical model, with which one could identify irrespective of their cultural background through compassion and shared, non-mediated, non-verbal experience of the human condition.

And hence Chapter 8, "Walking through Atmospheres," opens with a critical analysis of the modern obsession with health and of the metaphysical assumptions that impede genuine flourishing. The section draws on philosopher Byung-Chul Han's assertion that Western society's preoccupation with physical health and bodily efficiency has supplanted ancient notions of flourishing and spiritual salvation, rendering humans narcissistic and disconnected from nature. Han argues that this focus, coupled with an aversion to pain, has created a "palliative society," where a sterile, non-atmospheric self prioritizes productivity and aesthetic standards over true well-being. The chapter contrasts this contemporary sterility with ancient and alternative health perspectives, underscoring the value of pain and discomfort in fostering spiritual and psychological growth. Han's critique extends to modern Western medicine, which often isolates patients from their natural contexts, treating the body mechanistically and lacking a holistic approach.

Nietzsche's critique of his era, describing life as "comfortable survival," and his emphasis on transformation through pain, aligns with Han's diagnosis. Nietzsche's practices, which sought vitality through specific environmental conditions, exemplify a more integrated approach to health that acknowledges the atmospheric nature of well-being. This perspective is further supported by empirical research that links nature's interconnectedness with spiritual and eudaimonic well-being. The chapter explores health as an atmospheric phenomenon, emphasizing the environment's role in shaping our physical and mental states. Drawing on examples from the Białowieża Forest and Nietzsche's experiences, it argues that genuine health stems from a dynamic interaction with our surroundings, contrasting sharply with the lifeless interactions facilitated by digital technologies.

Philosophical reflections and meditative practices from the Thai Forest Tradition are examined to illustrate how nature can teach us to transcend egocentric perspectives and develop a more profound, holistic understanding of health. The chapter advocates for practices that enhance ecological sensitivity and eudaimonic well-being, such as walking through paths designed to integrate sensory stimuli with philosophical reflection. It concludes by proposing an alternative health model that combines modern medical advancements with a deeper, atmospheric understanding of well-being. This model critiques the potential alienation caused by over-reliance on technology and advocates for a balance that preserves our sensitivity to natural rhythms and to the "primitive present." Such "transgressive" metaphysics blurs the boundaries between the spiritual and corporeal, aiming for a holistic approach to health that promotes true flourishing.

Our reflections presented thus far are then reconceptualized in Chapter 9, "Toward the Metaphysic of the Garden," in which we propose an innovative model of health, founded on a garden-inspired metaphysics, surpassing the traditional *ego*logical perspective that views humans as entities separate from the world. In the section, we outline a vision of health as a communal good, anchored in a holistic metaphysics, where health is conceptualized through the lens of atmospheric ethics and the interconnectedness of all beings. Drawing on ancient and contemporary philosophical traditions, we examine the foundational principles of "garden metaphysics," stressing the dissolution of rigid boundaries and the significance of transgressiveness. We begin by clarifying the term "metaphysics," whose origins and various interpretations we trace throughout history, highlighting Aristotle's concept of metaphysics as "first philosophy" and inquiring into how later philosophers, such as the neo-Platonists, expanded its scope to encompass the suprasensory. This historical background sets the stage for an in-depth exploration of a garden-based metaphysical framework.

We argue that the "metaphysics of the garden" is characterized as a transgressive metaphysics that challenges rigid boundaries and conventional dualisms, such as self and non-self, soul and body, and nature and culture. Instead, it promotes a holistic view that acknowledges the interpenetration and interconnectedness of all entities. This approach aligns with deep ecological principles articulated by Arne Naess, which emphasize the intrinsic value of all life forms and the relational nature of existence. A thus construed "garden metaphysics" is contrasted with the modernist conception of the garden, particularly as critiqued by Zygmunt Bauman. Bauman perceives the garden as a controlled, artificial space shaped by human will and instrumental reason. In contrast, the proposed garden metaphysics emphasizes trust, affirmation, and harmony with nature's rhythms. It rejects the notion of the garden as a domain of dominance and control, instead viewing it as a realm of creative cooperation and mutual influence.

This new metaphysical perspective also incorporates the concepts of unconsumability and wildness, recognizing the value of pristine nature and the limitations of human knowledge. It advocates for a respectful and restrained interaction with nature, acknowledging the complexity and unpredictability of natural processes. This respect for wildness is deemed essential for genuine growth and flourishing, resonating with the ideas of thinkers like Henry David Thoreau. We conclude our reflections by framing the garden as the poetry of nature, where the nuanced interplay of relationships and dependencies can be best understood through art. Poetic sensitivity is crucial for appreciating the intimate, dynamic fabric of the world, transcending the rigid categorizations of science and practice. Thus, the metaphysics of the garden provides a foundation for a new understanding of health and well-being, rooted in the interconnectedness—and transgressiveness—of all life.

The reflections of the nine chapters ultimately bring us to "Conclusions," in which we collect our findings to formulate implications of our research for cultural practice. We emphasize that in order to truly implement the change in the present-day understanding of human health, healing-conceptualized-as-flourishing requires that conditions enabling patients to make self-discoveries and recognize their entanglement within the fabric of life should be institutionally fostered. In the context of the healing system grounded in transgressive metaphysics, patients can develop their own "intimate metaphysics" as a result of their engagement with nature. Nature, which often conceals itself, can only be unveiled through direct individual experience in a liminal space, where self intersects with environment. At the same time, we reiterate the dynamic characteristics of health, stressing the consequences of the fact that, transcending the confines of the body, health is fundamentally a *process*, but also, as various traditional practices demonstrate, it is a *communal* property. This approach transforms ethics into "atmospheric ethics," fostering a profound sense of community and interdependence. Ultimately, we propose that philosophy centered around biotopes, rather than specific philosophical schools, can transcend university walls and foster communal reflection, promoting a holistic approach to health and well-being. As such, our "philosophy of the garden" may serve as a catalyst for a true change, affecting the revision of the premises upon which new garden-based education, garden-based jurisprudence, or, like it is in Scandinavia, garden-based legislature can be founded to warrant lasting, eudaimonic, sense of communal and individual well-being.

Notes

1 The present research project was co-financed with the funds granted under the Research Excellence Initiative of the University of Silesia in Katowice, Poland.
2 Bertil Bengtsson, *Allemansrätten—Vad säger lagen?* (Stockholm: Naturvårdsverket, 2023).
3 See, for example, UN Women, "Persisting in the Pandemic: Youth Activism During COVID-19," December 2021, https://www.unwomen.org/en/news-stories/feature-story/2021/12/persisting-in-the-pandemic-youth-activism-during-covid-19; United Nations, "Young People Key to COVID Recovery and a Sustainable Future," United Nations Department of Economic and Social Affairs, https://www.un.org/en/desa/young-people-key-covid-recovery-and-sustainable-future; EAB, "Digital Student Activism: COVID-19," EAB, https://eab.com/resources/whitepaper/digital-student-activism-covid-19/; Juan Carlos Pineda Duque and Rodrigo Sandoval-Almazán, "Digital Activism in Times of COVID-19: Lessons from Latin America," *Social Sciences* 3, no. 1 (2023): 9, https://www.mdpi.com/2673-995X/3/1/9; Avery Hurt, "Gen Z: How Young People Are Changing Activism," *BBC Worklife*, August 3, 2022, https://www.bbc.com/worklife/article/20220803-gen-z-how-young-people-are-changing-activism; Michelle D. Olfert et al., "The Role of Social Media in Youth Activism During COVID-19," *Journal of Adolescent Health* 70, no. 2 (2022): 213–220, https://www.ncbi.nlm.nih.gov/pmc/articles/PMC10266065/; Global Youth Mobilization, *Global Youth Mobilization Interim Impact Report: February 2022*, https://globalyouthmobilization.org/wp-content/uploads/2022/03/Global-Youth-Mobilization-Interim-Impact-Report-February-2022.pdf.
4 Leszek Kołakowski, "Wielkie i małe kłopoty humanistów" [The Great and Small Problems of the Humanists], in *Kultura i fetysze* [Culture and Fetishes] (Warszawa: Państwowe Wydawnictwo Naukowe, 1989). The quoted passage has been translated by Paweł Jędrzejko.
5 The Anthropocene—a term originally employed by the Soviet scientists as early as in the 1960s, which became key in the late 20th century, when the atmospheric chemist Paul Crutzen and ecologist Eugene Stoermer presented their studies on the environmental effects of human activity. See Angela Zottola and Claudio de Majo, "The Anthropocene: genesis of a term and popularization in the press," *Text & Talk* 42, no. 4 (2022): 453–473. https://doi.org/10.1515/text-2020-0080.
6 A beautiful example of such a vision—involving figurations of the garden, cultivation of the land, experience of the ocean, the passage of time and the stages of human life, where the "earlier" stages are, allegedly, in closer attunement to the rhythms of nature, and finally metaphorically reinforcing the idea of the bygone innocence with the introduction of the white unicorn—is the German musical group

Enigma's official video for their single "Return to Innocence." The song, released on 4 January 1994 by Virgin, features a sample of the Taiwanese Amis chant "Weeding and Paddyfield Song," performed by Difang and Igay Duana. Its English lyrics, recited by Sandra—"That's not the beginning of the end, that's the return to yourself, the return to innocence"—are conducive to the philosophical interpretation of the song as an exhortation toward the return to the natural cycles of life. See: https://www.youtube.com/watch?v=Rk_sAHh9s08 (accessed: May 3, 2024).

7 See, for instance, Eric Baratay, *Le point de vue animal. Une autre version de l'histoire* ["The Animal Point of View: Another Version of the Story"] (Paris: Le Seuil Edition, 2012).

8 E. Richard Atleo (Umeek), *Tsawalk. A Nuu-chah-nulth Worldview* (Vancouver, BS: UBC Press, 2005). ISBN:9780774810845. The UBC Press introduces the book thus: "Western philosophy has long held scientific rationalism in a place of honour. Reason, that particularly exalted human quality, has become steadily distanced from the metaphysical aspects of existence, such as spirit, faith, and intuition. In Tsawalk, hereditary chief Umeek introduces us to an alternative indigenous worldview—an ontology drawn from the Nuu-chah-nulth origin stories. Umeek develops a theory of 'Tsawalk,' meaning 'one,' that views the nature of existence as an integrated and orderly whole, and thereby recognizes the intrinsic relationship between the physical and spiritual. By retelling and analyzing the origin stories of Son of Raven and Son of Mucus, Umeek demonstrates how Tsawalk provides a viable theoretical alternative that both complements and expands the view of reality presented by Western science. Tsawalk, he argues, allows both Western and indigenous views to be combined in order to advance our understanding of the universe. In addition, he shows how various fundamental aspects of Nuu-chah-nulth society are based upon Tsawalk, and what implications it has today for both Native and non-Native peoples. [...]." https://www.ubcpress.ca/tsawalk (accessed: May 3, 2024).

9 See, for instance, the following edition: Leslie Marmon Silko, *Ceremony*. Introduction by Larry McMurtry (London, New York: Penguin Books, 2006).

10 Floyd Favel—a theatre theorist, director, essayist, and cultural leader from the Poundmaker Cree Nation, an Indigenous community of the Cree people in Saskatchewan, Canada.

11 Thomson Highway—an Indigenous Canadian playwright, novelist, author of children's literature, and musician of the Cree Nation, Canada.

12 See, for instance, Jiddu Krishnamurti and David Bohm, *The Ending of Time. Where Philosophy and Physics Meet* (San Francisco: Harper One, 2014).

13 Ludwig Wittgenstein, *Zettel*, trans. G.E.M. Anscombe, ed. G.E.M. Anscombe and G.H. von Wright (Oxford: Basil Blackwell, 1967), n138.

14 Walt Whitman, "Song of Myself," https://www.poetryfoundation.org/poems/45477/song-of-myself-1892-version (accessed: May 4, 2024).

15 Walt Whitman, "Song of Myself," https://www.poetryfoundation.org/poems/45477/song-of-myself-1892-version (accessed: May 4, 2024).

16 See Jacquelyn Allen-Collinson, "Breathing in Life: Phenomenological Perspectives on Sport and Exercise," in *Routledge Handbook of Qualitative Research in Sport and Exercise*, ed. Brett Smith and Andrew Charles Sparkes (London: Routledge, 2016), 315–330; Jacquelyn Allen-Collinson, "Intention and Epoché-in Tension: Autophenomenography, Bracketing and a Novel Approach to Researching Sporting Embodiment," *Qualitative Research in Sport, Exercise and Health* 3, no. 1 (2011): 48–62, https://doi.org/10.1080/19398441.2010.541484.

17 Tony E. Adams and Stacy Holman Jones, "Telling Stories: Reflexivity, Queer Theory, and Autoethnography," *Cultural Studies ↔ Critical Methodologies* 11, no. 2 (2011): 108–116, https://doi.org/10.1177/1532708611401329.

18 Drick Boyd, "Autoethnography as a Tool for Transformative Learning About White Privilege," *Journal of Transformative Education* 6 (2011): 212–225, https://doi.org/10.1177/1541344608326899.

19 Cali Prince (née Vandyk-Dunlevy), "The Unexpected Effects of Autoethnographic Embodied Texts and Performances," in *Proceedings of the 2022 International Symposium on Autoethnography and Narrative*, ed. Silvia M. Benard Calva, Lisa P. Z. Spinazola, Rishi Raj, Christina L. Ivey, David F. Purnell, Cody M. Clemens, Marlen Harrison et al. (International Association of Autoethnography and Narrative Inquiry, 2022), 107–116.

20 See Przemysław Rura, "Doświadczenia zmysłowe w życiu codziennym: Autoetnografia wielozmysłowa" [Sensory Experience in Everyday Life: Multisensory Autoethnography], *Kultura i Społeczeństwo* 61, no. 3 (2017): 149–160, https://doi.org/10.35757/KiS.2017.61.3.9.

Part One

Back to the Garden (Toward Theory)

1 Outcasts of the Garden

Uncertainty as Inspiration

Since ancient times, philosophers have endeavored to identify the origins of philosophical inquiry and all reflective thought, exploring the catalysts that compel us to pose significant questions of a global nature. Among the myriad responses proposed by thinkers from antiquity to the present, three stand out: wonder, doubt, and the experience of liminality. It appears that the contemporary world presents us with a powerful mix of these three drivers with particular intensity. Facing periods of heightened uncertainty and encountering liminal situations unique to our era, we, like our predecessors, pose essential questions driven by wonder, often tinged with doubt, and occasionally marked by terror. However, a distinguishing feature of today's philosophical inquiries is their close connection to the global context of the issues at hand, which fosters holistic and sustained reflection.

Undoubtedly, despite the numerous tragedies it precipitated, the recent pandemic—particularly at its peak—has served as a profoundly potent catalyst for such thinking. Dominated by the sense of an invisible threat—which was itself intensified by compulsory travel restrictions, mandatory limits on direct human interactions, and the palpable dominance of monadic social structures—our experience of the pandemic, with its holistic dimensions, forced us to confront the existence of global problems that directly impact us and to recognize the agency instilled by the recentivist[1] fear encompassing these threats. The sense of helplessness, exacerbated by the prolonged period of isolation, appears to have fueled a breakthrough in the search for the nuanced origins of the pandemic-induced circumstances. This inquiry aims not merely to diagnose the causes of the pandemic but to prompt a comprehensive reassessment of our life model—our way of thinking about the world and what we hold most valuable—from the broadest perspective. Thus, the "pandemic fear" has underscored the significance of addressing fundamental questions.

One could risk the statement that the world as we know it in its familiar, "tame," form has started to slip from our control—and, admittedly, this is not the first time it happens. Such breakthroughs occur in culture and science rather frequently, but it seems that it is only today that we have eventually come to terms with uncertainty, learning to live with it continuously. The overlap of the meanings of "uncertainty" and "insecurity" has now become

DOI: 10.4324/9781003328469-3

palpable. On the one hand, we came to realize that we exist in a space of uncertain fate, unclear future, and of doubts that multiply incessantly. On the other, we discover that in such a space we must inevitably face the lack of security and a sense of disorientation. Uniquely, however, today, the broadly understood uncertainty has become embedded in our worldview: we are learning to function in our world without yearning for certainty. Consequently, we find ourselves in need to re-evaluate our thinking and redefine the meaning of our attitudes toward reality. It appears that in our case minor adjustments will not suffice; rather, fundamental changes are necessary. In the area of our interest, these changes should pertain to the relationship between humans and nature: our revisions should concern the understanding of nature on a global scale, the comprehension of humanity as a whole, and the dominant concept of health. Regarding the latter, it is essential to reconsider the very definition of health and its epistemological ramifications, especially that "the uncertainty associated with many of the predictions regarding the evolution of the pandemic poses questions about the limitations of quantitative models in health sciences and about the epistemological limits of multivariable analysis."[2] Still, doubts concerning quantitative models, the need for a qualitative shift, and epistemological challenges are not the only issues of our concern. In fact, these seem to be just the tip of the iceberg that the problem of health emerges to be: its understanding and, above all, its ontological complexity, depend on the understanding of the essence of humanity and upon the meaning we attribute to the human-nature relationship. This, clearly, indicates that there is no escape from fundamental philosophical questions.

The present that we are experiencing is undoubtedly the time of crisis—neither the first nor, sadly, the last such a situation in our lifetimes. Drawing an analogy with Popper's observation that science is in a state of permanent revolution, one could state that what we permanently face is *a continuum of crises*. However, while numerous crises peaking in recent decades had a (more or less) local dimension, it is the *global* nature of the present-day crises that terrifies us more than ever before, thus forcing us to ask adequately *global* questions.[3] If, however, we consider the term "crisis" more broadly and avoid monopolizing its content with commonly negative connotations, the term "crisis"—motivated by the Greek word κρίσις—will reveal itself in its original meaning, as "a moment conducive to heightened intellectual productivity." Thus, if "crisis" is understood etymologically—with as little bias as possible—the noun will emerge as synonymous with expressions such as "turning point," "critical period," or "change of the current course." Adopting this perspective, the time of the pandemic can be used to positive ends—as the time of breakthrough, whose critical potential must be skillfully harnessed. In this sense, the pandemic, like other challenging global crises, painfully reminds us that we have been "exiled from the garden" far longer than we might have suspected. Therefore, although it is important to learn from mistakes, it is also essential to cultivate one's propensity for turning problems into challenges in order to find the strength to inspire new reflections.

In this context, it becomes clear why we wish to propose that in the sections below the lexeme "crisis" should be understood in its most fundamental sense, colored with as little value judgment as humanly possible: a crisis is a pivotal, decisive period, in which previous lines of thought may be changed "just a little." Above all, our previous "grooves" of thinking about health should collapse: health can no longer be reduced merely to "a normal state" or the "lack of disease"—the recent global crisis has brought us closer to the recognition of the fact that health is a *process* in which one flourishes. The pandemic made us realize that to think of oneself as "healthy" in a more comprehensive perspective, it is not enough to merely "not experience illness": we have finally noticed, on a mass scale, that integral to health are the sense of connection with the world (involving the processual understanding of growth and development) and a deep sense of community. The time of the pandemic has proven to be the time of critical, groundbreaking, reflection on health, humanity, community, and nature.

Uncertainty—Crisis—Criticism

The term "crisis" (Gr. κρίσις) is etymologically linked to "criticism" (Gr. κρίνω),[4] underpinning their intrinsic connection. The most straightforward—and precise—definition of a critical stance, therefore, is as an attitude that vigilantly sustains a state of crisis. For some time, "crisis" has been a predominant term used to characterize our surrounding reality—or more precisely, the reality in which we operate. To accurately describe the world, it is useful to pair "crisis" with contextually appropriate adjectives such as "economic," "financial," "political," "identity-related," "axiological," and so forth. "Crisis" and "criticism" are dialectically intertwined—both positively, as each term encompasses functions of differentiation and resolution, and negatively, for instance, when "criticism" acts as an *antidote* to "crisis." Conversely, "crisis" can counteract the deficiencies of "criticism," or its critical shortfall. Therefore, maintaining a lucid critical attitude necessitates the continuous rejuvenation of the potential of critical thinking.

Critical thinking itself is fundamentally rooted in the capacity to discern accurately: the aptitude for evaluation and adjudication, and the predisposition toward anti-dogmatic skepticism, which fosters the expression of doubts. These components engender the vigilance necessary for the perpetual reevaluation of core questions. Adopting a critical perspective facilitates cognitive distance with respect to the problems under scrutiny, including one's personal beliefs, and enhances awareness of one's cognitive limitations, the underlying reasons for adopting certain perspectives, and the foundational principles guiding actions.[5] Consequently, critical thinking encourages openness; it derives from the broadening of the horizons of imagination and is the function of the ability to consider multiple variables. Single, narrowly focused scientific disciplines are insufficient for accurately diagnosing problems, defining objectives, or formulating specific, long-term social development strategies. This inadequacy underscores the relevance of philosophical inquiry and the value of cultivating a self-aware, well-balanced, and consciously adopted critical stance for the sake of practical applications.

Adopting a preliminary typology of criticism,[6] it is essential to recognize that its most rudimentary form is that of *separationist criticism*, manifest in acts of establishing fundamental distinctions. These foundational distinctions serve us as the basis upon which more nuanced divisions and more complex conceptual frameworks can be structured. Therefore, bearing in mind that in the context of our discussion the crucial area for re-evaluation is that encompassing the understanding of human beings in relation to nature, and consequently, the conception of health, it appears imperative to critically re-examine the fundamental dichotomies that have historically underpinned our thinking: man versus nature, soul versus body, nature versus culture, state versus function, and the physical, psychological, and social dimensions of our existence, affecting the reading of immanence versus transcendence, and I versus you. All of these distinctions warrant scrutiny, and even more: the very use of perfectionist language that presupposes idealized notions of "completeness" and "fulfillment" should be recognized as lacking in critical rigor.

Rethinking Health

To optimize our current critical self-awareness, it is advisable that we re-evaluate our prevailing definition of health. The preamble to the WHO constitution provides the following definition of health: "Health is a state of complete physical, mental, and social well-being and not merely the absence of disease or infirmity."[7] For decades, this definition has sparked debates concerning its accuracy and contemporaneity. Previously, health was often characterized as a "sane mind in a sound body," an expression that inherently embodies a strong sense of dualism. Recognizing the context in which the WHO definition was established (1946), it is important to acknowledge that the debate surrounding its relevance addresses only a subset of the larger issue.[8] Broadly speaking, the most significant critiques levied against this definition can be categorized into several groups:

a The problem of completeness. Health is defined as "complete physical, mental, and social well-being." This idealized requirement has been aptly termed an "utopian definition" (Garner),[9] or "euphoric definition" (Wood),[10] highlighting its overly ambitious nature.
b The issue of incorporating the social dimension, which raises concerns about the comprehensiveness and applicability of the definition in diverse sociocultural contexts.
c The challenge related to evolving health dynamics, where individuals with chronic diseases now enjoy longer lifespans, yet the definition categorizes these individuals as perpetually ill.

d The difficulty in operationalizing this definition, which complicates the measurement and practical application in public health policies and practices.
e The ambiguity surrounding the concept of well-being, which lacks a universally accepted interpretation and measurement criteria.

It is important to note that since the establishment of this definition, efforts have been made to implement specific amendments, though these cannot be exhaustively detailed here. However, it is essential to outline a few directional proposals. For instance, there has been a suggestion to shift the focus from purely social well-being to "psychosocial" health (e.g., RAND Health Insurance Experiment), which integrates mental and social well-being, and which adjustment acknowledges the influence of external social factors on individual health.[11] Another significant initiative is the Ottawa Charter, which underscores the importance of social and personal resources, as well as physical capacity. The proposal by Machteld Huber et al. is also of note:

> Just as environmental scientists describe the health of the earth as the capacity of a complex system to maintain a stable environment within a relatively narrow range, we propose the formulation of health as the ability to adapt and to self manage. This could be a starting point for a similarly fresh, 21st-century way of conceptualizing human health with a set of dynamic features and dimensions that can be measured.[12]

It is certainly enlightening to juxtapose this proposal with the critical discourses offered by those who argue that the WHO definition remains valid;[13] it is also worthwhile to consider the perspectives of some scholars who emphasize the necessity of incorporating the spiritual dimension into our understanding of health.[14] The above notwithstanding, it is essential to address three fundamental issues previously identified. The first concerns the claim to perfectionism or completeness, the second revolves around the imprecision of the definition of the concept of well-being, and the third pertains to the application of separationist criticism in both its positive and negative forms.

Regarding the first issue, the idea of "complete well-being" and its absolutist assertion are rather trivial; the second issue necessitates (a more or less arbitrary) elucidation; the third issue, in turn, involves:

i positive separationist criticism, which is evident in the differentiation among three domains (physical, mental, and social), and
ii negative separationist criticism, which involves not defining health merely as the absence of disease or infirmity.

While the second reservation appears entirely justifiable, as there are no substantial objections in this regard, the most often (cognitively) overlooked aspect of separationist criticism is that of the applicability of the *positive* separationist criticism, which is manifested in the distinct separation of three or more dimensions of human existence. The challenge lies not only in the number of these dimensions, but primarily in their cohesion and integration into a harmonious whole. It is possible to observe that there is a propensity to intensify and, consequently, multiply divisions in thinking about humans and their health, stemming from the advancements in separationist criticism. Yet, the recognition of finer distinctions often lacks a comprehensive, holistic perspective. Human health should not be viewed merely as coexisting dimensions—physical, mental, social, spiritual, among others—but rather as the effect of their integration. Moreover, the focus on "completeness" has overshadowed the importance of "unification" ("wholeness"). Health should be understood as a comprehensively conceived structure, and not necessarily as a concept confined to the individual alone. Therefore, we advocate for an approach to health that underscores its holistic nature, its supra-individual aspects, involving a connection between humans and their environment (nature), and its atmospheric dimension.

Before delving into these issues in greater detail, it is pertinent to revisit the insights of Hans-Georg Gadamer, articulated in his book, now somewhat overlooked, *Über die Verborgenheit der Gesundheit* [The Enigma of Health],[15] especially that many concepts introduced by Gadamer not only resonate with our views, but will also be further elaborated upon in this discussion. Gadamer astutely observes that "[t]he healthy individual is not simply someone who has been 'made' healthy,"[16] thereby asserting that health is inherently natural and can at most be restored, not manufactured. Thus, in his view, the art of healing does not consist in the invention of novel remedies, but in the reactivation—or resetting—of disturbed natural processes.

There is little doubt that the onset of modernity coincided with a significant shift in the world's historical consciousness, manifest in the acknowledgment of human agency. As Gadamer explains it, "[t]he mathematical-quantitative isolation of laws and regularities in the natural order aims at identifying specific causal contexts that afford human interventions replicable under precise conditions."[17] Yet, he argues that this "isolationist" approach should go hand in hand with a globalist and holistic perspective. In this vein, the philosopher underscores the critical importance of the concept of "equilibrium"—a principle with roots not only in Hippocratic thought but also in Pythagorean teachings:

> [...] in fact it is not merely the phenomenon of human health which almost cries out to be understood in terms of the natural condition of equilibrium. The concept of equilibrium also readily offers itself for our understanding of nature in general. The Greek concept of nature consisted in the discovery that the totality is an ordered structure which allows all the processes of nature to repeat themselves and to pass away in determinate configurations. Nature is therefore something which as

> it were holds to its own course, and does so in and of itself. This is the fundamental idea of Ionian cosmology in which all the original cosmogonic conceptions came to fulfilment: in the end the whole mighty harmonious balance of interacting events determines all things as a form of natural justice.[18]

A fundamental issue with the contemporary understanding of health is the diminished perspective on nature as an integrated whole, which has been supplanted by narrowly focused, specialized viewpoints. This shift is evident in the field of medicine, where patients' diagnoses are often fragmented across various medical specialties, each providing a different lens of examination. Undoubtedly, there is a pronounced need for a holistic approach that encompasses a more comprehensive balance:

> Modern natural science is not primarily a science of nature conceived as a self-maintaining and self-restoring totality. Our science is based not on the experience of life but on that of making and producing, not on the experience of equilibrium but on that of projective construction. This science is essentially—going far beyond the domain of that special science which bears the name—a kind of mechanics: it is *mechane*, that is, the artificial production of effects which would not come about simply of themselves.[19]

The foundational perspective for today should, therefore, prioritize the holistic nature of organisms, incorporating external influences—such as climate, temperature, seasons, types of activity, nutrition, and social interactions—which, collectively, shape human existence. This understanding aligns with the ancient Greek emphasis on the interplay of "balance" and "whole," further enriched by the concept of "harmony." Reflecting on this notion's origins, Aristotle discusses Pythagorean doctrine, stating that numerical measures are the foundational principles of all things, leading to the assertion that "the entire cosmos is harmony and number" ("τὸν ὅλον οὐρανὸν ἁρμονίαν εἶναι καὶ ἀριθμόν," Metaph., 986 a). Moreover, Aristotle connects harmony with the dynamic between opposing forces, echoing Alcmaeon's view. In fragment B 4 (according to Diels-Kranz), this Pythagorean-leaning physician posits that health is the equilibrium of opposing forces within the body (ἰσονομία τῶν δυνάμεων), whereas disease results from the dominance (μοναρχία) of any single force. Thus, the health-disease relationship is expressed as the opposition of ἰσονομία—μοναρχία. This conceptualization highlights the significance of balance and proper measure in preventing the excesses of any single force, which could lead to its tyrannical dominance. Meanwhile, Heraclitus suggests that the universal λόγος, while omnipresent, remains elusive, and is the objective of philosophical inquiry. He articulates it in fragment B 123 (Diels-Kranz) stating that "nature loves to hide" ("φύσις κρύπτεσθαι φιλεῖ") thereby asserting that "invisible harmony is stronger than visible" ("ἁρμονίη ἀφανὴς φανερῆς κρείττων", B 54). Here, ἁρμονία (harmony) signifies an orderly arrangement derived from the integration, or "fitting-together," of disparate elements, suggesting that this less apparent order is superior to the visible one, both ontologically and epistemically: the unseen underpins the seen, and the unapparent explains the apparent. Heraclitus's notion of superiority is not merely a statement of ontological robustness but also of explanatory capability. This dichotomy between the visible and the invisible parallels the distinction between the awake and the sleeping, the latter being those who overlook the underlying unity (harmony) of nature and focus solely on the discord of particulars.

> τοῖς ἐγρηγορόσιν ἕνα καὶ κοινὸν κόσμον εἶναι, τῶν δὲ κοιμωμένων ἕκαστον εἰς ἴδιον ἀποστρέφεσθαι." (B 89)—"those who are awake have a world that is one and in common, but that each of those who are asleep turns aside into his own particular world.
>
> (Laks/Most, III, Heraclitus R56)

This passage can be interpreted as advocating a particular cognitive approach, while simultaneously critiquing prevailing tendencies that divide humans on the grounds of their capacity to understand the complexity of reality, which also presupposes a critique of the dominance (μοναρχία) of particular worldviews.

Hans-Georg Gadamer persistently advocates for the harmonization of the drive toward the unknown with a re-engagement with the "balance-sustaining rhythm of the natural order."[20] He poses a critical inquiry: "Here we are confronted with the still unanswered question: what is health? We know, roughly, what illness is. It is that 'revolt' or rebellion which takes place when something starts to dysfunction."[21] The terms "revolt" or "rebellion" may not fully capture the essence of Gadamer's argument. Illness can be understood as a deviation from the natural rhythms of life; this "falling out," this "dysregulation," implies a network of relationships both within and external to an individual that may become disrupted. Conversely, health, conceptualized as the antithesis of this deviation, unambiguously introduces us to a holistic and comprehensive perspective. The pandemic, a globally traumatic event, vividly exemplifies this perspective on illness: the physical affliction of contracting the virus, along with the disruption of natural human interactions and the enforced isolation within homes, rendered the collective "falling out" of the "balance-sustaining rhythm of the natural order" palpable. Thus, our *global* experience of the *pan*demic energized—and motivated—the incentives to adopt a broadened, holistic approach to health.

Further in his analysis Hans-Georg Gadamer intricately connects the notion of balance and harmony to the concepts of "limit" and "measure," previously discussed in the context of "crisis" and "criticism." The concept of measure manifests in dual modalities: (i) as a scientific procedure serving to quantify and calculate specific facets of our organism, and (ii) as "belonging to the totality of our being-in-the-world."[22] This latter dimension pertains to a broadly defined therapeutic approach. To further elucidate this point, we turn once more to Gadamer, who states:

> Thus the mystery of health remains concealed. This concealment belongs to the preservation of good health and this consists in forgetfulness. [...] The limits of what can be mastered do not constitute the entire significance of what supports and sustains our lives. This is true for each individual, for society as a whole, for the relationship between all the different societies in the world, and for our relationship to nature. We all find ourselves constantly travelling on the same path. There are limits to what we can know and there are limits to what we can do. Knowledge—and not only the knowledge of science—is power and enables us to control things. Humankind must assert itself against nature as well. [...] But it is also and especially in the nature of human beings, in all they know and do, to sustain a relation of harmony with nature. This belongs to the most ancient wisdom of the stoics.[23]

Modernity heralds significant shifts in our conceptualization of the world, prominently through the lens of Descartian dualism (*res cogitans—res extensa*) and its ramifications. According to this framework, each substance is not only distinct but also self-sufficient, independent of anything else for its existence. Hans-Georg Gadamer articulates this point, noting that "[t]he self-certainty of self-consciousness is the unshakeable foundation of all certainty and so in the eyes of modern science of true knowledge."[24] In this modern perspective, truthfulness is supplanted by certainty, which ascends as the primary aim of human cognitive endeavors. Revisiting this notion from a historical-problematic viewpoint reveals the deleterious impact of separationist criticism, manifest as metaphysical dualism. This dualism, further compounded by an unyielding quest for certainty, poses a profound philosophical challenge in the aftermath of Descartes, its repercussions continuing to influence contemporary thought profoundly.

Both historical and problematic dimensions converge in what might be termed the "inspiration of the garden." In this context, we consider Leibniz's experience during a walk in Rosental, Leipzig. As Gadamer narrates, Leibniz, confronted by the richness and diversity of life, grappled with the indispensability of Galileo's physics, while simultaneously being compelled to embrace Aristotle's concept of entelechy. This dichotomy emphasizes a persistent challenge: establishing the garden as a paradigm within which it would be possible to transcend such dualism. Moreover, the garden metaphor advocates for the subversion of the fallibilistic pursuit of certainty as the ultimate aim of knowledge. The impulse for certainty breeds fear. To diminish this fear one must come to terms with the superfluity of the demand for certainty. Abolishing dualism, therefore, signifies a return to holistic thinking and the adoption of a pre-dualistic worldview, emblematically represented by the garden.

Garden—Agora—Paradise

In assessing human health through its relationship with the environment, it is instructive to revisit the attitudes entrenched in Western cultural heritage. Our analysis begins with the preserved thoughts of ancient Greek philosophers, whose discourses provide an early model of atmospheric health: the ancient Greeks, as has already been emphasized, viewed humans as fundamentally interconnected with both nature and community. This connection was exemplified by the agora—a space for genuine interaction, where individuals engaged in discourse and shared critical reflections, thereby fostering a communal bond. Yet today, in light of the pandemic, which has exacerbated the monadization of society and the algorithmization of communication, a thereto related social vacuum has become clearly visible. Logically speaking, addressing—and solving—this problem requires the (re)establishment of a "new agora" as a foundational element for community (re)construction. If we assume such a perspective, we will see why communities, including educational ones, should not be seen as mere networks of isolated—or more precisely (dis)connected—individuals communicating online: the social dimension of the Greek notion of universal connectivity is complemented by the concept of the Epicurean Garden—the quintessential representation of a space where humans grow and thrive in harmony with nature. Community bonds are forged through the collective recognition of shared values and wealth, and the acknowledgment of individual potential for growth. It is within—and owing to—these relationships that we can truly flourish in the garden of nature.

In his book *A Philosophy of Gardens*, David E. Cooper writes thus: "[...] because gardens may be 'seriously'—which does not mean 'solemnly'—apprehended, the question of why we garden and why we live in and with gardens is itself a serious one: one, that is to say, for philosophical reflection."[25] By pointing to the garden as a distinguished idea and a real place of meeting opposing tendencies, we do not aim for the most precise capture of its essence, but attempt to demonstrate that the garden—partly neglected by philosophy—is a fascinating and cognitively promising place conducive to the cultivation new ideas—or at least the space in which "mental crop rotation" could be exercised. It is worth to start with an insight into the core meaning of the garden as such in order to, in the next steps, rethink health, the human relationship to nature, and well-being in the perspective of the garden's extended implications. As Dan O'Brien notes, "[...] gardening is a human activity that engages with core philosophical questions concerning, among other things, human wellbeing, wisdom, the nature of time, political power and ideals, home, aesthetic experience, metaphysics, and religion."[26] The subject of philosophical reflection and philosophy itself find their reflections in the "garden" and "gardening," yet a reflection reinforced with an important supplement concerning the understanding of philosophy, to which we should now dedicate at least a few remarks.

A few decades ago, in the space of philosophical reflection, (yet another) partial reformulation of the status of philosophy could be observed, when Pierre Hadot demonstrated that ancient philosophers (although some other thinkers as well) understood their activity not so much as a collection of doctrines, philosophical systems, or simply theoretical reflections on reality, but rather as a particular *art of living*. For Hadot, the formula *exercices spirituels*, borrowed from the title of Ignatius of Loyola's work (*Exercitia spiritualia*) became key for such an understanding of philosophy. However, the Loyolan "spiritual exercises" are rooted in Greek thought, without which connection it would be impossible to understand this particular form of reflective activity. We will return to this tradition in the chapter on the journey to Greece, but before it happens, it is necessary to further define Hadot's unique understanding of philosophy, to which he drew the attention of academic philosophers already in his 1981 book *Exercices spirituels et philosophie antique*.[27]

Among the greatest admirers and continuators of the French philosopher's thought were such important thinkers as Michel Foucault and Arnold Ira Davidson.[28] The former interpreted Hadot's *Exercices spirituels* as *pratiques de soi*, a form of "work on oneself," the concept of which he further develops in the third volume of his *The History of Sexuality*, titled *The Care of the Self*.[29] The model of philosophizing both thinkers offer focuses not on *knowing* the external reality and encapsulating it in theory, but on knowing and changing *oneself*. Philosophy is thus an art of living,[30] and philosophizing is the transformation of the philosopher. It should be noted that in such a model of philosophy, the motivation to undertake the philosophical effort changes. Firstly, it is not driven by the need for cognition or acquisition of knowledge, but rather by the aspiration to change oneself, arising from the dissatisfaction with one's own state, from unfulfillment and lack. Such philosophy becomes a response to human challenges: its practice involves the transformation of one's way of life. Most importantly, philosophy understood in this way is primarily an *action*, an *activity*, a force of conversion. Hadot observes:

> In all philosophical schools, the goal pursued in these exercises is self-realization and improvement. All schools agree that man, before his philosophical conversion, is in a state of unhappy disquiet. Consumed by worries, torn by passions, he does not live a genuine life, nor is he truly himself. All schools also agree that man can be delivered from this state. He can accede to genuine life, improve himself, transform himself, and attain a state of perfection. It is precisely for this that spiritual exercises are intended. Their goal is a kind of self-formation, or *paideia*, which is to teach us to live, not in conformity with human prejudices and social conventions—for social life is itself a product of the passions—but in conformity with the nature of man, which is none other than reason. Each in its own way, all schools believed in the freedom of the will, thanks to which man has the possibility to modify, improve, and realize himself.[31]

In Hadot's approach, these spiritual exercises are linked to practices and habits that define the path of spiritual development, which renders philosophy an activity oriented toward oneself and one's relationships with reality.

At this point we may allow ourselves to take a small step forward in revisioning the understanding of philosophy and construe it not only as *action* and *self-transformation* but—more pictorially—as *gardening*. In the literature of the subject, it is possible to find studies indicating that the aesthetic dimension of gardening practice can be considered a part of the set of philosophical spiritual exercises, while gardening itself helps to understand oneself through exposure to a variety of temporal experiences.[32] Ourselves, we perceive philosophy itself as a form of gardening in several key dimensions, because, in our view, spiritual exercises may be construed as practices and habits of *cultivating oneself* in tune with the rhythm of nature. In such a context, philosophy becomes self-creation by way of particular gardening practices: it is not an abstract act of theory-building, but a process of factual self-change, in which the object of cultivation—the garden—is the garden of the self. Hadot conceives the entire history of the West as the history of perfecting the conversational techniques serving the purpose of transforming human reality. Importantly, Hadot himself notes a significant difference between his own interpretation of philosophy and Foucault's position in this respect:

> It is quite true that, in this connection, the ancients did speak of an "art of living." It seems to me, however, that the description M. Foucault gives of what I had termed "spiritual exercises," and which he prefers to call "techniques of the self," is precisely focused far too much on the "self," or at least on a specific conception of the self.[33]

This brings us to the key issue defining the position of particular interest to us. Philosophy conceived as an art of living *cannot* focus on oneself alone. Rather than concentrate solely on the subject shaping only him- or herself, it is, therefore, about *transcending oneself*. Thus, its focus is on the order of the whole, where the entire reality is a garden to be cultivated.

> The psychic content of these exercises seems to me to be something else entirely. In my view, the feeling of belonging to a whole is an essential element: belonging, that is, both to the whole constituted by the human community, and to that constituted by the cosmic whole.[34]

According to Hadot, the proper attitude is that of self-transcendence: an attitude of thinking and acting in tune with universal reason. This divergence between Hadot's and Foucault's views is significant as it speaks volumes about their respective diagnoses of modernity and about the potential of the heritage of the Antiquity, which may now be creatively reinterpreted. Foucault does not treat ancient ways of self-care as a distinctive trait of the "age of light," nor does he perceive them as a model to be copied, or transplanted, into contemporary contexts. Instead, he understands them as emancipatory strategies, involving the shedding of one's originary elitism, and aiming at the subjectification of the individual. In the light of the multitude of lifestyles, in the dazzling world of production, marketing, and promotion of countless models of human functioning, Foucault sees the need for a "conversion-to-oneself," a gesture called for in ancient sources. Yet, such a strategy does not lend itself to being interpreted in terms of the Christian *metanoia* nor is it Platonic *epistrophe*; his conversion-to-oneself is not a *radical conversion*: it does not involve turning away from the world toward the reality of some transcendent order. In fact, Foucault argues that his concern is not about severing contact with this world or with self; rather, converting-to-oneself is tantamount to keeping one's self constantly in sight. Turning one's gaze away from others (the world) and focusing on oneself, one does not have to break existing ties, yet such a conversion *toward* self is an expression of one's subjectification and, thereby, emancipation.

Unlike Foucault, Pierre Hadot promotes the perspective of spiritual exercises as means toward "reclaiming oneself" by way of the realization of one's belonging in the wholeness of nature—or even one's unity with nature. The philosopher writes:

> To summarize: what Foucault calls "practices of the self" do indeed correspond, for the Platonists as well as for the Stoics, to a movement of conversion toward the self. One frees oneself from exteriority, from personal attachment to exterior objects, and from the pleasures they may provide. One observes oneself, to determine whether one has made progress in this exercise. One seeks to be one's own master, to possess oneself, and find one's happiness in freedom and inner independence. I concur on all these points. I do think, however, that this movement of interiorization is inseparably linked to another movement, whereby one rises to a higher psychic level, at which one encounters another kind of exteriorization, another relationship with "the exterior." This is a new way of being-in-the-world, which consists in becoming aware of oneself as a part of nature, and a portion of universal reason. At this point, one no longer lives in the usual, conventional human world, but in the world of nature. As we have seen above, one is then practicing "physics" as a spiritual exercise.
>
> In this way, one identifies oneself with an "Other": nature, or universal reason, as it is present within each individual. This implies a radical transformation of perspective, and contains a universalist, cosmic dimension, upon which, it seems to me, Michel Foucault did not sufficiently insist. Interiorization is a going beyond oneself; it is universalization.[35]

The passage above highlights the importance of two complementary movements: (1) interiorization, expressed in shedding the dictate of externality and focusing on self-control and inner independence, and (2) exteriorization, construed as a new way of transcending oneself, as a different relationship with externality. This second movement—rejected by Foucault—is particularly important from the perspective of our proposal: it consists in "becoming aware of oneself as a part of nature, and a portion of universal reason." Hadot dubs such an understanding of man's relationship to nature as "practicing 'physics' as a spiritual exercise." Ultimately, interiorization is a universalization, opening oneself to the global (cosmic) dimension, to an order that transcends all, or, to put it simply, to nature. It can be said that this bipolar movement is simultaneously one of radical subjectification, but also of, not any less radical, objectification. *Delving into oneself* is inextricable from *thinking about oneself*—but in Hadot, such a self-oriented reflection only construes oneself as a part of nature.

In this very sense, practicing "physics" becomes a spiritual exercise and thereby, as already suggested, we may treat gardening likewise. Furthermore, this observation leads to four theses: (1) practicing gardening becomes a spiritual exercise; (2) philosophy tantamount to self-creation as a particular form of gardening practice; (3) nature can be conceived as a great garden, and man may be construed both as an element of that garden and as its co-creator; (4) the garden of nature as a whole can be conceived of as transcending classic oppositions. A closer look at these formulations may shade light upon their consequences.

It is clear by now that one could call upon many examples that illustrate the demonstrability of the concept of gardening as a unique, contemporary, form of a spiritual exercise. Researchers primarily point to the dialogic dimension of the relationship between the gardener and nature, as well as the temporal aspect of their mutual engagement.[36] Gardening thus manifests itself as dialogic, and thus impossible to reduce to one-sided contemplation. Mara Miller analyzes many temporal dimensions in the perspective of gardening (including subjective time, objective or shared time, cyclical time[37]), but those aspects that highlight gardening as a spiritual exercise are particularly worth attention.[38] Certainly, the training in patience—or in resistance to the dictate of the immediacy of the results of practice—comes to the forefront. The effects of our activities are visible with a delay, in line with the laws of nature, and in respectful recognition of the impossibility of reversing the natural process. Moreover, in that context, one can speak of training oneself to capture permanence in variability, as well as of practicing the awareness of cyclicality and repeatability. The vegetative cycles of plants, depending on the changing seasons, dictate the times of sowing and of collecting harvest, thus epitomizing the circularity of the rhythms of nature, to which we must become attuned.

Of course, thinking of philosophy-as-gardening does not equal reducing it to practical dimensions only. On the contrary, such an understanding ties its theoretical aspects with the practical ends, as actions in the world (garden) must be linked both to the knowledge of its basic laws and to the acquired skills of adapting oneself to the rhythms of life. From a meta-philosophical point of view, one can draw an outline of comprehensive reflection, in which unique care for the whole, and a sense moral commitment, simultaneously come to the forefront. Such a "philosophy of the garden" is concerned not only about *knowing* and *understanding* the world, but, first and foremost, about *keeping it alive* and creating conditions for it to flourish. In this context, the position of the observer (contemplator) fuses with that of the co-creator of life processes.

The vision of nature as a garden is one of the most obvious motifs in human imagination. In our case, however, we shift emphasis to the understanding of the fact that nature is a-whole-imbued-with-universal-reason. This whole, of which we are a part, does not depend solely on us; in fact, even we ourselves do not depend exclusively on ourselves. Gardeners in the garden of nature co-create nature; they strive to understand her in order to ultimately contemplate the results of their mutual engagement. The purpose of the garden is growth, blossoming, and flourishing—all of which testifies to the dynamics of the world-as-a-holistic-process.

Conceived in this way, the garden of nature transcends the classic oppositions, which forcefully, and often harmfully, affect our ways of thinking about reality. It should be noted that garden imaginaries evidently tend to blur particular boundaries, filling in deep conceptual chasms, and—as a result—discouraging thinking in terms of opposites. Interestingly, and somewhat paradoxically, the noun "garden" derives from the Proto-Indo-European root *gher-, meaning "to grasp, enclose." Although, etymologically, it was conceived as an enclosed or fenced place, the understanding we have adopted in these reflection, render the garden "unfenced" and "limitless." As it expresses the entirety of reality, it is not only inseparable from the larger whole but also blurs internal boundaries and distinctions *within itself*. This second aspect is suggestively expressed by Mara Miller:

> Organized and systematic, gardens operate at the border between the explicit and tacit, between communal or "universal" and individual, between objective and subjective, between the conscious and the unconscious or prereflectively conscious, between deliberate and taken-for-granted. They occupy, or perhaps create, a vast of transitional territory between the amorphous, unrepeatable, incommunicable chaos of irreduceably idiosyncratic experience and the readily communicated and comfortable norms of socially generated experience.[39]

The liminal nature of gardens finds its expression not in the figure of the fence, but in its "operating on boundaries," in its occupancy of the territory that could not be designated as located on either side of the threshold. The garden is simultaneously a reality and an idea—a metaphor, a symbol, and a model. Perhaps the garden is best characterized by its "in-betweenness": an-in-between-reality, a reality-in-contact, in-mediation. And yet, the liminality of the garden does not bridge temporality and transcendence (as Plato, or Eric Voegelin, would have it): instead, it is an area catalyzing the processes of the horizontal mixing of substantial particularisms.

In broadest terms, the garden epitomizes the intersection of art and nature, or rather *is* art expressed as nature and nature expressed in the form of art.[40] In this context, it is certainly worth recalling Rosario Assunto's studies on the philosophy of the garden, whose findings capture the profundity of this connection with particular precision. Assunto's point of departure is the premise that our relationship with nature stems from a sense of unity with it, and that the discovery of nature-within-ourselves leads us to the desire to care for it. As a result, the world is tended to: it becomes a garden in our care, in which we ourselves are neither completely *subject to nature*, nor are in the position to dominate or consume it. From Assunto's texts emerges an image of an individual who, finding himself in a physical garden, embraces the potential of a *metaphysical* experience. As a result—discovering that wherever he dwells, he always dwells in a garden—he becomes the practicioner of the garden philosophy. In light of Assunto's vision, one would be inclined to hypothesize that embracing the garden, everyone may find—and regain—their paradise lost (the idea to which we will yet return in our reflections).

The garden, although defined by its thresholds, is a space where externality and internality blur—as do the differences between living, creating, and contemplating, between practice and aesthetics, and, to some extent, also between human and nature, between subject and object. In his groundbreaking *Ontologia e teleologia del giardino*, Assunto reflects that the garden is nature shaped by man in such a fashion as to express his spirit.[41] By this virtue, the garden becomes a model of the relationships within the world—and, in particular, a model for human actions: following its principles, man should see himself as an extended human, and by that token, perceive himself as an extended garden. For the Italian philosopher, attempts to restore actual, physical gardens reflect the need for the renewal of man-as-a-garden, which calls for a change of attitude, for a revolution, which, in essence, aims to return to what once was, yearns for a re-reconciliation. A revolution in thinking is usually as much of an "overturn" as it is a "return"; to effect a revolution, one often needs to go back to concepts once proclaimed or revert to a state that has been abandoned.

The above notwithstanding, we should not lose sight of the fact that even in the garden, inevitable human violence manifests itself in more or less authoritarian choices: gardening, after all, involves making decisions as to what is to be allowed to grow, and what is to be weeded out (negated). The garden thus becomes a battlefield, in which man, entering the world of nature with the sense of entitlement, is the dominant player. Human violence severs the fundamental bond unifying man with the wild and untamed world of nature. Contrary to our naïve projections, we do not leave it unblemished, and our care comes at the price of "selective exclusion"—for example, the exclusion of plants that we choose to classify as weeds. Alluding to St. Augustine, one could state that God created nature, but man created weeds—or, more specifically, man selected certain plants and *named them weeds*. Yet, irrespective of human intrusions, the world of nature reveals itself to us as a struggle, a fierce dispute, which, *per se*, does not exclude the existence of a higher order. The Greeks, linking together the categories of war, logos, and unity, may have realized this truth better than anyone else.

Still, we should not allow these general remarks about the essence of gardening obscure the importance of other human actions. Even without in-depth analyses of the phenomenon, we do realize that people feel the urge to be *in* and *with* the garden; they need the connection with nature—and the recent pandemic put this need in a very bright spotlight. Jane Perrone's article "How Coronavirus Changed Gardening Forever" published in the *Financial Times* in September 2020 offers an eye-opening testimony to the latter claim. The author demonstrates that the pandemic revealed the need for the communion with nature (or, in fact, with greenery in general) as fundamental: "Gardening was listed as the second most popular lockdown activity people planned to do after watching TV, according to a survey by GlobalData market research in May, ahead of cooking, reading, and exercising."[42] Contact with nature proved necessary in its purely physical aspect, but—as Perrone points out—it was also needed in the psychological dimension. In this sense, the pandemic, which indeed changed gardening, also allowed us to palpably recognize what is truly important to us.

Empirical research in recent years highlights both the detrimental effects of severing contact with nature and the positive health benefits associated with gardening. Richard Louv demonstrates that contact with nature is essential for the healthy development of children, as well as the physical and mental well-being of adults.[43] His strong recommendation is evident in the subtitle: "saving our children from nature-deficit disorder." To counteract the digital isolation of children from nature, it is crucial to practice constructive boredom in nature and, importantly, to emphasize the significance of time spent in nature during childhood. Other studies indicate that breaking ties with nature leads to an "extinction of experience."[44] Surveys among students in Japan revealed that both current and childhood frequency of contact with various forms of nature positively influenced students' emotional connection with nature.

The positive effects of maintaining contact with nature for young people need to be supplemented with data concerning individuals more advanced in years. Daoyan Wang and Thomas MacMillan provide evidence that gardening is a beneficial activity for the elderly. They analyze 22 articles on the benefits of gardening, which include improved health, quality of life, physical strength, fitness, cognitive abilities, and socialization.[45] Particularly noteworthy are the results of Masashi Soga, Kevin J. Gaston, and Yuichi Yamaura's meta-analysis examining the effects of gardening, including horticultural therapy, on health.[46] The project encompassed 22 case studies discussed in 21 scientific articles, revealing that most researchers reported positive effects of gardening, with none reporting negative effects.

> To our knowledge, this meta-analysis is the first to provide a quantitative synthesis of the evidence that gardening is beneficial for human health. Overall, the results suggest that participating in gardening activities has a significant positive impact on health. Indeed, the positive association with gardening was observed for a wide range of health outcomes, such as reductions in depression and anxiety symptoms, stress, mood disturbance, and BMI, as well as increases in quality of life, sense of community, physical activity levels, and cognitive function.[47]

Moreover, the studies demonstrated that even short-term gardening activities have an immediate positive impact on health. Gardening also embodies a strong element of equality, as no differences were found in studies based on socioeconomic status. In summarizing their research, the authors write: "Although our meta-analysis presents a consistent result, untangling the causal relationships between gardening and improved health outcomes is not an easy task. There are several possible, but not mutually exclusive, pathways through which gardening promotes health."[48] They distinguish four non-exclusive causes or pathways:

a. direct experience with nature; this refers to cognitive regeneration from nature.
b. encouragement to undertake physical exercises.
c. gardens allow for interpersonal interaction, strengthening social bonds, and a sense of community.
d. the opportunity to provide oneself with a healthy diet.[49]

The "interpersonal interaction" referenced in the third point offers mental space that has the potential to integrate the functions of both the garden and the agora. In the subsequent sections of this book, we discuss specific examples of these interactions from various locations worldwide. At this juncture, however, it is important to outline a general model that encapsulates the concepts of the agora and the garden, focusing on the idea of relational entanglement. Understanding humans likely involves the recognition of the intricate relationships and dependencies among the temporarily distinguished, and variable, objects. While the agora embodies a complex network of interpersonal relations, the garden represents the reciprocal embrace of nature and culture, with each constituting its own network of relations.

The key point in this section is to emphasize the idea that the relationship between humans and nature is an attempt to regain a lost paradise—and idea rooted in the belief that humans are part of—and responsible—for nature. In the dissertation *Il giardinaggio come operazione filosofica*, Rosario Assunto metaphorically interprets the biblical story of paradise as a narrative of the disturbance of the world order. Emphasizing its relevance, the philosopher asserts that the source of this disturbance—manifest as a radical loss of the beauty of the world conceived as a garden—is the conception of knowledge as power, which subordinates nature. The consumption of the fruit and the expulsion from paradise signify the end of a particular relationship between humans and nature, envisioned as a garden.[50] In Assunto's view, the replacement of the contemplation of the world's garden by pragmatic—or positivistic—narratives (a transformation almost universally orchestrated with the dull sound of bulldozers destroying gardens) occurred in the mid-20th century. This transformation is, in fact, a form of nomadism, representing humanity's flight from itself. The disappearance of gardens in the last century appears to be a faint echo of the modern shift in thinking, the consequence of which was our expulsion. Metaphorically speaking, this also signifies our exile from Arcadia, which symbolizes the enduring bond between humanity and nature conceived as an inseparable whole.

In the 17th century, the humanist and theorist who later became Pope Clement IX created an aphorism as beautiful as it is ambiguous: *Et in Arcadia Ego* (loosely translated: "and/even in Arcadia I"). This phrase has triggered numerous interpretations across literature and painting (notably in the works of Nicolas Poussin), typically revolving around themes of paradise and death, exile and return, forgetfulness and longing. Many of us deeply feel the consequences of the exile from paradise (Arcadia)—and many of us, simultaneously, sense an awakened desire to return to the happy garden, or reclaim what was lost. This Latin adage is incomplete and always requires some sort of exegesis. Goethe, for example, relates the "ego" to himself, proclaiming: "I too have been in Arcadia." Another interpretation, referring to King George III, suggests that Death speaks from the grave: "Even in Paradise am I," signifying that even there, one cannot escape death.

It appears that the aphorism, due to, or perhaps owing to, its incompleteness, serves as an excellent pre-text to express one's relationship to paradise (Arcadia)—the quintessential garden. Expelled from it, we feel an unquenchable desire to return. What drives this desire may differ from individual to individual, but two major motivations have been presented above: nostalgia and death. Today, however, one more important factor complements the timeless two: the *entanglement*. It is not only the desire to return exacerbated by the hardships of exile (nostalgia), and it is not only the inevitability of death, but also the realization of the existence of a network of innumerable relationships between humans and nature that should prompt us to think of the latter and us within it as a Garden, where the multitude of dependencies can only be revealed in the perspective of wholeness and unity. Let us return to thinking of nature as Arcadia, a garden of harmonies, life rhythms, and compassionate care: *and/even in Arcadia I*.

Notes

1 Recentivism (from Latin *recens*, meaning "present") is a contemporary philosophical movement associated with the work of the Polish philosopher, Józef Bańka. The central argument of his seminal monograph *Ja teraz. U źródeł filozofii człowieka współczesnego* [I Now: At the Sources of Contemporary Human Philosophy] (Katowice: Wydawnictwo "Śląsk", 1983). In Bańka's philosophy, the significance of human life lies in the *immediate*. Phenomena are only true in the present moment—whatever exists, exists now.
2 Róbson Ramos Dos Reis, João Carlos Onofre Pinto, Bruno Nobre, Andreas Gonçalves Lind, and Ricardo Barroso Batista, "Thinking the Pandemic: Philosophical Perspectives," *Revista Portuguesa de Filosofia* 77, no. 2/3 (2021): 478–479, https://doi.org/10.17990/rpf/2021_77_2_0478.
3 In a book published over ten years ago, Paul Crosthwaite argues that two characteristics define contemporary crises: their diversity (financial, economic, environmental, geopolitical, terrorist, public health) and their global status. See Paul Crosthwaite, ed., *Criticism, Crisis, and Contemporary Narrative: Textual Horizons in an Age of Global Risk* (New York: Routledge, 2011).
4 It was Paul de Man who, suggestively, demonstrated this relation already in 1983. See Paul de Man, "Criticism and Crisis," in *Blindness and Insight: Essays in the Rhetoric of Contemporary Criticism*, rev. ed. (Minneapolis: University of Minnesota Press, 1983 [1967]), 9–34.
5 Dariusz Kubok, *Krytycyzm, sceptycyzm i zetetycyzm we wczesnej filozofii greckiej* (Katowice: Wydawnictwo Uniwersytetu Śląskiego, 2021), 28.
6 Dariusz Kubok, *Krytycyzm, sceptycyzm i zetetycyzm we wczesnej filozofii greckiej*...[Criticism, Skepticism, and Zeteticism in Early Greek Philosophy], 65–74; Dariusz Kubok, "Comments on the Sources of Greek Philosophical Criticism," in *Forms of Criticism in Philosophy and Science*, ed. Dariusz Kubok (Katowice: Wydawnictwo Uniwersytetu Śląskiego, 2015), 14–17.
7 World Health Organization. "The Preamble to the WHO Constitution." (accessed: June 29, 2023). https://www.who.int/about/governance/constitution.
8 See, for instance, James S. Larson, "The World Health Organization's Definition of Health: Social versus Spiritual Health," *Social Indicators Research* 38, no. 2 (June 1996): 181–192.
9 Leslie Garner, *The NHS: Your Money or Your Life* (New York: Penguin Books, 1979),14.
10 Philip H. N. Wood, "Health and Disease and Its Importance for Models Relevant to Health Research," in *Scientific Approach to Health and Healthcare*, ed. B. Z. Nizetic et al. (Copenhagen: World Health Organization, 1986), 57.
11 John E. Ware et al., "Choosing Measures of Health Status for Individuals in General Populations," *American Journal of Public Health* 71 (1981): 620–625, https://doi.org/10.2105/AJPH.71.6.620.
12 Machteld Huber et al., "Health: How Should We Define It?" *British Medical Journal* 343, no. 7817 (2011): 237. https://doi.org/10.1136/bmj.d4163.
13 Trevor Shilton, Michael Sparks, David McQueen, Marie-Claude Lamarre, and Suzanne Jackson, "How should we define health? Proposal for new definition of health," *British Medical Journal* 343 (2011): 435. https://doi.org/10.1136/bmj.d5359.435.
14 See for example, James S. Larson, "The World Health Organization's Definition of Health: Social versus Spiritual Health"..., 181–192.
15 Hans-Georg Gadamer, *Über die Verborgenheit der Gesundheit* (Berlin: Suhrkamp Verlag, 2010); the English version: Hans-Georg Gadamer, *The Enigma of Health. The Art of Healing in a Scientific Age*, translated by Jason Gaiger and Nicholas Walker (Cambridge: Polity Press, 1996), ISBN 9780804726924.
16 Hans-Georg Gadamer, Über die Verborgenheit der Gesundheit..., 40.

17 Hans-Georg Gadamer, Über die Verborgenheit der Gesundheit..., 40.
18 Hans-Georg Gadamer, Über die Verborgenheit der Gesundheit..., 41.
19 Hans-Georg Gadamer, *Über die Verborgenheit der Gesundheit*..., 43.
20 Hans-Georg Gadamer, *Über die Verborgenheit der Gesundheit*..., 86.
21 Hans-Georg Gadamer, *Über die Verborgenheit der Gesundheit*..., 95.
22 Hans-Georg Gadamer, *Über die Verborgenheit der Gesundheit*..., 98.
23 Hans-Georg Gadamer, *Über die Verborgenheit der Gesundheit*..., 132–133.
24 Hans-Georg Gadamer, *Über die Verborgenheit der Gesundheit*..., 141.
25 David Edward Cooper, *A Philosophy of Gardens* (Oxford: Clarendon Press, 2010), 6.
26 *Gardening Philosophy for Everyone: Cultivating Wisdom*, ed. Dan O'Brien (Chichester, West Sussex: John Wiley & Sons Ltd., 2010), 1.
27 Pierre Hadot, *Exercices spirituels et philosophie antique* (Paris: Études Augustiniennes, 1981).
28 Arnold Ira Davidson, "*Spiritual Exercises* and Ancient Philosophy: An Introduction to Pierre Hadot," *Critical Inquiry* 16, no. 3 (1990): 475–482, https://www.journals.uchicago.edu/doi/10.1086/448542.
29 See Michel Foucault, *Histoire de la sexualité: 1: La volonté de savoir* (Paris: Gallimard, 1976); Michel Foucault, *Histoire de la sexualité: 2: L'usage des plaisirs* (Paris: Gallimard, 1984); Michel Foucault, *Histoire de la sexualité: 3: Le souci de soi* (Paris: Gallimard, 1984).
30 Alexander Nehamas, *The Art of Living: Socratic Reflections from Plato to Foucault*, 1st ed. (Berkeley: University of California Press, 2000).
31 Pierre Hadot and Arnold Ira Davidson, *Philosophy as a Way of Life: Spiritual Exercises from Socrates to Foucault* (Malden, MA: Blackwell, 1995), 102.
32 See Marianna Favara-Kurkowski, "Reclaiming Time Aesthetically: Hadot, Spiritual Exercises and Gardening," *Eidos. A Journal for Philosophy of Culture* 5, no. 2 (2021): 7–21, https://doi.org/10.14394/eidos.jpc.2021.0002.
33 Pierre Hadot and Arnold Ira Davidson, *Philosophy as a Way of Life: Spiritual Exercises from Socrates to Foucault*..., 206–207.
34 Pierre Hadot and Arnold Ira Davidson, *Philosophy as a Way of Life: Spiritual Exercises from Socrates to Foucault*..., 208.
35 Pierre Hadot and Arnold Ira Davidson, *Philosophy as a Way of Life: Spiritual Exercises from Socrates to Foucault*..., 211.
36 "When nature is involved as nature in our appreciation of the activities we carry out in the garden, we can call the form of temporal experience afforded dialogic, that is, a form of dialogue between the gardener and the natural processes, versus the monologic form of observation. In other words, being involved with a dialogic temporality means that we are directed by what nature demands every time we have to do something in the garden." Marianna Favara-Kurkowski, "Reclaiming Time Aesthetically: Hadot, Spiritual Exercises and Gardening"..., 16.
37 Mara Miller, *The Garden as an Art* (Albany, NY: State University of New York Press, 1993); see also *Gardening Philosophy for Everyone: Cultivating Wisdom*, ed. Dan O'Brien..., 178–191.
38 In this context, Marianna Favara-Kurkowski makes the following observation: "Gardening, however, reformulates the conditions for considering spiritual exercises in a modern key without appealing to a metaphysical system that is now inoperative. However, the aesthetically mediated dialogue with nature, so rare, especially for homo urbanus, preserves a sense of unity and cosmic consciousness, albeit faded." Marianna Favara-Kurkowski, "Reclaiming Time Aesthetically: Hadot, Spiritual Exercises and Gardening"..., 18. Using the example of gardening, the author attempts to explain the essence of spiritual exercises without recourse to the metaphysical dimension. In our view, she does not sufficiently

realize that what she seeks to exclude is not so much metaphysics *per se*, but rather a *certain type* of metaphysics—one close to Hadot. One of the intentions of our work is to show, on the one hand, an alternative version thereof as a metaphysics of the garden, and on the other hand, to indicate that in any activity conceived as a whole, one will invariably find the rudiments of some metaphysical assumptions or assertions.

39 Mara Miller, *The Garden as an Art…*, 4.

40 See, for instance, Angela Ferraiolo, "Fictional Characters as Adaptive Systems: The Malaspina and de Graaf Gardens," in *Proceedings of EVA London 2020* (London: BCS Learning & Development Ltd., July 2020), 313–321, https://www.scienceopen.com/hosted-document?doi=10.14236/ewic/EVA2020.57.

41 Rosario Assunto, *Ontologia e teleologia del giardino* (Milan: Guerini e Associati, 1994), 19–38.

42 Jane Perrone, "How Coronavirus Changed Gardening Forever," *Financial Times*, September 11, 2020, (accessed: June 10, 2024) https://www.ft.com/content/c3abc0bb-7ade-4ae1-8d2b-d3fa6be53416.

43 Richard Louv, *Last Child in the Woods: Saving Our Children from Nature-Deficit Disorder* (Chapel Hill, NC: Algonquin Books of Chapel Hill, 2005).

44 See for instance: Masashi Soga, Kevin J. Gaston, and Yuichi Yamaura, "Gardening is Beneficial for Health: A Meta-analysis," *Preventive Medicine Reports* 5 (2017): 92–99, https://doi.org/10.1016/j.pmedr.2016.11.007.

45 Daoyan Wang and Thomas MacMillan, "The Benefits of Gardening for Older Adults: A Systematic Review of the Literature," *Activities, Adaptation & Aging* 37 (2013): 153–181, https://doi.org/10.1080/01924788.2013.784942.

46 Masashi Soga, Kevin J. Gaston, and Yuichi Yamaura, "Gardening is Beneficial for Health: A Meta-analysis"…, 92–99.

47 Masashi Soga, Kevin J. Gaston, and Yuichi Yamaura, "Gardening is Beneficial for Health: A Meta-analysis"…, 97.

48 Masashi Soga, Kevin J. Gaston, and Yuichi Yamaura, "Gardening is Beneficial for Health: A Meta-analysis"…, 97.

49 Masashi Soga, Kevin J. Gaston, and Yuichi Yamaura, "Gardening is Beneficial for Health: A Meta-analysis"…, 97.

50 Rosario Assunto, "Il giardinaggio come operazione filosofica," in *Paradisi ritrovati. Esperienze e proposte per il governo del paesaggio e del giardino. Con un'introduzione di Rosario Assunto*, ed. Mariapia Cunico and Domenico Luciani (Milan: Guerini e Associati, 1991), 7–12.

2 From the Garden of Eden to *Project House*

One of the most hilarious passages in J.R.R. Tolkien's *The Lord of the Rings* describes the scene, in which Gandalf the Grey catches Samwise Gamgee eavesdropping on his secret conversation with Frodo Baggins:

> "Well, well, bless my beard!" said Gandalf. "Sam Gamgee is it? Now what may you be doing?"
>
> "Lor bless you, Mr. Gandalf, sir!" said Sam. "Nothing! Leastways I was just trimming the grass-border under the window, if you follow me." He picked up his shears and exhibited them as evidence.
>
> "I don't," said Gandalf grimly. "It is some time since I last heard the sound of your shears. How long have you been eavesdropping?"
>
> "Eavesdropping, sir? I don't follow you, begging your pardon. There ain't no eaves at Bag End, and that's a fact."
>
> "Don't be a fool! What have you heard, and why did you listen?" Gandalf's eyes flashed and his brows stuck out like bristles.
>
> "Mr. Frodo, sir!" cried Sam quaking. "Don't let him hurt me, sir! Don't let him turn me into anything unnatural! My old dad would take on so. I meant no harm, on my honour, sir!"[1]

Of course, Sam meant no harm. But aware that he was caught red-handed, he is scared that Gandalf might do *him* harm, turning him into something questioning the "order of nature" that the young gardener comprehends. The inverted commas are intended: as the sections below clearly demonstrate, the fluctuating projections of what is "natural" and what is "unnatural" have proven to be heavily dependent on the dominant philosophical conceptions of the subsequent periods in the cultural history of the garden.

The (Un)Natural in Arcadia?

Consider some obvious examples from the history of Western painting. For instance, when Hans Memling (born ca. 1435 in the German town of Seligenstadt) painted his stellar triptych of *The Last Judgment*,[2] he rendered the garden of Eden—epitome of the pre-lapsarian "nature"—as entirely

DOI: 10.4324/9781003328469-4

Figure 2.1 Hans Memling (ca. 1435–1494), Bruges, *The Last Judgment*, triptych (1467–1473), oil and tempera techniques on wooden panel, gilding. Muzeum Narodowe w Gdańsku, inventory no.: MNG/SD/413/M. Reproduced by permission of the National Museum in Gdańsk, Poland.

"comprehensible" for women and men of the time, because he transferred the responsibility for its visualization entirely to his viewers:

In fact, the array of the Saints—whose bodies are young and healthy, and who are welcomed at the gates of heaven by Saint Peter and the angelic train—climb beautifully crafted crystal stairs to pass into the garden through an entrance fashioned to resemble an intricately sculpted gate of a rich Gothic cathedral (which reminds the viewers that the path to Heavens leads through the Church). The garden, concealed behind the gate, remains invisible; the only elements of "nature" featuring in the panel are the rocks, upon which the bottom landing of the stairs rest, a patch of low (perhaps trimmed) grass, red lilies, and human bodies (all of which, in Memling's triptych, are Caucasian). Man-made architecture dominates the whole section of the painting and the "logic" of the ascent of the Saints follows the "obvious" principles of gravity, and—despite the suggestion of equality-in-nudity—rather strict social order (the Pope leading the Saints). Rigid symmetry of the architectural marvel of the gateway—both vertical and horizontal—emphasizes the organization of the "natural," i.e., feudal, order even further.

Unlike heaven, Memling's hell bears no traits of human intervention. Resembling the crater of an active volcano, enwrapped in storm clouds, erupting with fire, the entrance to hell seems as natural as one can possibly imagine. With no man-made structures present, the uncanny of the image rests upon the human fear of the lack of control (the damned give in to the forces of gravity and fall) and upon the presence of dark-skinned demons, whose bodies are either human-animal hybrids, or bear "unnatural" features of hermaphroditic fusions (male torso, female genitalia).

A fearful image, beyond doubt—one, which could inspire Samwise Gamgee's horror in the face of Gandalf's anger—especially that, quite evidently, J.R.R. Tolkien's narrative is Christian to the core: to redeem "humankind," a weak Hobbit, all too human himself, must carry the ring of power—a cross-like, unbearable burden, under which he falls more than once—to the heart of Mount Doom, and eventually defeats the forces of evil at the cost of his own life, sacrificed for the good of the Middle Earth. Such a vision is subcutaneously based on the ethics relying upon the dichotomy of "good" and "evil" as reflected in the binary of the "natural" and the "unnatural," which is plainly analogous to that manifest in Memling's painting. In Peter Jackson's cinematic interpretation of the novel, the spaces of righteousness—the Shire, Gondor, Rohan, and Rivendell—are not only constructed as bright, harmonious, and inhabited by wise, noble, and ridiculously gorgeous people (suffice it to mention that the central characters were impersonated by actors representing the pinnacle of the preferred Western standard of beauty: Cate Blanchett, Orlando Bloom, Sean Bean, Viggo Mortensen, Sarah McLeod, Peter McKenzie, Dominic Monaghan, Liv Tyler, or Hugo Weaving), but also as spaces of stunning natural allure and architectural elegance. In Jackson's vision, Gondor—a white, vertically structured city upon a hill, is a monument to human skill, surrounded by awe-inspiring beauty of nature. Rohan, echoing Nordic traditions, harmoniously blends the man-made (mostly wooden) structures with their natural environs. The architecture of the Shire leaves no doubt: Hobbits are literally *glebae adscripti*: not only do they live in houses that barely disrupt the landscape, but also, as farmers, they embody the purity of the uncorrupted morals of the "natural" countryside, a concept vividly present in the texts of the Western Age of Reason. Rivendell, however, is the most intriguing of all of the kingdoms of the Middle Earth. Modeled after the Swiss Lauterbrunnen Valley (in Tolkien) and after Kaitoke Regional Park, enhanced with special effects (in Jackson), visually, Rivendell is almost identical to the landscapes featured in the canvas of the pre-Raphaelites, and just as striking is the correspondence between the cinematic Elves and the pre-Raphaelite portraitures of women and men alike. More importantly perhaps, the architecture of the Elvish palaces (both in Rivendell and in Lothlórien) deconstructs the binary opposition between the interior and the exterior, the crafted and the biological, thus stressing the importance of the Elves as synecdochal analogons of undivided unity of being and ultimate attunement to the natural rhythms.

Conversely, Mordor, whose very name sounds ominous, surrounded by swamps and high mountains, and separated from the Middle Earth by an impenetrable wall, is represented as a curious fusion of a desolate caldera, which still exudes toxic fumes, and an industrial complex. Mordor is dark, and is inhabited by Orcs, Goblins, and the Uruk-hai: all tortured, all distorted, humanoid, yet inhumanly demonic creatures, whose bodies suggest decay and putrefaction, and whose deaths, even if accompanied by evident signs of agony, do not evoke sympathy. The reader of the novel, like the viewer of the film, may experience highly emotional states when Boromir of Gondor or Théoden of Rohan heroically perish, but any kind of empathy with the "unnatural" creatures of Mordor is actively discouraged. After all, "[t]he divine is beauty, wisdom, goodness, and the like; and by these the wing of the soul is nourished, and grows apace; but when fed upon evil and foulness and the opposite of good, wastes and falls away."[3] Notably, when Saruman the Wise chooses to join forces with Sauron, his own Isengard, once lush green, becomes barren: the industry, heedless of nature, causes the destruction of the harmonious relationship between its inhabitants (now ousted by the Orcs and the Uruk-hai) and their environs, resulting, eventually, in the uprising of the Ents, who, upon winning the battle against their oppressor, take over—and heal—the wasteland of human pride.

Curiously, the representation of hell in the controversial polyptych *Visions of the Underworld* (*Blessed and Cursed*), painted by Hieronimus Bosch (1450–1516) and dated between 1490 and 1516, while generally observing the principles of "natural" physics (the force of gravity), resembles the wastelands of Mordor:

The fires below, not unlike the fires heating the furnaces of Mordor and Isengard, emit unsettling, "unnatural" light, trickling upward into impenetrable darkness. The cursed, devoid of hope, tortured by demons, dwell in the barren space, among sharp rocks and dead trees, while the blessed, carried by

Figure 2.2 Hieronimus Bosch, *Four Visions of the Hereafter*, ca. between 1490–1516, Catalogue no. 182–184. (*Fall of the Damned*, oil on oak, 88.8 × 39.6 cm; *Hell*, oil on oak, 88.8 × 39.6 cm; *Earthly Paradise*, oil on oak, 88.5 × 39.8 cm; *Ascent into Heaven*, oil on oak, 88.8 × 39.9 cm); Venice, Palazzo Grimani, Gallerie dell'Accademia di Venezia, Italy. Reproduced with permission of the Ministry of Culture.

the angels, pierce through the darkness of the night and ascend into a tunnel of light, promising eternal bliss in the garden of Eden. Uncharacteristically, in his *Visions of the Underworld*, unlike in his most famous representations of heaven and hell, Bosch, operating in dim chiaroscuros, resorts to what we could call "a painterly rhetoric of the barren wasteland by night," which wasteland only differs from those we encounter in nature because the latter are not populated by uncanny figures or flooded with "infernal" light. In contrast, Bosch's representations of heaven, hell, and earth in his numerous renditions of the Last Judgment (such as his 1486 triptych residing at the Groeningemuseum in Bruges, Belgium[4]) rely on his evident juxtaposition of spaces in which words/objects correspond to interpretable categories (such as purpose) and those, in which objects and their senses are utterly divorced:

Bosch's confrontation of the "natural" semiotic order of heaven with the infernal semiotics of chaos may suggest that in the artist's philosophical interpretation of the world the category of "the natural," tantamount to the category of "the comprehensible," stands in contrast with the category of the uncanny/incomprehensible/unfamiliar/*unheimlich*, which, in this context, is synonymous to "the unnatural," or, in other words, to hell. Hell is thus an absurd space of incomprehension, devoid of principles, impossible to tame, and therefore terrifying. Yet, quite evidently, Bosch distrusts clear-cut binaries.

Unlike *The Last Judgment* from the Groeningemuseum, Bosch's most famous painting, *The Garden of Earthly Delights*, although its power relies, essentially, upon the same contrastive device, offers the viewer a vision of an

Figure 2.3 Hieronimus Bosch, *The Last Judgment*, ca. 1486. Inventory number: 0000.GRO0208.I Reproduced by permission of the the Groeningemuseum in Bruges, Belgium. Attribution: Public Domain Mark 1.0 Universal.

altogether revolutionary deconstruction of traditional Christian binaries, whose components fall into a tantalizing freeplay owing to minor—or perhaps only small-sized—details:

In the panel representing hell, like in his vision of the Last Judgment, the semiotic divorce of recognizable objects from their "natural" epistemological relations defines the inferno as the space of utmost incomprehension. Feces, anal torment, demonic music, Mordor-like lighting, slimy amphibians, uncanny hybrids—all add to the sense of dread. Heaven, however, although "landscaped" to accommodate the topology based on vertical and horizontal symmetry, although inhabited by the Creator, Adam, and Eve, and populated by a plethora of (more or less) fantastic animals and plants of matchless beauty, contradicts the dogmatically established concepts of Eden. *Et in Arcadia Ego*… Indeed, there *is* death in the garden: like in nature, predators hunt their prey. Revising, or even reversing, the biblical idea of "perfect harmony," the painter appears to question, or perhaps mock, the utopian visions of the pre-lapsarian heavenly garden, where the "wolf and the lamb shall feed together, and the lion shall eat straw like the bullock […]."[5]

Figure 2.4 Hieronymus Bosch, *The Garden of Earthly Delights*, 1490–1500, grisaille/oil on oak panels, 205.5 cm × 384.9 cm (81 in × 152 in). Inventory number P002823. Reproduced by permission of Museo Nacional del Prado, Madrid, ©Photographic Archive Museo Nacional del Prado.

Furthermore, the Earth, also symmetrically oriented, and represented in the central panel as the proper "garden of delights," initially strikes one as the space of utmost joy, happiness, and freedom from oppressive semiotic orders based on binary oppositions. In the garden of earthly delights, humans may enjoy physical love without shame or restriction, and nudity is innocent; fantastic plants, disproportionately towering over petite humans and animals, epitomize the flourishing of the world *in-statu-nascendi*. Human habitations are not architectural masterpieces, but eggshells, flower buds, or fruit, from which individuals or groups emerge, or which they enter. Human-animal and human-plant hybrids share the garden on par with non-hybrid humans—all races thrive in each other's presence, and none is superior to any other. The garden thus nullifies the opposition between "natural" and "unnatural": in its early stage, nature, clearly, is still free from stifling definitions: grape- or berry-headed beings, like human-headed beings, enjoy the same status as fish-tailed beings, and none of them is "unnatural," or even "less natural" than others. They may grow out of flowers rather than be born; the botanical, the zoological, the human, the inanimate—each may transmogrify into the other: the language determining perceptions and relations is not yet categorically fixed, and therefore even human anus, as Bosch would have it, may exude a legitimately floral aroma.

And yet, when seen through details, Bosch's narrative becomes as unsettling as it may seem liberating. In the garden, whose inhabitants celebrate newly created life, the indulgence in the "delights" sometimes involves the pleasure of inflicting suffering on others. An inch-by-inch inspection of the garden reveals that the painting certainly does not idealize the newly created Earth: murder, stealth, or rape *always-already* hide in the pre-lapsarian universe, introducing a rift to the otherwise seamless whole. Disorder, miscomprehension, uncertainty, irrational instincts, uncontrollable drives cannot be relegated to the margins of experience. The "perfect" is not "natural," "the natural" is not "perfect," and most certainly—not safe. Regardless of Bosch's original intentions, and irrespective of his contemporaries' interpretations of the painting, the *Garden* undermines the mythological fundaments of the one of the most important (if not *the most important*) grand narratives of the Western world, simultaneously questioning naïve idealism, uncritical dogmatism, and easy distinctions between the "natural" and the "unnatural," which opposition Samwise Gamgee will eventually learn to question, without quite dismissing it.

Even a brief analysis of the visions of the garden in the works of the two—very different—painters indicates that even in early modern Europe, the perception of "nature" as a dialectial phenomenon was by no means unproblematic. Some would see the correspondence between "good" and "comprehensible" as an exponent of the "natural order," whereby the physical, wild, nature, because it is "untamed," questions this order's essential premises and, by extension, the status quo. As a result, paradoxically, nature is projected as incomprehensible and evil, feeding the imaginaries furnishing hell. Others, like Bosch, would extend the dominant vision of the inferno to encompass the ultimate failure of language with its world-making categories: hell, indeed, is a desolate place, but it is not devoid of man-made objects. Far from it: these man-made things, whose semiotic attribution went awry, are the essence of hell. In Memling, the Saints, who follow the rules, enjoy God's grace, manifest in the resurrection of their healthy, young bodies, free from the afflictions of old age or disease, and unadulterated; in Bosch—the prelapsarian garden pulses with hybrid life. These distinctions bring to mind Aristotle's dispute with the Pythagoreans from Book One of *De Anima*:

> [Plato's] view, in company with most theories about the soul, involves the following absurdity: they all join the soul to a body, or place it in a body, without adding any specification of the reason of their union, or of the bodily conditions required for it. Yet such explanation can scarcely be omitted; for some community of nature is presupposed by the fact that the one acts and the other is acted upon, the one moves and the other is moved; interaction always implies a special nature in the two interagents. All, however, that these thinkers do is to describe the specific characteristics of the soul; they do not try to determine anything about the body which is to contain it, as if it were possible, as in the Pythagorean myths, that any soul could be clothed upon with any body—an absurd view, for each body seems to have a form and shape of its own. It is as absurd as to say that the art of carpentry could embody itself in flutes; each art must use its tools, each soul its body.[6]

Legitimizing this position for centuries to come, Thomas Aquinas comments upon Aristotle's view thus:

> [...] the soul of a fly for instance might perchance enter the body of an elephant. This cannot in fact happen; for the body of each particular thing, and especially of living things, has its own form and species and type of movement: hence there are great differences between the bodies of a worm, a dog, an elephant and a gnat. When they say that any soul can enter any body, it is as if one were to say that the art of weaving could enter flutes, or that the art of the coppersmith could enter a weaver's loom. If it was in the power of these arts to enter bodies or instruments they would not do so indiscriminately, but the art of playing the flute would enter flutes, and not lyres, while the art of playing stringed instruments would enter stringed instruments and not flutes. In the same way, if there is a body for every soul, any soul does not enter any body; rather the soul shapes the body fit for itself; it does not enter a ready-made body. Plato and the others who speak only about the soul are too superficial; they fail to define which body answers to which soul, and the precise mode of existence of each in union with the other.[7]

According to this logic, Bosch's visions of hell is entirely legitimate—both flutes and string instruments, serve the art of torture. But his vision of the original fluidity of the "palingenetic" potential of bodies, which, in his *Garden of Earthly Delights*, have not yet settled into solid categorial distinctions, rather than simply "metempsychotic" (Pythagorean) potential of "souls," reverberates with Meister Eckhart's call for the move toward "the poverty in spirit," or a return to the pre-lapsarian "indifferentiation" as the *sine-qua-non* condition of one's ultimate dissolution in transcendence, where a "fly" and—by extension—also an "elephant," become superfluous categories:

> Now we say that God, inasmuch as He is "God," is not the supreme goal of creatures, for the same lofty status is possessed by the least of creatures in God. And if it were the case that a fly had reason and could intellectually plumb the eternal abysm of God's being out of which it came, we would have to say that God with all that makes Him "God" would be unable to fulfil and satisfy that fly! Therefore let us pray to God that we may be free of God that we may gain the truth and enjoy it eternally, there where the highest angel, the fly, and the soul are equal [...].[8]

Yet, more importantly, it is Aristotle's vision of "besouled bodies," redeemed for Christianity by Aquinas, that affected early modern Europe more pronouncedly than its alternatives, and therefore it is also less than surprising that Meister Eckhart found himself in conflict with the authorities of the Church: an institution emphasizing the rational, the orderly, and above all, the hierarchical, in its teaching of spirituality. And despite the shock of the discovery of the dual American continent, and irrespective of the radical revisions of the dominant paradigms brought about by the Reformation, the long early modern epoch, spanning from the late 15th to the late 18th century, eventually, veers away from the visions of nature underlying Bosch's representation of the pre-lapsarian universe. Yet, the study of his *Garden* allows one to better comprehend why, at its most productive stages, the definition of health, based primarily on the humoral theory, had to extend beyond "the absence of disease."

At Bosch's time, the European philosophy of health—championed by such figures Hippocrates and Galen, who posited that good health could only be achieved through balance among the body's four humors—was build on the premise of the equilibrium: a state of overall harmony within the body and its external environment. Such a holistic view, integrating mental, spiritual, and physical well-being, would emphasize that health was also influenced by an individual's temperament, lifestyle, diet, or astrological factors. As such, it connects the early modern idea of health with theories evolved in ancient Greek philosophy—theories embracing the idea of universal connectivity, and rooted in the Epicurean Garden, the space where humans-in-nature thrive.[9]

Needless to say, as European Renaissance came to its close in the late 16th century, the onset of the Baroque marked a multidimensional cultural shift, which would powerfully affect the philosophical (and theosophical) visions of the human-and-the-universe relation. The most significant drivers of change were, of course, the Protestant Reformation, the ensuing Counter-Reformation, aimed to reaffirm the authority of the Roman Catholic Church, and the Thirty Years' War (1618–1648), which devastated much of Central Europe. Equally importantly, the twilight of Renaissance saw significant economic changes: on the one hand, colonial expansion resulted in the influx of wealth from the New World and energized the growth of new trade networks, and, on the other, the ensuing rise of capitalism would increase the overall level of affluence in Europe. In effect, toward the end of the 16th century power structures shifted toward absolute monarchies.

Unsurprisingly, art and architecture did not remain unaffected: the initial early modern focus on harmony, proportion, and balance in artistic interpretations of reality shifted to the emphasis on contrast, movement, and emotional intensity. Partly—in effect of the "generational" reaction to the restrained, and often idealized style of the Renaissance, and partly—in response to the demands of the Counter-Reformation politics, with the Roman Catholic Church sponsoring art capable of evoking religious fervor through awe. And yet, even such policies could not avert the inevitable transformations of the global intellectual landscape: the Scientific Revolution, championed by such scholars as Galileo and Newton, lastingly altered the perceptions of the universe, testifying to the need of certainty in the time of anxieties, tensions, and doubt.

Yet, in the context of our debate, one should primarily bear in mind that the religious and economic underpinnings of the Baroque would significantly affect the mechanisms of patronage: during the heyday of Renaissance, culture would primarily be sponsored by wealthy merchant families; in the Baroque, royalty and aristocracy became the primary patrons of the arts, which came to serve as vehicles for the expression royal authority and control, conveying the message of power and divine right (as epitomized by grandiose royal palaces, like Versailles or Peterhoff, masterpieces of architecture widely imitated by ambitious aristocrats). The same projection informed many other grand-scale projects—including the design of gardens.[10] Commissioning their gardens (symmetrical in layouts, featuring geometrically shaped hedges, elaborate water features, and statuary embodying classical themes), aristocratic owners would demonstrate their power over people and over land, at the same time—theatrically[11]—calling for the return to the "unadulterated" values of the antiquity, the rebuilding of the Arcadia, and embracing "nature."

Walls and *Ha-Has*: Man-Made Arcadias

Exploring the conceptual intricacies of the 17th- and 18th-century gardening in their monograph *Geometry, Winding Paths and the Mansions of Spirit*, David Jarrett, Tadeusz Rachwał, and Tadeusz Sławek offer insights that draw one's attention to the perseverance of egotism in human perspectivist reception, or subjective projection, of the epistemological relations binding the universe. Embedded in human constructions of the notion of "the natural," egotism goes hand in hand with the naïvely blatant pretense that it is not the case. The three scholars' analyses demonstrate that 17th- and 18th-century gardens, usually fenced or walled in, offer a very good point of departure for such argumentation. It seems trivially obvious: even if such protective (or even prohibitive) measures were neatly camouflaged as *ha-has*,[12] the garden remained closed to wild animals, trespassers, and individuals of "different tastes" alike, though (reluctantly) not to the labor:

> The landscape garden or park occupies the territory of [formerly open-access] locality thus becoming the locality of the rich who reorganize the space according to their own taste which they see as universal, though not accessible to those whose tastes are different. If [...] arduous travel is eradicated from the garden, it is simultaneously introduced as a hidden artery which, though invisible from the territory of the garden, connects the garden with the outside world, with the sphere of production for which there is no space in the garden itself. This artery, perhaps paradoxically, brings producers *out*, and products *into* one's estate. Labour, along with its deliverers, becomes thus and intrusive element within the landscape, a necessary evil of sorts which also spoils the landscape of one's life. [...] Though unwelcome, labour must thus remain within the space of visibility, at least as an object of some supervision, even of one's own work. The pleasure of one's walk in the garden itself testifies to one's fruitful labour of designing, for instance. Since idleness in its pure form is unwelcome to the eighteenth-century English eye, it is the endowment of labour with pleasure that also translates into walking onto a fruitful activity of enriching, or composing one's self in contact with "nature." Interestingly, one of the reasons for the superiority of the English constitution over the French, according to [Edmund] Burke, is some biological compulsion for relaxation within the English.[13]

Even if the fencing of such a garden were a *ha-ha*, the lack of obstruction to the "natural" vista does not contradict the exclusiveness of such a space: the 18th-century aristocratic garden, by its very assumption, produces "outcasts." And although an open view could suggest the philosophical need of the owner of the property (or of the landscape designer) to assert the limitlessness of the garden in recognition of the limitlessness of nature *sensu largo*, such a man-made vista remains, in fact, a tribute to the proprietor's ego. Even though, as Burke declares, such a garden caters to the "biological compulsion for relaxation," and allows its proprietor to become involved with the noble labor of "enriching, or composing one's self" in contact with "nature," its true healing capacity remains questionable. Walled-in and egocentric, a landscaped garden contains only carefully selected components; those unwelcome (whether animate or inanimate) are weeded out, largely in accordance with the Dominion Mandate, while contact with thus manufactured "nature," paradoxically, intensifies the detrimental consequences of the owner's egotistic self-projections. Therefore, irrespective of how many copies of Greek statues would be installed in such a space, such a garden must, inevitably, fall short of what the Epicurean Garden, the "garden-agora," requires.

These shortcomings become clear in Jarrett, Rachwał, and Sławek's analysis of the implications of the philosophical insights of William Shenstone, an English poet and one of the earliest practitioners of landscape gardening. Shenstone, they observe,

> quite explicitly states in his *Uncollected Thoughts on Gardening* (1764) that "landskip, or picturesque gardening […] consists in pleasing the imagination by scenes of grandeur, beauty, or variety." The objects selected by an "able gardiner" should not be very striking, but such as "serve to connect ideas, that convey reflexions of the pleasing kind."[14]

Furthermore, while analyzing the eroticism of Jean-Honoré Fragonard's famous painting *The Swing* (1766), the scholars assert that

> Since […] wealth is inevitably linked with economic display of property, the pleasure of the "Secret Satisfaction" (as Addison used to describe it) must have been allotted its special space, where it could be enjoyed after the glory of the perspectival vista and public openness with which wealth is connected. These spaces of the "secret Delight" (yet another phrase borrowed from Addison) are a permanent part of the garden design in the eighteenth century. Grottoes, kiosks, pagodas, gazebos, side-paths are all instruments of separation and seclusion which give the eye the liberty to look without being seen.[15]

The garden, if designed properly, i.e., in a fashion that would combine the allegedly all-inclusive openness and unobstructed perspective, "pleases the imagination" of the visitor with objects "serving to connect ideas" that not only "convey reflexions of the pleasing kind," but also allow the owner to communicate the message concerning their own status (wealth/power) without either brazenly asserting it by way of direct demonstration of any form or rhetorically encoding it in words. The enjoyment of one's own recognition in the eyes of others must be arrived at "in secret," almost voyeuristically: the (in)visible observer spying on the (un)aware visitor, who (un)willingly admires (un)concealed wealth, inspiring in them reflections of "the pleasing kind," himself or herself "delights" in the awe he or she sees in the eyes of the Other.

Moreover, in the space of our reflections, Fragonard's painting, whose cultural contexts Jarrett, Rachwał, and Sławek insightfully interpret, triggers yet another doubt. Its frivolous contents—the "theatrics" of courtship (the young gentleman attempting to catch the sight of the young lady's upskirt as she enjoys the pleasures of the swing, the lady's shoe flying off her foot) is complemented by the *mise-en-scene* of "the wild." The image, outwardly celebrating "nature," epitomizes the "theatricality" of the outdoors: "nature" is little more than a painterly decoration in the background of the merriment shared by the humans (dominating the scene) and the neoclassical sculptures, curiously observing the fun. Geometrically arranged, the "wild" becomes "tame"—a synecdoche of "nature":

Even more obvious in this respect are some of the most famous painting by Thomas Gainsborough, of which the portrait of Mr. Robert and Mrs. Frances Andrews (often described as the "triple portrait" of Mr. Andrews, his wife, and his land) and *The Morning Walk*—the portrait of Mr. and Mrs. William Hallett (1875)—stand out as particularly telling. While the "triple portrait," celebrating the landowner's dominance over the docile nature (arable land) *and* his spouse (rather uncomfortable, seated on a bench curiously installed at a vantage point affording a broad vista of Mr. Andrews's

Figure 2.5 Jean-Honoré Fragonard (1732–1806), *The Happy Accidents of the Swing* (*Les hasards heureux de l'escarpolette*), 1767–1768, oil on canvas, 56 cm × 46 cm. Reproduced by permission. ©Wallace Collection, London, UK/ Bridgeman Images.

property), requires little analysis, *The Morning Walk* could easily be interpreted as a picturesque attempt to combine the comfort of the painter's studio with the demands of the outdoors. The description offered by the National Gallery in London reads as follows:

> Gainsborough painted this portrait of William Hallett (1764–1842) and Elizabeth Stephen (1764–1833) shortly before their marriage on 30 July 1785. The couple are shown arm-in-arm on a morning walk with a Pomeranian sheepdog. Gainsborough has suggested the dog's thick soft coat with sparse, light brushstrokes which he also uses for Mrs Hallet's dress. [...] The couple may be wearing their wedding clothes. Elizabeth's gown is of fine ivory silk, caught at the waist with a black silk band. Her breast is covered with a frilled muslin kerchief with a knot of pale grape-green ribbon under it, which is repeated in the extravagant bow on her hat. The three downy ostrich plumes crowning her hat resemble the colour and texture of her powdered hair. The gauzy silk stole draped over her arms echoes the contrived carelessness of her hairstyle and dress. Gainsborough's slanting brushstrokes in the sky and foliage give the impression of movement, as though there is a light breeze. William Hallett wears a black silk velvet frock-suit, his hair is powdered and he carries a round black hat [...].[16]

To a critical observer, it is clear that the extravagance of the aspiring couple's attires clashes with the ominous clouds, gathering over the strolling figures. Most evidently, in such weather the morning walk would be far from pleasurable, and, even if attempted, it would certainly have required a rather different choice of clothes. But because nature, whose representation in Gainsboroughs painting already foresages the onset of the dominance of the Romantic aesthetics of the sublime, is nothing more than a theatrical backdrop, the picturesque "Walk" would not raise any eybrows among the members of the Hallet's millieu. Especially that many of their contemporaries would second William Gilpin's assertion that:

> Nature is always great in design; but is seldom so correct in composition. She is an admirable colorist; and can harmonize her tints with infinite variety, and inimitable beauty; but is seldom so correct in composition, as to produce a harmonious whole.[17]

Therefore, as Gilpin continues elsewhere, rather than expose yourself to the uncouth crudeness of nature, one should seek to improve it: "Turn the lawn into a piece of broken ground: plant rugged oaks instead of flowering shrubs [...] in a word, instead of making the whole *smooth*, make it *rough*, and you make it also picturesque."[18] Following the same principle, both the landscape gardeners and the painter create gardens of imagination, in which the walk is

Figure 2.6 Thomas Gainsborough, *Mr and Mrs William Hallett* (*The Morning Walk*), 1785. Oil on canvas, 236.2 × 179.1 cm. N-6209-00-000037. NG6209. Bought with a contribution from The Art Fund (Sir Robert Witt Fund), 1954. Reproduced by permission of the National Gallery, London, license number: 3798.

both pleasurable and safe—and, above all, utterly unnecessary as a physical exercise. Jarrett, Rachwał, and Sławek continue thus:

> Interestingly, walking in pre-enclosure landscape was something quite different from walking after the acceleration of enclosure, which also enabled the construction of more and more landscape gardens. In an open-field village, as John Barrel notices, the road system was "a labyrinth, whose secret cannot be learned without a guide, and which it is positively dangerous to enter without one." From the point of view of the locality, walking was not travelling, it actually "set boundaries and did not break them, it moved in its own circles and did not move through, it remained at home and did not travel." Walking, as it were, defined by locality, kept it in its place with the roads directed inside rather than outside. This rural home, which Goldsmith recollects as the ideal of natural life in his *The Deserted Village*, was not "adorn'd for pleasure" by a rich foreigner to the locality, but was constituted by local movements to labour, to church, to one's house.[19]

And so, safe within their home, Mr. and Ms. William Hallet take their morning walk along a picturesque path, an analogon of which would be described by William Mason over a century later. In his poem "To a Gravel Walk," Mason appears to act as the epigone of the neoclassical aesthetics when he declares:

> Here, as thou leads't my step through lawn or grove,
> Liberal though limited, restrained though free,
> Fearless of dew, or dirt, or dust, I rove,
> And own those comforts all derived from thee!
> Take then, smooth path, this tribute of my love,
> Thou emplem pure of legal liberty![20]

Heedless of the weather, "fearless of dew, or dirt, or dust," the couple embraces "nature" as construed by the intellectuals, who would build their understanding of the world upon the fundament of egotism rooted in the anthropocentric visions of the Universe, determined by the Newtonian-Descartian dualist metaphysics. The philosophy of landscape gardening, manifest in painting, is inextricably bound to the aesthetics of beauty: "comprehensible," neatly composed, within the bounds of imagination, avoiding aporias. Yet, such "ethico-horto-aesthetics,"[21] albeit nominally it should lead one toward the recognition of powers beyond human control (however they are conceived of), becomes a vehicle for the communication of a particular *episteme* and a source of a particular language, which has particular consequences:

> This double point of ethico-horto-aesthetic is subsumed under a larger argument on the strength of which man's aesthetic activities are related to his/her cognitive and perceptual abilities. Shenstone's garden is a place where man learns a lesson concerning his/her station in the universe; one is tempted to say that he garden is for Shenstone what the lands of Lilliput and Brobdingnag were for Gulliver. A walk in the garden can open immesurable distance from which the human body and mind appear very different from what they used to look like indoors. The problem is one of time and the impossibility of immediate perception. Shenstone complains that man cannot intuit the wholeness of nature "at one survey" and, implicitly, that the inevitable sequentiality and fragmentariness of perception brings about the drama of time. There would be no semiotics, no science of signs, were man able to "comprehend the universe at one survey." In such a situation, man could impose his/her *techne* "upon" nature because both man and nature would exist somehow outside the (to use Bergson's terminology) quantitative time, outside the "vulgar sense of time" Heidegger meditates upon in his *Being and Time*. The necessary limits, aporias of human art and thought originate in man's temporality which pushes

> reflection and action from the position of "upon" to that of "beneath" or "inside." [...] The garden for Shenstone is somehow pre-positioned, i.e. it exists in a temporal dimension different from the human temporality and therefore man's art must remain concealed, must "take its glory" (Shenstone's words) to hide its operations. The garden's positioning and topography have been already laid out before the place received the name of a "garden."[22]

The assumption of the impossibility of the *immediate* perception, which Romanticism will dismiss with, triggers the imposition of *techne* upon nature, or the violence of which we wrote earlier, and which goes hand in hand with another consequence of the dominance of the *episteme* born out of the dualist paradigm (legitimized by the authority of science of the time): the superficial decentralization of man. In the post-Cartesian *episteme*, the human-hiding-in nature is, in fact, an act to conceal the abhorrence of the lack of control over nature itself: nature which "likes to hide." It is, therefore, possible to risk the statement that, along with the Jansenist revision of the concept of *Deus Absconditus*, *Natura* also became *abscondita*. Although the Age-of-Reason explosion of natural sciences would suggest otherwise, the renewed recognition of the actual *presence* of nature seems to have required the breakthrough of the Romantic turn and a transgression of the principles of the horto-ethico-aesthetics rooted in the Enlightenment into the Transcendentalist, or existentialist,[23] aesthetics of the sublime. If the former evolved from the traditions of Western idealism and mysticism, the latter, quite possibly, was born out of the recognition of the value of direct, unmediated, experience, which—albeit escaping categorization—was acknowledged as instrumentally affecting our comprehension of our position in the universe, and thereby, inevitably, our functioning within our environment, our understanding of our bodies, and the revival of our interest in the holistic idea of what it means to be healthy.

Toward the Transgression of Duality

Apparently, in the history of the nature-culture binary, the paradigm in which the "natural" would be opposed to the "unnatural" would first yield to one juxtaposing "natural" and "artificial" (or, better still, "artful"), and then to a model in which nature and culture would collapse into one. The above notwithstanding, because in the Western context the trajectory of the post-lapsarian ideas of "tending the garden" may be construed as an ascending spiral, our visions of the universe oscillate between the very non-duality that characterized Bosch's (transgressive) vision of the universe (smuggled into his *Garden of Earthly Delights*) and the duality-based metaphysics, informing so much of the intellectual history of the Occident. Between caesuras in the history of culture, dominant narratives of gardening seem to have been revolving from strictly anthropocentric concepts rooted in the biblical

Dominion Mandate[24] (manifest, for instance, in landscape gardening or industrial cultivation) to those interpreting the latter in terms of stewardship[25] (visible, for instance, in the Romantic era development of legal regulations serving to protect "natural" areas), to those motivated by the urge for the return to the pre-lapsarian unity with the Divine (Transcendentalist and mystical traditions), to more recent ecosophical perceptions of the world (ranging from deterministic insights historically evolving from Darwinism, Freudianism, or Marxism, to poststructural, posthumanist, transhumanist discourses of race, class, and gender, deep ecology, and animal studies), to Indigenous methodologies or Eastern thought (particularly various form of Buddhism), only to return to the ever-revised concepts of the human domination, affecting the commonplace, androcentric understanding of the garden, determined by old binarities (ownerless/owned, open/closed, wild/tame, weed/plant).

Still, our overview of the selected canonical works of art seems to demonstrate that despite the cultural strength of the binarity-based paradigms, the subcutaneous sense of the essential unity perseveres: the Dutch, English, and French landscape paintings, as much as the Hudson River School legacy, seem to testify to the generation's heightened sensitivity to the complexity of the phenomenon of nature, whose contemplation—emphasizing its sublimity—might be interpreted as a form of a "spiritual exercise." Even in paintings embracing biblical themes, the shift from the centrality of the human toward the centrality of nature *sensu largo* is visible. An interesting illustration of the above tendency is Thomas Cole's *Expulsion from the Garden of Eden*, painted in 1828, which, although upholding essential binary opposition underlying the judeochristian metanarrative, departs very far from Hans Memling's vision of heaven and hell:

In Cole's painting, the central position of the symmetrically organized image is occupied by a portal, envisioned as a natural arch (as opposed to Memling's gothic gateway), to the right of which extends a pristine landscape of lush green—fertile and flourishing. To the left, in the foreground (suggestive of the viewer's vantage point), the landscape—likewise natural—is rocky and barren. The lifeless tree trunks, the ill-boding clouds, dangerous predators, and the impenetrable abyss separating the space of thriving from the space into which humankind is exiled, are all semi-concealed in a shadow; the "outside" of the vibrant "garden of Eden" is overwhelmingly dominated by the monochrome, implying the lack of distinction, and ultimatelty—death. Importantly, the founder of the Hudson River School relegates both God and the humans into secondary positions: the divine is reduced to the supernatural flare-up in the gateway; the anguished humans, insignificant, are almost indistinguishable from the dreary landscape. Departing from the androcentrism of the "traditional" representations of the Fall, the painting prompts the contemplating viewer to expand the "canonically" biblical reading of the scene and to contemplate the "original sin" of the modern era: having succeeded in subduing the Earth, humankind, paradoxically, ends up in the

Figure 2.7 Thomas Cole, American (born in England), 1801–1848 *Expulsion from the Garden of Eden*, 1828. Oil on canvas, 100.96 × 138.43 cm (39 3/4 × 54 1/2 in.) Museum of Fine Arts, Boston. Gift of Martha C. Karolik for the M. and M. Karolik Collection of American Paintings, 1815–1865, 47.1188. Reproduced by permission. Photograph © 2024, Museum of Fine Arts, Boston. Media License #24-50456-CC.

space they themselves rendered barren. Having ousted nature to the margins of their weltanschauung, with the industrial revolution considerably contributing to the transformation of the formerly agrarian, nature-dependent world into a more or less well-oiled, capitalist, machine, humankind finds out that, cut off from the rhythms of nature, unable to recognize the arbitrariness of the Mandate, it cannot thrive. To flourish, Cole seems to nostalgically suggest, the human must be one-with-the-garden. This idea seems to pervade most of the Romantic reflection, and culminates in the emergence of Transcendentalist (idealist) thought, but is also visible, for instance, in the philosophy underlying some of the most famous paintings of Joseph Mallord William Turner, whose brushstrokes cancel out the distinction between the human and the non-human, the man-made and the natural, the wild and the tame:

Turner's *Rain, Steam and Speed* seems to illustrate the latter point rather well. The "man-made" and the "natural" blend together: the bridge, spanning the two banks of the river which a lonely boat struggles to cross, the train (a fascinating human invention that forever changed the landscape, social relations, and the sense of time), the hare on the tracks, all enwrapped

Figure 2.8 Joseph Mallord William Turner, details from: *Rain, Steam, and Speed — The Great Western Railway.* 1844. Oil on canvas, 91 × 121.8 cm. N-0538-00-000030. NG538. Turner Bequest, 1856. Reproduced by permission of the National Gallery, London, license number: 3799.

in the golden haze, lose their outlines. The painter's technique, contributing chiefly to the atmospheric tonality of his expression, draws one's attention to the shift of emphasis from the way the artist "speaks" and from how the painting makes one feel to the *affects* that the setting itself—in this case, the landscape of blunted distinctions—evokes in the viewer. In fact, the "atmospheric" dimension of Romanticism, defining much of the art and writing of the period, might be considered to be precisely one of the leitmotifs of culture that revolve in an ascending spiral, periodically shifting from the (productive) center to the (hibernating) margin of the cultural polysystem, only to reappear in a new form when the productivity of the dominant epistemic formula is exhausted.

The modern era saw such an oscillation of philosophical positions defining ethico-horto-aesthetics between paradigms based on unity and those emphasizing clear-cut binarity, between the sense of "pre-lapsarian" oneness and "post-lapsarian" divisions, between wholeness and fragmentation, between beauty and the sublime, and between the perceptions of nature as opposed to culture, and those construing culture as an inextricable part of nature. It is, therefore, hardly surprising that the twilight of the Age of Reason witnessed a repositioning of accents: the emphasis on the cultivation of the "garden"—designed, arranged, requiring labor—shifted to the protection of spots of natural beauty: "national parks" and "natural reserves." And even though human agency remains important (after all, the choice of non-intervening with the natural processes remains, in fact, an intervention), the emphasis on the value of nature understood in terms of the sublime, gave rise to new formulas of outdoor recreation: the "biological compulsion for relaxation," once motivating landscape gardening, came to be realized on mountain trails. Mountaineering, hiking, or sport sailing quickly become fashionable on both sides of the Atlantic, and organizations promoting outdoor activities in nature gain on importance. Romanticism, distrustful of the supremacy of reason and cautious of the power of language, exploring the liminality of non-mediated experience, and embracing aporias, veers away from the episteme of the Enlightenment, returning to the idea of essential oneness of being in a renewed incarnation: either as non-distinctive, absurd, and unmotivated essence, or as inseparable from existence, perceived outside of language, devoid of distinctions, non-verbal, non-categorial, and transparent.

This epistemic continuity—and transcultural interconnectedness—of the ever-present metaphysical insight that transgresses duality is illustrated by a conversation between Ireneusz Kania, a stellar Polish translator, and Czesław Miłosz, Nobel Prize-winning poet, who delve into Emanuel Swedenborg's 18th-century vision of the universe in the preface to the Polish edition of Signe Toksvig's biography of the Swedish philosopher.[26] The relevant passages of their dialogue, quoted *in extenso*, could be rendered in English thus:

Czesław Miłosz: To discuss Swedenborg, one must overcome certain taboos. When Swedenborg's name is mentioned, people involved in literature frown; to them Swedenborg represents some form of mysticism, things contrary to reason; they do not understand the significance of the connection the philosopher has had with the history of literature over the past two hundred years. Swedenborg exerted an extraordinarily strong impact on the literatures of various countries, even though writers often failed to mention his name or credit the sources of their borrowings. It is well known that, in a sense, William Blake was a disciple of Swedenborg in terms of his symbolic ideas. And of course, the entire symbolism of the 19th century has strong ties with Swedenborg, most notably Baudelaire's famous sonnet "Correspondences." Moreover, there is this relatively recent discovery confirming Swedenborg's influence on Walt Whitman.

Ireneusz Kania: Yes, and besides Whitman, of course, the entire 19th-century American literary milieu under Swedenborg's influence, including figures such as Hawthorne, Melville, and the entire circle centered around Emerson. In Whitman, one can clearly see references to the Swedenborgian "new man," liberated from the fear of posthumous punishment and the sense of guilt, which Swedenborg rejected. On might say that certain motifs—Swedenborg's unique anthropological optimism and, globally, his anthropocentrism—formed the foundation of Whitman's entire ideology—even the myth of the "American man" and the age of Newness. The most tangible influence of Swedenborg's work is evident in the very title of Whitman's main work—*Leaves of Grass*. Scholars have wondered where this strange title came from, as we should expect something like *Blades of Grass*, which would sound "more natural" in English. However, the title has a profound meaning, recently unearthed with reference to two dictionaries by American researchers, Hindmarsh (1794) and Nicholson (1800), who compiled indices of symbolic terminology in Swedenborg's major theological and exegetical works. In these dictionaries, we read, among others, that in Swedenborg "grass" means "what is alive in man," and "leaves" mean "truths." Thus, the title of Whitman's poem would essentially mean "Truths of What Is Alive in Man." It is, essentially, a coded Swedenborg. This goes to prove just how deeply Swedenborg's influence penetrated even the linguistic fabric of his poetry. And in the poem itself, we read:

A vast similitude interlocks all,
All spheres, grown, ungrown, small, large, suns, moons, planets,
[All distances of place however wide,
All distances of time, all inanimate forms,]
All souls, all living bodies though they be ever so different, or in different worlds,
[All gaseous, watery, vegetable, mineral processes, the fishes,
the brutes,
All nations, colors, barbarisms, civilizations, languages,
All identities that have existed or may exist on this globe, or any globe,
All lives and deaths, all of the past, present, future,]
This vast similitude spans them, and always has spann'd,
And shall forever span them and compactly hold and enclose them.
[...]
To me the converging objects of the universe perpetually flow,
All are written to me, and I must get what the writing means.

Here emerges the Swedenborgian concept of "correspondences" and the idea that the world, the entire cosmos, is essentially a language—the language of God—a collection of signs through which God reveals himself outwardly and which man must read if he wants to restore his condition before the Fall, before his severance from God.

Miłosz: You know, reading Swedenborg, I thought he was one of the happiest people in history.

Kania: He described himself as such.

Miłosz: ... Because he had his vision of another world, so for him death was like stepping from one room to another.

Kania: Essentially—awakening, an awakening to normal life.

Miłosz: Yes; he was a man entirely free from the fear of death. In a sense, we find the same thing in Whitman! Whitman was not afraid of death. So perhaps some echo of Swedenborg could be noted here.

Kania: I think so; this Swedenborgian optimism, this acceptance of the world, maybe even a certain deafness to evil—because, in Swedenborg, the problem of evil was not at all sharply formulated. In his reflection, one cannot see the terrible rift that we observe in many other thinkers—or writers. According to Swedenborg, evil originates in man, resulting from the separation of man from God, and it can be overcome essentially through reintegration with God.

Miłosz: I was astonished to see that Swedenborg, the naturalist, who was so involved with nature, had no sense of suffering in nature. None.

Kania: Yes. But this is probably related to the fact that—paradoxically, because he was a scientist, a naturalist, who knew and described nature very thoroughly—Swedenborg did not seem to have held a great sentiment for nature.

Miłosz: He did not, because in his view, nature was primarily a "garden of symbols."

Kania: Yes, primarily. Another matter is that he was an avid gardener, planting flowers and fruit trees. So perhaps in some way, he did like nature. But he did not see in it the rift, darkness, fracture, horror.

Miłosz: No, he did not see any of it. He did not see suffering in nature. [...]

Kania: [But] I wanted to [ask about] your personal connections to Swedenborg. I was intrigued that your latest volume of poetry is called *Second Space*. This title comes from *The Land of Ulro*, from the part where you talk about Blake. After twenty-five years (because *The Land of Ulro* was published in 1977) the idea somehow returns to you. Evidently, you still find Swedenborg intriguing [...].

Miłosz: [...] I do not declare myself as a believer in Swedenborg's spiritual space; on the contrary, I believe that we are in a situation characterized by the fact that we cannot imagine any spiritual space.

Kania: Agreed, but is this evidence that it does not exist, or does it only indicate our further fall? The fact that we are losing certain abilities?

Miłosz: As you rightly pointed out (on another occasion), this "second space" of Swedenborg is roughly the same as the Tibetan Buddhist *bar-do*.

Kania: "Intermediate space."

Miłosz: Roughly, because in Tibetan Buddhism, it is an intermediate space one enters before the next incarnation.

Kania: Or before nirvana, because there are two paths.

Miłosz: Yes. In Swedenborg, it looks different, but in any case, it is a space created by the mind.

Kania: That is, this space in Swedenborg, analogous to *bar-do*, is the so-called spiritual world—a lower, intermediate sphere preceding the heavenly worlds, which people may enter only after some preparation...

Miłosz: Or fall into hell!

Kania: Yes—or fall! In this sense, it is the exact analogue of the Buddhist *bar-do*.

Miłosz: I have encountered these ideas reading Daisetsu Teitaro Suzuki, who called Swedenborg the "Buddha of the West."

Kania: Ha! Balzac called him the "Buddha of the North"! Suzuki translated Swedenborg—you know that, right?—and widely promoted him in Japan [...] because indeed, the analogies with Buddhism are very clear here. However, regarding this space, the following question emerges: should we understand this space as something absolutely subjective, or does it have some other status? After all, Swedenborg brought back specific, precise information from his excursions to other worlds. [...] [And his theory of *proprium*, or the human *ego*, is also very important; he] understood it as the Hindus do, [...] as something, in a certain sense, unreal, illusory.

Miłosz: And very harmful.

Kania: Yes, but on the other hand, also necessary for humans, because only thanks to the *proprium*—as he writes somewhere in *Arcana Coelestia*—normally being an illusion for man, something that misleads him (like the Indian maya), can one, upon the realization of its illusory nature, accept as his own what God wants to give him. And God wants to give him the world, to make what He created available to him, because it is all for man. And man recognizes the world in its true, somewhat unreal nature only when he uses the *proprium*. In other words, Swedenborg, acknowledging the illusory nature of the human ego, simultaneously assigns it an important operational role. This ego is simply necessary for us to contact the empirical world, as it is its correlate. [...] The point is that Swedenborg wanted to build a homogeneous, total vision of the Universe. Essentially, Einstein referred to the same motif, trying, for many years, to construct the so-called unified field theory. He did not succeed. And it has not been attained to this day. But attempts are continuously being made, and maybe they will eventually lead to a new, homogeneous vision of the world, which Swedenborg aimed for.[27]

Blake and American Transcendentalists, Melville, Hawthorne, Baudelaire—the most influential writers and thinkers of Romanticism—remaining "in conversation" with Emanuel Swedenborg, the "transgressor" of the rationalist-empiricist-mechanicist-utilitarian paradigm of the 18th century, develop their distinctive positions while seeking answers to questions that the proponents of the principal philosophical theories of the Age of Reason failed to successfully tackle. Importantly, however, Swedenborg's search for a holistic theory of the Universe seems unique inasmuch as it departed from the dominant trends of the time; amplified by the onset of the Romantic era, the return of the idea of the "whole" to the center of the cultural poly-system should be perceived as representative of a broader phenomenon. Although the interlocutors—Miłosz and Kania—do not suggest it explicitly, their conversation, expanding to embrace notions of different cultural provenance, seems to indicate more than the ebbs and flows of "akin intutions" within the culture of the West: at the level of the fundamental human experience, the sense of the whole appears to be omnipresent across cultures. After all, the 19th-century crisis of faith in the West brought about attempts to seek alternatives in the cultures of the East: Emerson's insights into the *Bhagavad Gita*, Melville's exploration of Zoroastrianism, Hinduism, and, eventually Buddhism (via Schopenhauer), Longfellow's return to Native American traditions—all seem to testify to the recognition of the key importance of the immediate—or, linguistically un-mediated—experience: the experience that the paradigms of the Enlightenment failed to grasp.

Beyond doubt, the resurgence of some of the intuitions that drove Swedenborg's reflection may be considered to characterize the "Zeitgeist" of Western Romanticism, conducive to the rethinking of the relationship between the human and the Universe beyond distinctions, which reflection, to a lesser or greater extent, seems to color the literatures and arts eversince. In particular, the "Transcendentalist-idealist-mystical" and the "existentialist" veins seem to be especially pronounced in the late modern and then post-modern (post-World War II) reflection, but despite their "experiential" common denominator, the fundamental difference between them finds an interesting illustration in the correspondence between Herman Melville and Nathaniel Hawthorne, in which the author of *Moby-Dick*, to whose reflections on the pantheist "unity" we return in subsequent chapters of this book, assumes the role of a tongue-in-cheek critic of serene, idealist, positions of some of his contemporaries:

> In reading some of Goethe's sayings, so worshipped by his votaries, I came across this, "Live in the all." That is to say, your separate identity is but a wretched one,—good; but get out of yourself, spread and expand yourself, and bring to yourself the tinglings of life that are felt in the flowers and the woods, that are felt in the planets Saturn and Venus, and the Fixed Stars. What nonsense! Here is a fellow with a raging toothache. "My dear boy," Goethe says to him, "you are sorely afflicted with that tooth; but you must live in the all, and then you will

be happy!" As with all great genius, there is an immense deal of flummery in Goethe, and in proportion to my own contact with him, a monstrous deal of it in me.

H. Melville.

P.S. "Amen!" saith Hawthorne.

N.B. This "all" feeling, though, there is some truth in. You must often have felt it, lying on the grass on a warm summer's day. Your legs seem to send out shoots into the earth. Your hair feels like leaves upon your head. This is the all feeling. But what plays the mischief with the truth is that men will insist upon the universal application of a temporary feeling or opinion.[28]

To Melville, whose "quarrel with God"[29] propelled his lifelong philosophical struggle to find a framework of ethical reference that would not require theodicy, the sense of unity that he describes in the letter above using language largely reminiscent of that of Whitman's, leads, in fact, to a reflection polarly different from that central to Emerson's thought: Melville's understanding of "unity" seems to stress the "indifferentiation of essence," which is visible not only in his sea-locked novels (e.g., Wellingborough Redburn's experience of helplessness while witnessing the slow death of the destitute family in Liverpool, the description of White Jacket's fall from the yardarm, Pip's "madness," or the impenetrable whiteness of the whale), but also—and perhaps particularly so—in his later poetry (e.g., *Rammon*, "Buddha," or "Pebbles"), where the pantheistic "All" of Goethe's or Emerson's "Oversoul"—both concepts equivalent to "nature"—find a counterproposition in the non-theistic rhetoric of "liquidity," "flow," "annulment," "annihilation"—denoting a terrifying dissolution in absurd, non-divine, homogenous essence, that offers no other "harmony" than ineffability. As Hawthorne observes in his journal,

> Melville, as he always does, began to reason of Providence and futurity, and of everything that lies beyond human ken, and informed me that he had "pretty much made up his mind to be annihilated"; but still he does not seem to rest in that anticipation; and, I think, will never rest until he gets hold of a definite belief. It is strange how he persists—and has persisted ever since I knew him, and probably long before—in wondering to-and-fro over these deserts, as dismal and monotonous as the sand hills amid which we were sitting. He can neither believe, nor be comfortable in his unbelief; and he is too honest and courageous not to try to do one or the other. If he were a religious man, he would be one of the most truly religious and reverential; he has a very high and noble nature, and better worth immortality than most of us.[30]

Although, toward the end of his life the writer was eventually able to come to terms with the "liquid" universe (deifying the "inhuman sea," epitomizing the universal fluidity of essence),[31] he has remained critical of the

Transcendentalist interpretation of the experience of the "dissolution of self" in the "All/Oversoul": what Emerson's experience of transcendence renders uniformly transparent, Melville's experience of the "terror of the terrible" seems to render uniformly opaque; what to Emerson is "divine," to Melville is "inhuman," what to Emerson is mystical, to Melville is absurd and thus terrifying. Even though, as William Braswell demonstrates, Melville himself realized rather well that he might not have been too profoundly acquainted with the philosophical tenets of Transcendentalism,[32] he would strongly condemn Emerson's "reason" as deficient nonetheless. Reading *The Poet*, on the margins of his copy, Melville—who, in his already quoted letter to Hawthorne, declared: "I stand for the heart. To the dogs with the head! I had rather be a fool with a heart, than Jupiter Olympus with his head"—annotated the New England philosopher's observations on language as "fossil poetry" thus:

> This is admirable, as many other thoughts of Mr Emerson's are. His gross and astonishing errors & illusions spring from a self-conceit so intensely intellectual and calm that at first one hesitates to call it by its right name. Another species of Mr Emerson's errors, or rather blindness, proceeds from a defect in the region of the heart.[33]

As Braswell demonstrates, Melville's "anti-rational" criticism of Transcendentalism (and its sources) took an even more radical form in his novel *Pierre*:

> Certain philosophers have time and again pretended to have found it; but if they do not in the end discover their own delusion, other people soon discover it for themselves, and so those philosophers and their vain philosophy are let glide away into practical oblivion. Plato, and Spinoza, and Goethe and many more belong to this guild of self-impostors, with a preposterous rabble of Muggletonian Scots and Yankees, whose vile brogue still the more bestreaks the stripedness of their Greek and German Neoplatonical originals.[34]

No doubt, Melville, although his intuitions as a writer lead to eye-opening emulations of unmediated experience, is not a trained philosopher; his insights would, possibly justifiably, raise numerous eyebrows in the profession. And yet, if one seeks to demonstrate the persistence of the "grassroots" need to come to terms with the experience of "non-distinction" that sooner or later, more or less memorably, becomes everyone's share—the observation of which propels the search for a philosophy that would transcend the binary models dominant in Western metaphysics—Melville emerges as an "antitheoretical" doubter in "things systemic," including the transcendental order. The gaps in his academic education notwithstanding, his floating Harvard and Yale allowed the writer (who begins his career as a Romantic, and, arguably, ends it embracing elements of Naturalism) to create a body of work whose focus on the human condition came to be recognized as an indispensable part of the Western canon. Admittedly, the recognition only came after the calamity of

World War I: the phenomenon of the Melville Renaissance, arguably, was the function of the fact that the post-war readership came to the recognition of questions concerning the viability of the possible sources of ethics in the face of the utmost exhaustion of the pre-war values and the overwhelming sense of uncertainty. Unsurprisingly, along with the existentialist philosophies of participation of the previous century, also Emersonian/Swedenborgian thought returns to the cultural polysystem's productive center, coloring, among others, the naturalist reflection of modernist writers, poets, and artists—both in America and in Europe.[35] And both these discourses have returned to stay.

At Home with Nature

The trauma of the Great War brought about the intensification of determinist reflection, sprouting out, on the one hand, from the mid-19th-century Darwinist, Spencerian, and Marxist thought, and inspired by the turn-of-the-century Freudian/Jungian insights on the other. Directionless and "immobilized," the World War I generation, suspended between the disappointment with the failed promise of modernity and the dubiousness of the ethics legitimizing "the survival of the fittest" and other basic instincts as the sole principles explaining the nature of *conditio humana*, is "lost" in more than one sense of the word. Yet, it is the global experience of the World War II, with its industrial-scale production of death, its mortally efficient technology of destruction, and its institutionalization of social Darwinism in the name of *Lebensraum* and purity of race, along with the nuclear annihilation of Hiroshima and Nagasaki, that ultimately extinguished the last glowing embers of modernist optimism. Complementing the reflection of the academic existentialists of the 20th century, seeking to make sense of the tragedy beyond human ken, Kurt Vonnegut Jr. aptly captures the horror of the perverted war-time reality, turning a slaughterhouse in Dresden "inside-out" in his eponymous novel. Paradoxically, it is within the abhorrent building, in which animals would be put to death, that accidental people accidentally survive, while the whole world outside of it, in this inverted perspective, is exposed as a universal "slaughterhouse," ruled by no ethical principles other than that of chance, which further reinforces the sense of the loss and exacerbates the nauseating absurdity of pointless human existence.

Three decades later, an Icelandic artist, Hreinn Friðfinnsson, employs a similar concept to draw his viewers' attention to the only ethics, the practice of which could potentially avert the otherwise inevitable consequences of the escalation of greed-driven violence of the Cold War period. Yet, rather than resorting to (post)apocalyptic visions, which post-modern writers and artists would often present to the public, the sculptor promotes the atmosphere of friendship by reference to the intimate space of home—a space, whose importance, intuitively sensed by most if not all, requires no justification. The official website of the Moderna Museet in Stockholm offers the following description of Friðfinnsson 1974 *Husproject*:

The Origin

In the book *Íslenskur aðall* (Icelandic Aristocracy) by the author Þórbergur Þórðarson, which was first published in 1938, there is a little story of a man called Solon Gudmundsson—an "aristocrat." He lived in a fishing village in the North West of Iceland where he had a small house. He was a very hospitable man and his house served as a private hotel for many other "aristocrats" who did not have a fixed address or position in society, but chose freedom and mobility to be more fit for the study and enjoyment of life. Solon Gudmundsson was a man of many talents. Early in life he became familiar—and efficient—with most jobs on the sea and on land. For example, he was a good carpenter. He composed a very special and personal kind of poetry which he himself called "light jokes." He also made several interesting objects of no practical purpose except to help develop his and other people's consciousness. It is likely that he was considered mad by most people, but his madness was tolerated because he did not bother anybody, could look after himself and even help others, and because he and his acts were a welcome source of conversation in the village. When Solon Gudmundsson was already an old man, he sold his house and started building a new one, using mostly wood and corrugated iron, a widely used building material in Iceland.

The following is a shortened and freely translated extract from the book *Íslenskur aðall*:

He started building the house in the following manner: First he made a wooden construction in an ordinary way, then fixed corrugated iron sheets on the inside of the wooden frame and started living in the building at this stage. The building process was to continue depending on the economic situation and other circumstantial conditions. But Solon wanted to build this house completely in the reverse order to traditional architecture, that is to say putting the corrugated iron on the inside and finishing with wallpaper on the outside. When Solon was asked why he planned to build the house in this way, he answered with a faint smile: "Wallpaper is to please the eye, love, so it is reasonable to have it on the outside where more people can enjoy it." But Solon did not get very far with his project because some very concerned friends managed, with much difficulty, to persuade him to retire to an old people's home. There he was well cared for until he died on the 18th of October, 1931.

The House

In the summer of 1974, a small house was built in the same fashion as Solon Gudmundsson intended to do about half a century ago, that is to say an inside-out house. It was completed on the 21st of July. It is situated in an unpopulated area of Iceland, in a place from which no other man-made objects can be seen.

The existence of this house means that "outside" has shrunk to the size of a closed space formed by the walls and the roof of the house. The rest has become "inside."

This house harbours the whole world except itself.[36]

Solon Gudmundsson's "mad dream," when realized as artwork, becomes a poignant ethical statement. With the "outside" limited to the miniscule space demarcated by the walls and the roof, the universe—including all of its animate and inanimate components—is a natural part of the welcoming, warm, household. In it, every human is a kin, every plant is a houseplant, every animal is a family member, all the seas, rivers, mountains, fields, and forests form the household patio, all the celestial bodies provide light, and all the technology are household appurtnances. Such an extended house does not allow for expulsions: there are no thresholds to transgress, nor is there space, to which transgressors could be exiled. Thus, the house inside-out manifests itself as home to the undivided, albeit complex, universe, which—as every element of one's household—deserves compassionate, loving care.

Indeed, Thoreau was right when he wrote that

> [i]t is something to be able to paint a particular picture, or to carve a statue, and so to make a few objects beautiful; but it is far more glorious to carve and paint the very atmosphere and medium through which we look, which morally we can do. To affect the quality of the day, that is the highest of arts.[37]

Figure 2.9 Hreinn Friðfinnsson, *Husprojekt*, 1974. Purchase 1977. MOMB 26. From the archive of Moderna Museet/Stockholm.

Gudmundsson did, and, naïve as such a concept might sound at first, his gesture, through Friðfinnsson's metaphor, taps into the atmospheric sense of essential togetherness—which seems to gain substance when, for instance, one recalls our recent, shared, experience of the COVID-19. When our homes turned into our prisons, our instinctive urge to break through their prohibitive walls and roam the plain air became painfully evident. Reverberating in our memories with particular might, the stifling restrictions of the pandemic motivate revisions: to be healthy, we cannot be contained; yet, in order to be uncontained, we cannot "contain" ourselves within the limits of our body walls. Transgressing the binary opposition between the home and its outside by turning it inside-out, we extend ourselves to include in our self-definition the air we breathe, water we drink, rocks we walk on, fire that keeps us warm, and nutrients we need to live, and social relations within our household. With such a revision, we become extended humans, inhabitants of the extended home, living the "crazy" dream of Solon Gudmundsson, a simple man nurtured by the rhythms of demanding Scandinavian nature, whose intuitive understanding for the need of compassion required no fundament in "armchair" philosophy.

Such "grassroots philosophies," practiced among many local communities world-wide, foster the balanced relationship between the elements and the human—as well as non-human—subjectivities. Simultaneously, our need to know-for-sure notwithstanding, our tendency to seek for truths beyond sharp categories—whether through direct experience, contemplation, or deconstructive re-readings of the central metanarratives of culture—seems to never quite wane. Even when the dominant philosophical paradigms enforcing perceptions of the human-in-the-world through the binary (or dialectical) lens are at the heights of their productivity, literature, arts, and some elements of geographically conditioned cultural practice demonstrate that the "centrifugal," "transgressive," reflection is never dormant. This allows one to hypothesize that the non-binary, holistic approaches to the universe (whether quotidian, naïve, or methodologically deepened)—older than the "systems," and apparently engulfed in the "mythical" time—energize one's intuitive longing for the wholesome sensation of one's "legs sending out shoots into the earth" and one's "hair feeling like leaves upon one's head." A sensation of which requires no rationalization whatsoever, but which nonetheless makes us feel united with Gudmundssonian household, where everything is a part of the lovingly tended-to nature, and where every atom that belongs to me belongs to the garden.

Notes

1 J.R.R. Tolkien, *The Fellowship of the Ring, Being the First Part of the Lord of the Rings* (Harper Collins e-books), 83. j-r-r-tolkien-lord-of-the-rings-01-the-fellowship-of-the-ring-retail-pdf (accessed: August 28, 2023).

2 See the interactive guide on the triptych at: http://sadostateczny.mng.gda.pl/en (accessed: July 31, 2024).

3 Plato, *Phaedrus*, trans. Robin Waterfield (Oxford: Oxford University Press, 2003), 6 [246e].

4 See, for instance: https://en.wikipedia.org/wiki/The_Last_Judgment_%28Bosch,_Bruges%29#/media/File:Bosch_laatste_oordeel_drieluik.jpg (accessed: August 28, 2023).
5 Isaiah 65:25 (King James Version).
6 Aristotle, *De Anima*, Book One, Chapter Three https://psychclassics.yorku.ca/Aristotle/De-anima/de-anima1.htm (accessed: May 6, 2024).
7 See paragraph 131 in "Lectio Eight" in Thomas Aquinas, *Commentary on Aristotle's De Anima*, translated by Kenelm Foster, O. P. and Sylvester Humphries, O. P. (New Haven: Yale University Press, 1951), html edition by Joseph Kenny, O. P., https://isidore.co/aquinas/english/DeAnima.htm (accessed: May 6, 2024).
8 Meister Eckhart, "Sermon Eighty-Seven: *Beati pauperes spiritu, quia ipsorum est regnum coelorum*," in: *The Complete Mystical Works of Meister Eckhart*, translated and edited by Maurice O'C. Walshe. Revised with a Foreword by Bernard McGinn (New York: The Crossroad Publishing Company, 1979), 422.
9 Apart from Hippocrates, often referred to as the "Father of Medicine," and Galen, whose writings further developed Hippocratic humoral theory, the most important philosophers of health included Avicenna—whose *Canon of Medicine* was a mandatory reading in European universities, Paracelsus—who introduced chemical remedies to medicine, thereby challenging traditional Galenic methods and paving the path for modern pharmacology, as well as Thomas Sydenham, "English Hippocrates," who fostered observational approach to medicine, advocating for case-oriented treatments, attentive to individual needs of the patient.
10 See, for instance: Clemens Steenbergen, Wouter Reh, *Architecture and Landscape. The Design Experiment of the Great European Gardens and Landscapes* (Basel–Berlin–Boston: Birkhäuser Publishers, 2003).
11 See, for instance: Abbie Elizabeth Rufener, *Blurring the Lines: The Intermingling of Garden and Theater in Seventeenth Century France*. Theses and Dissertations, no. 1615 (Brigham Young University—Provo, 2008). https://scholarsarchive.byu.edu/etd/1615 (accessed: August 28, 2023).
12 The *ha-has*, or sunk fences, were a popular landscaping solution in 18th century Europe; a fence or wall installed in a ditch would clearly demarcate the limits of a given property and help ward off unwanted (human and non-human) visitors without obstructing the allegedly "natural" vista. See, for instance, Antoine Joseph Dezallier d'Argenville, *La théorie et la pratique du jardinage* (Paris: Jean Mariette, 1709). The English version of the work is the following: Antoine Joseph Dezallier d'Argenville, *The Theory and Practice of Gardening: Wherein is Fully Handled All that Relates to Fine Gardens, Commonly Called Pleasure-Gardens, as Parterres, Groves, Bowling-Greens, &c.*, trans. John James (London: Geo. James, 1712).
13 David Jarrett, Tadeusz Rachwał, Tadeusz Sławek, *Geometry, Winding Paths and the Mansions of Spirit* (Katowice: Wydawnictwo Uniwersytetu Śląskiego, 1997), 82.
14 David Jarrett, Tadeusz Rachwał, Tadeusz Sławek, *Geometry, Winding Paths and the Mansions of Spirit…*, 75.
15 David Jarrett, Tadeusz Rachwał, Tadeusz Sławek, *Geometry, Winding Paths and the Mansions of Spirit…*, 171.
16 See the description of Thomas Gainsborough, *Mr and Mrs William Hallett ("The Morning Walk")*, 1785, oil on canvas, 236.2 × 179.1 cm, National Gallery, London, https://www.nationalgallery.org.uk/paintings/thomas-gainsborough-mr-and-mrs-william-hallett-the-morning-walk. (accessed: December 12, 2023).
17 William Gilpin, *Observations on the River Wye and Several Parts of South Wales & Relative Chiefly to Picturesque Beauty; Made in the Summer of the Year 1770* (Richmond, Surrey: Richmond, 1973), 18. Quoted in David Jarrett, Tadeusz Rachwał, Tadeusz Sławek, *Geometry, Winding Paths and the Mansions of Spirit…*, 79.

18 William Gilpin, *Three Essays: On Picturesque Beauty; On Picturesque Travel; and on Sketching Landscape: To Which Is Added a Poem on Landscape Painting* (London: R.Blamire, 1792), 8. Quoted in David Jarrett, Tadeusz Rachwał, Tadeusz Sławek, *Geometry, Winding Paths and the Mansions of Spirit…*, 79.
19 David Jarrett, Tadeusz Rachwał, Tadeusz Sławek, *Geometry, Winding Paths and the Mansions of Spirit…*, 81.
20 William Mason, "To a Gravel Walk," quoted in David Jarrett, Tadeusz Rachwał, Tadeusz Sławek, *Geometry, Winding Paths and the Mansions of Spirit…*, 85.
21 A term coined by David Jarrett, Tadeusz Rachwał, and Tadeusz Sławek, see their *Geometry, Winding Paths and the Mansions of Spirit…*, 112–116.
22 David Jarrett, Tadeusz Rachwał, Tadeusz Sławek, *Geometry, Winding Paths and the Mansions of Spirit…*, 113.
23 Such as the aesthetics characterizing the perceptions of Søren Kierkegaard in Europe or Herman Melville in America.
24 "Be fruitful, and multiply, and replenish the earth, and subdue it: and have dominion over the fish of the sea, and over the fowl of the air, and over every living thing that moveth upon the earth." King James' Bible, Gen. 1:28.
25 —which, although axiologically different, also celebrates human agency.
26 Signe Toksvig, *Emanuel Swedenborg, Scientist and Mystic* (New York: Swedenborg Foundation, 1948).
27 Czesław Miłosz and Ireneusz Kania, "Ciągle poszukujemy klucza. O Swedenborgu rozmawiają Czesław Miłosz i Ireneusz Kania," in *Emanuel Swedenborg: uczony i mistyk*, by Signe Toksvig, trans. Ireneusz Kania (Kraków: Universitas, 2002), 1–9, 14–15, 20–21. The passage has been translated into English by Paweł Jędrzejko.
28 Herman Melville, Letter to Nathaniel Hawthorne, June [1?] 1851, in *Herman Melville, Correspondence*, ed. Lynn North (Evanston and Chicago: Northwestern University Press and the Newberry Library, 1991), 193–194.
29 See Lawrance Thompson, *Melville's Quarrel with God* (Princeton: Princeton University Press, 1952).
30 Nathaniel Hawthorne, *The English Notebooks*, based upon the original manuscripts in the Pierpont Morgan Library and edited by Randall Stewart (New York: Modern Language Association of America; London: Oxford University Press, 1941), 423–427.
31 For a more extensive insight into Melville's existentialism, see: Paweł Jędrzejko, *Płynność i egzystencja: Doświadczenie lądu i morza a myśl Hermana Melville'a* [Liquidity and Existence. The Experience of the Land and the Sea in Herman Melville's Thought] (Katowice–Sosnowiec–Zabrze: BananaArt.pl/ExMachina/MStudio, 2008); Paweł Jędrzejko, *Melville w kontekstach: Prolegomena do studiów melvillistycznych* [Melville in Contexts. Prolegomena to Melville Studies] (Katowice–Sosnowiec–Zabrze: BananaArt.pl/ExMachina/MStudio, 2007).
32 William Braswell, "Melville as a Critic of Emerson," *American Literature* 9, no. 3 (November 1937): 317, https://doi.org/10.2307/2919662.
33 Herman Melville quoted in William Braswell, "Melville as a Critic of Emerson"…, 331.
34 Herman Melville, *Pierre*, quoted in William Braswell, "Melville as a Critic of Emerson"…, 318.
35 —and later prompting Charles Ives to "translate" Emerson into unsettling music of the *Concord Sonata*.
36 Hreinn Friðfinnsson, *Husprojekt*, 1974. Moderna Museet, Stockholm. https://sis.modernamuseet.se/objects/192/house-project (accessed: December 10, 2023).
37 Henry David Thoreau, *Walden: A Fully Annotated Edition*, ed. Jeffrey S. Cramer (New Haven: Yale University Press, 2004), 88.

3 If Health Were a Cloud

Everything seems "normal," at least according to classical physics, which, for instance, states that a brick of a specific mass, when dropped from a certain height, will accelerate to a precise speed before hitting the ground, without deviating even by a fraction of a second. This aligns with Cartesian metaphysics, which slips a screen of coolness between the thinking mind and the physical world that extends in various directions. The world, of course, is a collection of objects in space, and the human body is just another part of this collection. But then, suddenly…

> Again you show yourselves, you wavering Forms,
> Revealed, as you once were, to clouded vision.
> Shall I attempt to hold you fast once more?
> Heart's willing still to suffer that illusion?
> You crowd so near! Well then, you shall endure,
> And rouse me, from your mist and cloud's confusion:
> My spirit feels so young again: it's shaken
> By magic breezes that your breathings waken.[1]

What happens to Johann Wolfgang Goethe (or Faust, his character?) on the first page of the famous poem? The poet submits to the atmosphere, and then "unsteady figures," shapeless or faintly outlined, crowd around him. And then—something intriguing transpires: Faust (or Goethe?) begins to interact with the "wavering forms," commanding these atmospheres to become shapes. Yet, although he, the subject—perhaps thanks to his conscious focus—issues the command himself, it becomes clear that giving a shape to an atmospheric phenomenon does not at all equal controlling it. We learn this lesson from meteorology, a science which has developed a taxonomy of clouds, but even though it has worked out methods allowing us, for instance, to predict (at least to some extent) what will happen with the cumulus currently building up over our heads, it cannot *program* the weather. Meteorological analogies seem appropriate in the context of the first lines of *Faust*: over the procession of misty figures "a spell is blowing," just as the wind blows over the earth; shapeless, invisible entities can act like a

DOI: 10.4324/9781003328469-5

squall—quickly and powerfully, in a series of strikes and respites. Atmospheric phenomena are not necessarily soft: their concreteness is expressed in their dynamics, not in their coagulation into an object.

What then is the cognitive value of reading this passage from *Faust*? Are we able to learn anything about the nature of human cognition through it, or should Goethe's description—far from scientific—be dismissed as pure fantasy and, as such, exempted from the realm of philosophical inquiry? The answers become less easy when one realizes that it is not only poets who address phenomena such as those introduced above. Analyzing Goethe, Hermann Schmitz, a contemporary phenomenologist, writes thus:

> [T]here is another kind of feeling, the mere perception of the atmosphere without being in the grip of it; sometimes the transition from this perception to being affectively involved occurs so smoothly that the gripping power in causing this involvement can be clearly traced. In this manner Goethe's Faust is involved when he enters Gretchen's champer as a lustful spy: What a feeling of peace, order, and contentment breathes round![2]

Atmospheres, as Schmitz describes them, settle upon us—and when it happens, the landscape of our internal emotions changes as well. Faust (henceforth we shall assume his persona as the lyrical "I," facing the invasion of the misty figures) experiences a crisis triggering uncertainty, and, if we were to read further into the poem, undergoes a self-transformation—a sharpening of his cognitive apparatus—which makes him capable of perceiving phenomena previously invisible, thus allowing him to leave the old Newtonian-Descartian world behind. In other words, he experiences a cognitive "reset," ensuing which comes a new recognition of the territory of life and a new form of criticism. His attention focuses anew, but before it happens, he must first lose his former, Descartian, focus. Faust must leave a part of his old self behind, which is possible (as one can surmise) owing to the process of self-development that he had undergone. When this happens, things change. The protagonist declares:

> A shudder takes me: tear on tear, entire,
> The firm heart feels weakened and remote:
> What I possess seems far away from me,
> And what is gone becomes reality.[3]

Faust becomes a new person. The core of his identity has undergone a shift—a shift from the mediated world, inhabited by present, tangible objects, into a world of memories, coming to him in waves of emotions. A memory (or a series of memories) moves to the fore, becoming truth (at least for the moment), while all that used to be true until then—recedes. Yet, the replacement of one truth with another cannot be described with reference to the Descartian binarism of the self. The self is not binary: it is not that *it exists* or *does not exist*. If anything at all, it is an atmospheric state of being, elusive and ephemeral, feeding on truths that are always temporary, amalgamating in the form of sensations (in line with the certain forms of ancient skepticism)—these truths are true only for a moment, and only under certain circumstances.

When the new truth emerges, Faust does not have to seek its sanction in philosophy or science: it is *his* truth ("truth to me"). Yet the impact of this new truth is physiological: "tear on tear" flows, a "shudder takes him." Doubtless, the verses are work of literary fiction, but such fictions are created to point to some truth. Millions of book lovers, generation after generation, read *Faust*—as they do all other outstanding works of literature—because in it they find a truth: at the very least, a truth about their emotions. On the first pages of *Faust*, that which is new does not simply *enter stage*, it rolls in, or slides into, the space of our perception as a ready *object*. It "heaves up"; it does take shape—but a shape not definite enough to become a *thing*. And yet, it affects the mind of the observer more than any object defined by tangible boundaries. If we were to adopt Epicurus's simple definition that truth is what exerts an impact, we could say that, by this virtue, that which is new *becomes* truth—and this definition of truth is certainly not worse than those claiming that what exists is what philosophers, scientists, or specialists editing the WHO documents have determined to exist. Furthermore, the Epicurean definition might also prove useful in medical practice.

Why should we not see health in this perspective? Why should we not perceive it as something that "heaves up," something that "acts," but never solidifies into one state—a rhythmic process in which we flourish or wither, as something that cannot be "arrested" in a definition. Health researchers may always find consolation in the thought that they may rest assured that their work will never be done—and that dull retirement will never be their share. Indeed, health, like atmosphere, is and will always remain a mystery: we can either acknowledge this fact or pretend to be able to fully comprehend it. Yet—mystery is *good*: acknowledging it, we also acknowledge that our health is co-dependent on our environment, thus pairing our epistemic humility regarding health with metaphysical humility. Thereby, the metaphysics of health we explore is skeptical in character and, as such, feeds on crisis: the best provider of doubt. And therefore, as is clear, we, the authors of this book, would never aspire to deliver an "authoritative" definition of health.

In such a lens, however, the arrival at some "full and complete" understanding of the phenomenon of health is no longer of the essence; instead, what is important is finding ways to approach it, comprehend more and more of it, more and more precisely with each iteration, circling around it, on board with the conceptual apparatus. In such a perspective, someone who wants to live a healthier life—someone usually called a patient—they must stop thinking about their health as if it were their car: a car that will run just fine if the mechanic's (doctor's) recommendations are followed. Especially, bearing in mind that in the contemporary Western medical practice the patients heed the advice of a whole *consillium* of mechanics/doctors, each of them specializing in a particular subfield of medicine, the expertise of each of them narrowed down to the functioning of a particular organ. Departing from such a perspective, we propose thinking about health from the vantage point of a gardener—or, more specifically, through the lens of the whole garden that bears fruit: health as a *lifestyle*, whose ultimate goal is flourishing. In this sense, we attempt to inscribe health into everyday life as a practical category, comprising life creation and contemplation: contemplation intentionally programmed to transcend egocentrism. To attain this, however, we need to begin with some redefinitions.

Redefining Our Frontiers: From Inside Out

Investigating various phenomena, philosophy sometimes gets entangled in hopeless issues, often hitting dead ends, or ending up building fanciful constructions only to pile further fanciful constructions upon the former. Occasionally, it delves into specific phenomena that seem "niche" at first, but—as it may turn out later—have universal significance. One such phenomenon, "pathocentric epistemic injustice," proves particularly consequential for our attempts to redefine health.

The term "pathocentric epistemic injustice," coined by Ian James Kidd and Havi Carel,[4] describes a situation familiar to many patients, who complain that they are "ignored, silenced, or dismissed"[5]—all because medical staff do not consider them to be credible sources of information: their subjective descriptions of the symptoms they observe are either ignored or deemed of little value. As a result, the lived experience to which they wish to testify turns out irrelevant: it is rejected as imprecise, vague, unintelligible. Consequently, deprived of medical feedback to their accounts, the patients themselves end up unable to "make sense of one's experiences, a state that Medina calls hermeneutical death."[6] The very concept of epistemic injustice is an instance of "people being harmed specifically in their capacity as knowers: individuals capable of knowing or providing knowledge."[7]

Patients are placed in the frustrating role of "half-people," impaired individuals, only because they suffer from some kind of a "defect." This "defect" could be their physical or mental disability, as cataloged in medical books—compendia of defects. But the patients' impairments truly escalate when these very patients are not listened to—when they are refused to be taken seriously. In such situations, the patient is relegated to the position of a child dependent on parents who "know better," and hold all the tools of control in their hands. This, ultimately, deprives patients of their sense of agency: they become disenfranchised objects (of medical procedures).

> When patients are asked "how they are," for instance, they are typically expected to limit their answers to providing factual information about somatic sensations, functionality, response to medication, or the

> appearance and severity of symptoms. Such a piecemeal approach yields a partial view of the full phenomenon: it recognizes and privileges the physiological dysfunction (disease), while hiding the experience of this dysfunction (illness). In effect, this approach glosses over and marginalizes the holistic, existential nature of the illness experience, reducing the patient's ability to see the wide-ranging impact of illness on their life.[8]

The inability to make sense may seem unimportant under the assumption that a person is a piece of matter brought to the hospital to be fixed. But what if the object of treatment is a mental illness, at the center of which is the impossibility of making sense of everyday experience? Acute destabilization of patients' structures of meaning resulting from their inability to understand the ailment itself only exacerbates the condition. This problem emerges with particular clarity in the light of Antoni Kępiński's acute observations:

> A feature of life is dialectics of changeability and conservation. Life consists of constant change, which is an effect of its metabolic character—living beings are so-called open systems, i.e., they live owing to continuous exchange (the metabolism) of energy and information with the environment, without which they cannot exist.[9]

In this description, particularly interesting in the context of the doctor-patient relationship is the concept of "informational metabolism," as a sub-function of metabolism in general. If we accept that health is harmony, it becomes clear that epistemic injustice is an important element hindering healing, understood not only as return to flourishing (as proposed in this book), but even as the removal of symptoms and the restoration of a state of normality, as it is most commonly understood today in Western medicine. In his *Discipline and Punish: The Birth of the Prison*,[10] Michel Foucault goes even further: he criticizes the very concept of "normality," placing prison, court, school, and hospital side by side as institutions, which—in the name of restoring people to a state of normality—exercise power over them. He notes the strange alliance of "normality" and "scientificness," where the former is supposedly grounded in the latter—an alliance whose aim is disciplining man in the modern world and, ultimately, depriving individuals of their freedom.

The ancient ideal of philosophical flourishing mandates distrust of normality, which is often defined by errors in thinking and prejudices rooted in a given society. Perhaps the madness of a given culture can be measured by the distance of a person, who participates in it, from his or her own felt-body. Perhaps this madness can be diagnosed on the scale of man's alienation from nature, of his or her withdrawal from the great project of life and, at the same time, his or her exile from their own home—which, according to the Greek idea of *oikeiôsis*, is the cosmos. And perhaps these two areas are closely related. One cannot tune into the process of life in any way other than through contact with what Schmitz calls the "primitive present," which, as he believes, contains a rich life "full of situations, from the meaningfulness of which no individual meanings can be isolated"[11]; meanwhile, the Western doctor's message, which must meet the requirements of completeness and separation, consists solely of isolated individual meanings. In the following chapters of this book, we identify the space conducive to such an attunement in the garden, because the "primitive present" is a rhythm that does not reverberate on one side of the human skin only: such a rhythm disregards "dermo-boundaries."

One could say that the goal of informational metabolism is care for one's sense of meaning, a sense often lacking among patients of modern hospitals. Depriving patients of this sense through medical procedures, only to restore it to them later, seems to be a counter-productive practice. Kępiński points out that in the human world it is primarily that which has a name that exists:[12] when physicians discuss disease in their jargon, they introduce new, incomprehensible, mysterious entities into the world of the (psychiatric) patient—fictional objects, that further destabilize this world. According to Kępiński, both psychiatry and psychology should be "simple and clear, because they deal with fundamental human matters."[13] He argues that the patient's own world order is also the order of their environment, and both these orders clash in the processes of informational metabolism, to eventually come together to form a comprehensive picture of the world.

> Kępiński, as a psychiatrist who has observed for dozens of years what he calls different models of reality employed by his patients, formulates an intriguing thesis: in delusional syndromes and delusional personalities, the typical rhythm of death and resurrection is being disrupted. A paranoid person fears the death of his or her models of reality, clings to them, and, as a result, new models cannot be created.[14]

In the context of epistemic injustice, the situation is even worse. A paranoid person has no chance of building and updating their own reality models, as the only model to which any value of the patient-doctor relation is assigned is the one imposed by the doctor. Such a model, besides the fact of its being incomprehensible to the patient, somehow "clogs" the space, in which the patient could update models of their own. Significant in Kępiński's description is the invocation of "the rhythm of death and resurrection," that is, an invocation of the cyclical processes taking place every year (and sometimes at shorter intervals) in the garden. Certain studies show that perhaps it would be best to place hospital beds or hammocks (as long as the particularity of a given illness does not prevent it) directly in the garden—especially that even the mere view of trees outside hospital room windows proves of consequence to the healing process:

> Statistically significant differences were found between the tree-view patients and brick wall-view patients on patient length of stay, pain medication use, and nurse notes. Patients with a view of trees were

> hospitalized shorter (7.96 days) than patients who had a view of the brick wall (8.7 days). Brick wall-view patients had more negative nurses notes (3.96 per patient, examples of negative notes—"upset and crying" or "needs much encouragement", examples of good notes—"in good spirits" and "moving well") than tree-view patients (1.13 per patient). Patients with nature window views received fewer analgesic doses 2–5 days after surgery. The tree-view patients more frequently received weaker pain medications like aspirin or acetaminophen while brick wall-view patients [...] needed stronger pain medications such as narcotics. No significant difference was found between the groups on doses of antianxiety medications. Tree-view patients tended to have lower scores for minor postsurgical complications. However, the difference was not statistically significant.[15]

Toward the end of this chapter, we introduce the concept of "attuning oneself to the process of life," which explains situations analogous to that described above in the Pennsylvania hospital, through the synchronization of the rhythms of an individual organism with the environment—a concept that is part of the Open-source Intelligence hypothesis, discussed in the above-quoted monograph. Before we explore it in greater detail, however, it is important that we acknowledge that the causes of epistemic (and pathocentric) injustice vary, and are numerous. From the point of view of our inquiry, the most interesting among the causes (as identified by Kidd and Carel) is the very definition of health and its underlying doctrine—a doctrine of significant practical consequences, influencing the way of thinking and talking about disease, and about people affected by it.[16]

Pathocentric epistemic injustice arises from the cognitive filter which captures, in human experience, that which is named "pathology"—a filter routinely implemented in the contemporary Western healthcare systems (hence Kidd's and Carel's term "pathocentric injustice"). Pathology appears as an objectively existing state, as if "detached" from the whole organism coexisting with its environment: a state calling for repair, or—better still—elimination. That which remains after such an elimination has been effected is precisely what is called health.

> Ideally, health is defined from a value-free, third-person perspective, independently of subjective and contextual values and other "extra-biological" factors, such as "intentional actions," "goals," and other aspects of human subjectivity.[17]

This "value-free" definition is, of course, justifiable: it is a consequence of the approach that stands behind the success of empirical sciences. The problem arises when the complex process of re-entry into harmony, called healing, is reduced to the implementation of a certain naturalistic procedure.

> As a consequence, both the conceptualization and treatment of health and illness become the primary, if not exclusive, responsibility of biomedical science. It is this epistemic and practical privileging of naturalistic resources that can, we argue, act as a source of pathocentric hermeneutical injustice.[18]

What the authors call "naturalistic or biomedical conception of health that privileges the concepts, theories, and terminology of natural science"[19] in medicine is actually the result of applying an even denser filter than the one singling out pathology in the health-disease area, namely the Newtonian-Descartian filter. It results in the patient's body being perceived and treated as separate from the psyche (the Descartian part), while various parameters, measured with the use specialized equipment (the Newtonian part), are elevated in their status from the level of the *description* to the level of the *truth* about the disease or health—which triggers particular treatment procedures. This approach deems as objective what is "displayed"—on the surface of the patient's body, on the computer screen showing an image of the body's internal organs, on a page populated with numbers spat out of a printer, etc. And although it often results in curing the disease—or solving the problem—such treatment often concerns only a symptom, the procedure itself forgoing a holistic investigation into the source of the disturbance in the patient's well-being. Needless to say, such a cure cannot be considered "a step forward" if we see it in the perspective of health-as-flourishing—also because such "healing" does not go hand in hand with the development of the patient's awareness.

The philosophically informed or "ontologically diverse" procedure of caring about one's well-being, based on what we propose to call "transgressive metaphysics" (we expand this concept in the further sections of this book), would appeal to a more profound kind of naturalism, whose elementary epistemological intuition—to use Heraclitean terminology—would involve the recognition of the fact that "nature loves to hide."[20] If such an approach were to be faithful to pre-Descartian philosophical traditions, it would necessarily link well-being to a much broader system of reference than that limited to a single organism in a hospital bed or on an operating table. In the perspective of the ancient philosophers of the West, as well as for Chinese philosophers, nature operates in the same way in the human body as it does in the environment, and even at the level of cosmic laws (as we have demonstrated in the previous chapters, invoking, among others, the metaphor of the human body as a garden). This is why, seen in such a context, the processes that contemporary medicine calls "healing" manifest themselves as inevitably consisting in the reactivation of a healthy pattern of nature of which the affected organism is a part.

However, it is not our goal to carry out an investigation of the essence of health using a reference system of one philosophical tradition only: such a choice, most often, imposes a specific metaphysical position, which would be inconsistent with the idea of the "ontologically diverse" approach we propose. Likewise, it is not our goal to pass judgments on the "truthfulness" of various types of metaphysics. Instead, the insights presented thus far serve us as a multifaceted background for creating a metaphysically neutral frame of reference: a philosophy of the garden, which, applied to reflections on well-being, abandons the lingering vision of health as a matter concerning a single organism, does away with the separation of two independent Descartian substances, dismantles the myth of the inadequacy of patients' subjective experiences, and rejects the mechanistic concept of nature along with the belief in the objectivity of scientific description.

This, however, begs a question: is the language derived from the practice of gardening—either that used for the purpose of the actual description of the phenomenon of the garden, or that transferred from this area to philosophical and life practice[21]—self-standingly suitable to address the living experience of what we call "illness" and "healing," and to enable communication with the environment that helps "patients" flourish? Is it, alone, capable of "capturing" the phenomenon called "health"—even if such an approximation were not to be ultimate, or even sharp?

Our initial diagnosis suggests that this is not necessarily the case. The language of the garden has clear but limited heuristic value when applied to the phenomenon of health/disease, or to flourishing. For example, Buddhist teachers representing the so-called Thai Forest Tradition use it to facilitate the flourishing of their students, but—although the language of gardening can provide directions for the practice of meditation among the beginners, or even help correct some of their errors in their approach to this practice—it is usually deployed for the purpose of *comparing* the practice of meditation to the life of a garden or, for instance, to the vegetative processes in a single tree. A typical example of such an application of the language of gardening may be drawn from the teachings of the legendary master Ajahn Chah:

> In meditation, you must continuously be attentive, just like when planting a seedling. If you plant a seedling in one place, then after three days you pull it up and plant it in another place, and after three more days, pull it up again and plant it somewhere else, it will just die and not grow up and bear any fruit. Meditation is the same. If you do a seven-day meditation retreat and after leaving it, for seven months you go around "soiling" the mind, and then come back and do another seven-day retreat where you don't speak and you keep to yourself, it's like the tree. Your meditation practice won't be able to grow and it will die without producing any real results.[22]

Admittedly, discussing the nuances of meditative experiences, Thai meditation teachers often replace such metaphorical idiom with technical terminology, composed of concepts developed in the language of early Buddhist texts—Pali. Similarly, in the context of our discussion on health, we, too, enrich the language of the garden, pairing it with another idiom—one derived from the holistic framework dubbed "new phenomenology," fostered by the already-quoted German philosopher, Hermann Schmitz. As the reader would have realized by now, the semantic package—or set of corresponding concepts, that we have chosen in this book for inquiries about "atmospheric health"—combines the languages of the garden and Schmitz's new phenomenology, in precisely adjusted proportions. We use the language of the garden to delineate the contours of the territory being described and to describe the environmental aspect of the phenomenon of our interest, while the language of new phenomenology fills in these contours and becomes the medium for subjective descriptions of individual people's lived experiences. And although the fusion of these languages is central to our understanding of the process of human flourishing, we do not employ the philosophical heritage of phenomenology with the view to creating a "systematic" framework of reference; rather, we treat it as a compound source of inspiration, consisting of Schmitz's *language*—especially his concept of "atmospheric health," featuring in the title of his book—and of his *insights* related to the nature of human experience of the world. And, as the reader already senses it, our very application of the concept of "atmospheric health" in the contexts of our search for a philosophy of the garden suggests that we attempt to remedy the "imprecision," of which doctors accuse patients when the latter describe their ailments, not by means of the enhancement of the sharpness of the fragmentary image, but by means of expanding the image itself to account for the "wholeness" of the human in their universe. And even though such an image may be characterized by indeterminacy, this fact does not have to hinder the renewed understanding of health in the context of sense-making.

Environment Keeps Us Alive: From Outside In

Atmospheres, like the figures in the opening paragraphs of *Faust*, come to us from the outside. They surround us, perhaps even seep into us, eliciting somatic reactions (such as crying) and psychological ones (such as longing). They differ in character from objects in space: Schmitz focuses on "diffuse, undefined, and affective states" filling the "lifeworld" (*Lebenswelt*),[23] emphasizing experiences ignored by health specialists (at least in the West), even though they form the basis of human experience of the world. He builds his method of reaching the elements of "spontaneous life experience"[24] by refraining from experiencing constructed intentionally. Beyond intentionally constructed experience remains the experience gained in a "felt" manner, expressed by means of an "embodied alphabet." It is to this alphabet that we resort throughout this book.

An important component of our lexicon is the concept of "atmosphere," which Schmitz defines as "unbound occupation of surfaceless space in the region of what is experienced as present."[25] The expression "surfaceless space" clearly signifies that the atmosphere is not an object—not even one with blurred boundaries. It is not an illusion, either. Its reality is confirmed by the fact that we experience it in the present moment. It is not an emotion, although the latter can accompany it—as Goethe so accurately demonstrated it on the example of Faust, who "enters Gretchen's chamber as a lustful spy," but is immediately captivated by the sense of peace and harmony, already present in her room. Other instances of a "gripping atmosphere" are, for instance, the feeling of meaninglessness or "a kind of boredom intermingled with nausea/disgust" which has no direction, or the "eerie mood of the evening when the pale coolness of the fading day makes everything appear distant."[26] These sensations occur in what Schmitz dubs "area-less spaces."

> The two most important types of area-less spaces are the space of the felt body and the space of emotions as atmospheres. I define the felt body of a person as the epitome of everything that he himself can perceive as belonging to himself, in the vicinity—not always within the boundaries—of his material body (*Körper*), without making use of the five senses of sight, hearing, touch, smell, taste and of the perceptive material body schema (the habitual image of a person's own material body) that has been obtained from his or her experiences, particularly those of sight and touch. These include first of all the purely corporeal stirrings such as fright, anxiety, pain, hunger, thirst, itching, sharp pains, disgust, vigor, fatigue; secondly, the corporeal stirrings which are the affective marks of emotions, e.g. of joy, sadness, anger, shame, fear, courage, compassion, contentedness, despair; thirdly, the perceived voluntary and involuntary movements such as walking, grasping, jumping, dancing, shivering, twitching, swallowing; and fourthly, the irreversible corporeal directions, sometimes occurring without movement, as when one gazes, sometimes tied to movements, as in breathing out and swallowing. All these corporeal events are area-less. One cannot perceive any areas on one's own felt body. One can see them and feel them only on one's own material body. The felt body is usually a surging of blurred islands, amongst which individual ones stand out through consistency, structure and function.[27]

Although, to our best knowledge, Schmitz has not analyzed the atmosphere of a hospital room in any of his texts, the experience of most hospital patients will confirm that the space of the hospital is dominated by an atmosphere of anxiety and uncertainty. This atmosphere, in turn, affects what Schmitz calls "self-ascription"—a sense of one's own identity that fosters one's sense of security. Self-ascription, "based on self-consciousness without identification,"[28] is rooted in what Schmitz calls the "primitive present," a preconceptual feeling of one's presence in the world, to which we usually have limited access, and which is sometimes described in the language of poetry, but not in the language of contemporary Western medicine. By depreciating the "primitive present" in patients through ignoring their subjective reports, doctors further disrupt their self-ascription, which is already weakened by the very situation of illness, but which can be ultimately shattered by, for example, a diagnosis of a terminal disease. The sense of identity in someone who must suddenly stop thinking of themselves as having years of life to live must, inevitably, change.

Treatment can be painful. Schmitz analyzes the feeling of pain and concludes that we can experience it directly, even before the process of identification occurs. Often, pain prevents self-ascription, as in the case of shock, which renders individuals immobile, and results in a situation in which "[t]he conscious subject has no access to itself."[29] Pain is accompanied by a series of strong emotions, which are also ignored in the reductionist model of medical procedure. However, in the model of health-as-flourishing, pain plays an important role. For Buddhists and Epicureans, for example, developing a healthy relationship with pain—which includes, among other things, the understanding of its necessary presence in human existence and the ability of separating pain from psychological suffering—is an important step on the path of self-development. But in order to attain it, one must first confront pain (and the emotions accompanying it) at the level of the "felt-body," as Schmitz describes it; only then an intellectual attempt can be made to incorporate pain into a meaningful life project. Needless to say, contemporary Western medicine does not see pain as an opportunity for development: it is only perceived as a pathological symptom that needs to be eliminated.

The goal of Schmitz's new phenomenology is to understand one's own life, and this is impossible without access to the spontaneous experiencing of it, effected in a series of successive contemplations. The language of specialists (not only physicians but also psychologists, or authors of texts popularizing science) is the language of the "psychologistic-reductionist-interjectionist paradigm."[30] Such a language, especially when treated as the only one capable of objectively describing experience, curtails an individual's capability to spontaneously experience the world.

The psychological character of language and the general "psychologization" of experience, to use a term proposed by Ángel Gordo and Jan De Vos,[31] have their origins in philosophy. According to Schmitz, the 5th century BC Greece witnessed a split of the world in people's perception. As a result of this split, a private, internal world emerged, encompassing experiences independent of the environment around—a previously unknown, sterile psyche, or soul, came into existence.[32] Since then, we have had to control this soul, simultaneously developing a need to conquer the world external to it.

The reductionist character of this paradigm lies, among others, in its breaking down both the soul and the external world into individual elements, of which the consequence is the loss of a holistic picture. The interjectionist aspect of it, in turn, involves the internalization of emotions into a closed-off inner space—into the bubble of the psyche. Schmitz illustrates this with the example of the juxtaposition of the *Iliad* and the *Odyssey*, demonstrating that the paradigm became entrenched in the Western tradition between the creation of the first and the second work. In the universe of the *Iliad*, humans are exposed to atmospheres created by the gods—to their love or hate. No mental entities are featured; what we call "mental" in the *Iliad* is actually the matter of the felt-body. Personal emancipation only begins with the heroes of the *Odyssey*: they start to control their facial expressions and recognize in themselves a *daimon*, which is nothing other than an internalized system of emotions. The harmful effect of this transformation, according to Schmitz, is the "introjection of feelings into a closed-off inner space."[33] What follows is a privatization of the emotional sphere, or the enclosure of a person in a "black box," similar to the ones with which airplanes are equipped. Twenty-some centuries later, Descartes, having inherited this concept, would give it a new life.

Importantly, however, as a result of the privatization of emotions, we lose the sense of collective situations and of our place in the universe. In the modern world, this manifests as depression, alienation, or compulsive need for ego gratification (sometimes misleadingly called self-realization) combined with the need to impress others. We seek superficial stimulation, losing sight of felt-body experiences. But that does not mean that the felt-body with its needs ceases to exist: humans still feel that they are like an archipelago of blurred islands amidst the seas, some more distinct than others. A person confined to a hospital bed probably senses this more clearly than someone running around the city taking care of their errands, yet the bottom line is that the islands communicate with each other and reach out to the outside world in search of the community of collective situations. But there, they encounter repression or neglect on the part of the psychologistic-reductionist-interjectionist health system.

Research on the human-nature relationship shows that even minimal exposure to natural patterns has a positive impact on human health and human sense of well-being. In the model we propose, human consciousness, which contracts and expands, meets a similar pattern in the natural environment, resulting in healthy synchrony. Viewed from this perspective, leading a healthy life requires that we attune our attention to the patterns of our environment—the wilder the patterns, the better.[34] In such synchrony with nature, our attention, constantly distracted by the demands of quotidian life keying it to artificial patterns, becomes restored.

Harmonious Synchronization

An example of a medical intervention that is characterized by balance rather than epistemic injustice is provided by Yanhua Zhang, the author of the article titled "Negotiating a Path to Efficacy at a Clinic of Traditional Chinese Medicine."[35] The observations of the scholar, who observed the work of a certain Chinese physician for ten months to understand the essence of the "stagnation of emotions," could be summarized thusly:

> The typical western biomedical model employs separated ontologies for body and mind. The Chinese art of healing is based on an ontology that understands life as a complex psycho-physical process. Description of

> phenomena in this system of reference is mental and physical at the same time. Physical description is applied to mental worlds in the form of expressions such: as stagnation of emotions, "vexed-heart emotion," "heart emotion not flowing smoothly." The Chinese healer removes the blocks by various methods such as herbal application but also by therapeutic talks about patient's thoughts and emotions.[36] The word "talk" in this context is more appropriate than the word "interview" that we use for doctor-patient conversations in Western medicine because, according to Zhang, the patients "actively engage the doctor in their perspectives, help define their own illnesses, and even negotiate their treatment."[37]

The mentioned negotiations are, of course, conducted within a particular cultural context. Both the doctor and the patient, unlike it is the case in such relationships in Western medicine, assume a certain geography of corporeality, namely, one that involves both sides of our body wall. The primary target to be influenced in the healing process is *qi*, and the energy of *qi* flows both in our bodies and in the environment. According to Chinese medicine, the human body is not "a skin bag" filled with internal organs: it is a layered, complex entity. *Qi*, or the energy layer, is one of these layers. Complementing it are the layer of *jing*, which manifests as what we can call the physical body, and the layer of *shen*—the level of spirit, associated, among others, with one's way of thinking and the worldview one holds. These three layers mutually influence each other and their harmonious interaction makes the free flow of energy possible. Sometimes, however, the energy goes stagnant, and this is the problem which the Chinese healer must address.

> According to this Chinese art of healing, the disease of stagnation can be healed only if the patients change their perspective on life, which requires from the healer what we in the Western world would consider as highly developed psychotherapeutic skills. A purely physical treatment would not work.[38]

Strong emotions also obstruct the flow of *qi*. Hence a traditional Chinese doctor has no other choice but to work to restore the harmony (and thereby patient's health) through acupuncture, herbal infusions, massage—but also through conversation. The repertoire of his methods is founded upon four basic ways of obtaining information, which can be arrived at by means of gazing, listening/smelling, questioning, and touching. His patient does not have to "somatize" their emotions, as the story of these affects is an essential—and necessary—condition of a precise diagnosis. From the perspective of anthropology, such a doctor simultaneously diagnoses various types of dysfunctions: physical, emotional, and social. Yet it is not his goal to name the disease, or identify its textbook designation; nor does the diagnosis itself only serve the doctor to develop a concept of the ailment at hand. His contribution to the interaction with the suffering person assumes taking actions that in the Western context would be considered psychotherapeutic. The doctor must choose his words carefully in conversation with the patient because words, like acupuncture needles, have the power to remove blockages in the flow of *qi*. The patient is encouraged to talk about their quotidian problems, and, unlike it is the case with Western medicine, the doctor influences the patient in such a way as to allow him or her to develop a thorough understanding of their affliction at more than just the physiological level.

Zhang's anthropological study includes quotes from the patient's medical record, including phrases such as "vexation of heart," "feeling wrong and having a tendency to weep," "lack of interest," or "pulse strung."[39] Such knowledge is not obtained as a result of interview structured by the physician; often, it is the patient to initiate the conversation and determine its topics, especially that, as Zhang observes, patients "seldom submit themselves completely to the authority of the doctor." Instead, they "actively engage the doctor in their perspectives, help define their own illness, and even negotiate their own treatment."[40] Such a medical procedure features more elements of ancient philosophical *therapeia* than does contemporary psychotherapy; the interview does not revolve around the doctor presenting an "objective" viewpoint on the patient's emotions, but consists in engaging the doctor and the patient in a meaningful conversation about the human condition that both of them share.[41] The doctor works toward explaining to the patient *what is* and *what is not* under their control—much like the ancient Greek philosopher Epictetus, for whom the ability to distinguish between these two aspects of experiencing the world was key to the process of flourishing. In his view, for instance, the stagnation of energy stems from the patient's disappointment caused by the failure to achieve goals they wished (or would like to) to attain. In the described form of medical intervention, known as *zhongyi* in Chinese, healing, primarily, hinges upon the patient's readiness to adopt a new perspective on life.

> With Western psychotherapy, "the patient is encouraged to express the feeling, to gratify the desire, and to remove supression".[42] With *zhongyi*, the patient is encouraged to contain (*baorong*), dissolve (*huajie*), and then transform (*zhuanhua*) the conflicts by emphasizing the connectedness of self and other, individual and society, human being and the myriad of things of the world. What is emphasized is the sense of unity and harmony with nature and the world.[43]

Such "unity" and such "harmony" are offered by the garden, epitomizing not only the "unity of man and nature," but also the harmony of the relations among people. Epicurus practiced philosophy with his students in nature. In the past, the Icelandic parliament used to meet in a deep ravine, among trees and rocks. In Chinese medicine, the therapy described above is called *tiao*, which corresponds to the English *attuning*, *adjusting*, or *balancing*.[44] Perhaps, on the basis of the observations offered in the sections above, it might be possible to surmise that finding themselves in an environment such as that of a garden, people find it easier to feel a stronger sense of community owing to the shared sensation of the natural rhythms, pulsing around and within them. This, in essence, is the fundament of the "open-source intelligence" hypothesis[45] according to which intelligence—and thereto associated sense of meaning of existence and direction of life, as well as the sense of well-being—are functions of the organism's immersion in the environment. This hypothesis is based on the recognition of the rhythmicity of human consciousness, which wakes and fades in micro-segments of time, and on the assumption of the possibility of attuning the rhythms of the organism to the rhythms of the environment. It also invokes the concept derived from the analysis of the Buddhist *sutta Satipaṭṭhāna*—the possibility of developing the ability of self-attunement to the force of one's attention, tending to free itself from entrapment within a perspectival view.[46] In our exploration, we propose a concept of a system of interdependencies that resembles an immune system, but extends beyond a single organism. This expanded system is accountable for what we term "atmospheric health," a function of the synchronization of one's individual rhythms with the rhythms of nature—rhythms notably absent in the digital world. It is on such grounds that we expand the notion of the garden as a notion synthesizing self-awareness, cognition, well-being, and health into a unified framework.

In contrast to the Japanese practice of "forest bathing" (*shinrin-yoku*) and other health models advocating for extensive time in nature, we adopt a philosophically augmented method of fostering well-being—one founded upon transgressive metaphysics. As the subsequent sections demonstrate, our approach, rooted in philosophy-driven lifestyles, places equal emphasis on thoughtful engagement and meditative immersion in nature. We view thoughts as forms of attention, affected both by our physicality and our surroundings. Consistent with the *autodidactos* tradition (detailed below), we consider the recognition of the importance of the interplay between these forces as vital in healing and indispensable in flourishing. The above notwithstanding, it is essential to acknowledge the limitations of the narratives we use, and to train one's attention in such a way as to be able to adeptly navigate the non-narrative dynamics of natural processes. In this sense, philosophy holds an immense potential, opening access to direct knowledge of the body, intention, and will. In the sections below, not unlike the patients of the Chinese heelers, we seek to narrate the experiences of our own immersion in the processes of life in locations, in which communally shared lifestyles stem from the communally shared "grassroots" practices of attunement to the rhythms of nature, in hope to approximate the atmospheres conducive to the reinforcement of the sense of individual and collective well-being.

Notes

1 Johann Wolfgang von Goethe, *Faust*, translated by A. S. Kline. *Poetry in Translation*. https://www.poetryintranslation.com/PITBR/German/FaustIProl.php (accessed: October 10, 2023). For an alternative translation consult the "classical" edition: Johann Wolfgang von Goethe, *Faust*, trans. and intro. Walter Kaufmann (New York: Anchor Books, 1962).
2 Hermann Schmitz, *New Phenomenology. A Brief Introduction* (Sesto San Giovanni: Mimesis International, 2019), 98.
3 Johann Wolfgang von Goethe, *Faust*, lines 29–32.
4 Ian James Kidd and Havi Carel, "Pathocentric Epistemic Injustice and Conceptions of Health," in *Overcoming Epistemic Injustice: Social and Psychological Perspectives*, ed. Benjamin R. Sherman and Stacey Goguen (London: Rowman and Littlefield, 2019), 153–168.
5 Ian James Kidd and Havi Carel, "Pathocentric Epistemic Injustice and Conceptions of Health"..., 153.
6 Ian James Kidd and Havi Carel, "Pathocentric Epistemic Injustice and Conceptions of Health"...,159.
7 Robert Chapman, Havi Carel, "Neurodiversity, epistemic injustice, and the good human life," *Journal of Social Philosophy*, Volume 53, Issue 4, Special Issue:Analyzing Social Wrongs, Winter 2022, 614. https://doi.org/10.1111/josp.12456.
8 Ian James Kidd and Havi Carel, "Pathocentric Epistemic Injustice and Conceptions of Health"...,157.
9 Antoni Kępiński, *Psychopatie* [Psychopathies] (Kraków: Wydawnictwo Literackie 2007), 6.
10 Michel Foucault, *Discipline and Punish: The Birth of the Prison* (New York: Vintage Books, 1995).
11 Hermann Schmitz, *New Phenomenology. A Brief Introduction* (Sesto San Giovanni: Mimesis International, 2019), 75.
12 Antoni Kępiński, *Psychopatie...*, 9.
13 Antoni Kępiński, *Psychopatie...*, 9.
14 Marcin Fabjański, *Embodied Nature And Health. How to Attune to the Open-source Intelligence* (London, New York 2022), 70.
15 Roger Ulrich, "View Through a Window May Influence Recovery from Surgery," *Science* (New York, N.Y.). 224. 420–421. https://doi.org/10.1126/science.6143402.
16 Ian James Kidd and Havi Carel, "Pathocentric Epistemic Injustice and Conceptions of Health"..., 153–168.
17 Lennart Nordenfelt, *Action, Ability and Health: Essays in the Philosophy of Action and Welfare* (Dordrecht: Springer, 2014), 25; see also: Ian James Kidd and Havi Carel, "Pathocentric Epistemic Injustice and Conceptions of Health"..., 163.
18 Ian James Kidd and Havi Carel, "Pathocentric Epistemic Injustice and Conceptions of Health"..., 163.
19 Ian James Kidd and Havi Carel, "Pathocentric Epistemic Injustice and Conceptions of Health"..., 163.
20 Pierre Hadot, *The Veil of Isis. An Essay on the History of the Idea of Nature* (Cambridge, MA: Harvard University Press, 2006).
21 Three examples of this are presented in Part Two of this book.
22 Ajahn Chah, *A Tree in a Forest. A Collection of Ajahn Chah Similies* (Taoyuan: Yuan Kuang Publishing House, 1994), 69.
23 Hermann Schmitz, *New Phenomenology. A Brief Introduction* (Sesto San Giovanni: Mimesis International, 2019), 11.
24 Hermann Schmitz, *New Phenomenology. A Brief Introduction...*, 11.
25 Hermann Schmitz, *New Phenomenology. A Brief Introduction...*, 139.
26 Hermann Schmitz, *New Phenomenology. A Brief Introduction...*, 97.
27 Hermann Schmitz, *New Phenomenology. A Brief Introduction...*, 62.

28 Hermann Schmitz, *New Phenomenology. A Brief Introduction...*, 62.
29 Hermann Schmitz, *New Phenomenology. A Brief Introduction...*, 62.
30 Hermann Schmitz, *New Phenomenology. A Brief Introduction...*, 16.
31 Ángel Gordo and Jan De Vos, "Psychologism, Psychologising and Depsychologisation," *Annual Review of Critical Psychology*, 8: 3–7.
32 Hermann Schmitz, "Atmospheric Spaces," *Ambiances: Environnement sensible, architecture et espace urbain, Redécouvertes*, 2016, translation by Margret Vince, https://doi.org/10.4000/ambiances.711, 1 (accessed: April 19, 2019).
33 Hermann Schmitz, "Atmospheric Spaces"..., 6.
34 Interestingly, such—and similar—recognitions have already energized academic programs. For instance, the course "Re-Wilding Harvard," or the "Natural Neighbors Program" realized in Martha's Vinyard, explore the concept of rewilding through interdisciplinary research and public engagement. The program aims to integrate insights from biology, ecology, cultural studies, and the humanities to create a comprehensive understanding of rewilding. Traditionally, rewilding refers to the process of restoring ecosystems to their natural state by allowing nature to take care of itself, which can involve the reintroduction of native species, the removal of human-made structures, and the reduction of human intervention to enable natural processes to flourish. Such programs expands on this definition by incorporating cultural, social, and philosophical dimensions to understand how rewilding can also apply to human cultures and societies. By doing so, initiatives of this kind seek to address not only ecological restoration but also the cultural and social implications of rewilding, offering a more holistic approach to conservation and sustainability.
35 Yanhua Zhang, "Negotiating a path to efficacy at a clinic of traditional Chinese medicine," *Culture, Medicine and Psychiatry*, 31(1), 73–100. https://doi.org/10.1007/s11013-006-9039-6.
36 Yanhua Zhang, "Negotiating a path to efficacy at a clinic of traditional Chinese medicine"..., 73–100.
37 Yanhua Zhang, "Negotiating a path to efficacy at a clinic of traditional Chinese medicine"..., 78; see also Marcin Fabjański and Lorenza Tiberio, "Presence-Centered Flourishing: A Proposal of Alternative Strategies to Promote Sustainable Living," *Journal of Educational, Cultural and Psychological Studies (ECPS)*, no. 27 (2023): 129, ISBN 978-88-5513-096-7, https://doi.org/10.7358/ecps-2023-027-f, 129).
38 Marcin Fabjański and Lorenza Tiberio, "Presence-Centered Flourishing: A Proposal of Alternative Strategies to Promote Sustainable Living"..., 129.
39 Yanhua Zhang, "Negotiating a path to efficacy at a clinic of traditional Chinese medicine"..., 78.
40 Yanhua Zhang, "Negotiating a path to efficacy at a clinic of traditional Chinese medicine"..., 78.
41 Yanhua Zhang, "Negotiating a path to efficacy at a clinic of traditional Chinese medicine"..., 83.
42 Wen-shing Tseng, Q. Y. Lu, and P. Y. Yin, "Psychotherapy for the Chinese: Cultural Considerations," in *Chinese Society and Mental Health*, ed. T. Lin, W. Tseng, and E. Yeh (Hong Kong: Oxford University Press, 1995), 292.
43 Yanhua Zhang, "Negotiating a path to efficacy at a clinic of traditional Chinese medicine"..., 94.
44 Yanhua Zhang, "Negotiating a path to efficacy at a clinic of traditional Chinese medicine"..., 77.
45 Marcin Fabjański, *Embodied Nature And Health. How to Attune to the Open-source Intelligence* (London, New York 2022).
46 Marcin Fabjański, *Embodied Nature And Health. How to Attune to the Open-source Intelligence...*, 156.

Part Two

Atmospheric Health Practices (Autoethnologies)

4 Sharing the (Common)Wealth of the Garden

As every year throughout the last two decades, I feel an intriguing excitement while planning my next trip to Greece. It is not just about what objects to take with me on the plane—or, more recently, in the car, which offers more freedom—but mainly about places to see or to more thoroughly explore. Now, I *know* that making these trips to Greece, I do *not* want to be a tourist; I probably *cannot* be one anymore. In fact, I do not plan on places-to-visit, but rather on places in which uniquely-to-be. I want to understand these spaces from the inside, I want to immerse in their particularities, local color, hues, flavors, people, and nature. For a long time now, I have chosen not stay in hotels; instead, I rent apartments with the Greek families or right next door—more often in the mountains than by the sea. As a result, it sometimes happens that I encounter goats more often than I do people. It does not bother me in the least; I avoid the clamor of tourist chatter as much as I possibly can.

My excitement, stirring whenever I plan my stays in Greece, must be somehow linked with my conscious anticipation of combining the familiar with the unfamiliar. Indeed, over the years spent in Hellas, one may acquire particular habits and learn particular behaviors—learn to understand the uniqueness of particular places, to see potential dangers, and to understand hopes. At the same time, it seems clear that opening up to other spaces, exploring the new, allowing yourself to be surprised—is always worthwhile. And thus, in my case, a mix of the known with the unknown, of the omnipresent with the endemic, of the global with the local, is what fills the agenda of each next travel project. With such an approach in mind, all my trips *to* Greece, *in* Greece, and *for* Greece—fuse into one great journey. And I cannot disagree with Lawrence Durrell, who wrote: "Other countries may offer you discoveries in manners or lore or landscape; Greece offers you something harder—the discovery of yourself."[1]

A Greek Journey: A Return to the Unknown

I do not create catalogs labeled with subsequent years, where I would gather sets of memories keyed to specific dates; I do not draw trajectories of individual journeys locked in given time-spans. All these journeys are one

DOI: 10.4324/9781003328469-7

journey, broken down into a multitude of experiences, excitements, emotions, and sometimes disappointments. Dozens of times, I have traveled to Greece by planes, buses, or cars (my own or rented, though the rentals, for a moment anyway, would also feel "mine"). Locally, I have moved on foot or traveled by car, by bus, by train, truck, ferry, ship, sailboat, kaiki, rowboat, motorboat—or swimming; I have visited all the main regions of the country, although, of course, I have not been everywhere: I have experienced Greece from Epirus to the Dodecanese, from Thrace to the Peloponnese, and I have spent time on almost 60 of the nearly 200 inhabited islands; I have explored both the wilderness devoid of human presence and cities like Athens, populated by millions. Therefore, when I start planning another trip to Greece, I feel that a temporary break in my continuous journey is about to come to its end. But despite the diversity of these experiences, I make it a point to keep my original fascination with Greece, for which I fell at first sight. Alive with nostalgia, motivated by curiosity, I am also driven with routine. I try to be both: a discoverer and an Odysseus at the same time.

I write about my "one journey," or rather present excerpts from it, attempting to substantiate the theoretical discourse from the first part of our book a little further. I observe modern Greece as her *aficionado*, and sometimes as her critic; I also see her as the author of numerous publications on antiquity, on Greek culture, on ancient philosophy, and on metamorphoses of great ideas. Each time, however, I make an effort to look at Greece in such a way as to see the reflection of these grand narratives in the smallest details—to recognize vestiges of seriousness and kindness of the past long gone in the face of a shepherd waving at me, or to hear the echo of ancient symposia in the voice of a friend, with whom I drink *raki* in the evening. Greece—whose philosophy, reflection, sensation of the rhythms of nature fill her lands to the brim—entered my life by osmosis. All I wish to do now is to vigilantly listen, carefully observe, and then—I know that I will hear and see more and more. In such a mindful state, one can single out voices of particular significance today: mingled with the chirping of cicadas are the voices of the great sages, who still have important things to say.

Every time I plan another visit to Greece, I learn something new, welcoming the uncertainty of the unfamiliar. I believe that experiencing uncertainty is one of the most valuable aspects of traveling. As Paul Theroux famously wrote, "Tourists don't know where they've been, travelers don't know where they're going."[2] His adage does not only emphasize the traveler's openness to the unknown, but also pinpoints the central difference between "the traveler" and "the tourist." In fact, I think that the essence of this distinction lies in one's ability (or lack thereof) to accept uncertainty, to welcome ambiguity, to reconciliate the known with the unfamiliar. The modern tourist expects (and requires) that the actual conditions of their trip will match the commercial offer—that the photographs in the folders "tell the truth," and that the travel agency's narrative will correspond to the lived experience. A traveling tourist requires that everything go according to plan: all components of the tour must be predictable and accounted for to such an extent that the conformity of the "contents of the trip" to the "contents of the advertisement" often turns out to be more important than the very locations visited and phenomena explored. On the contrary, in the experience of the traveler's journey—or in expeditions embarked upon with the loved ones—what matters most are freedom, agency, creativity, and, above all, the shared consent to accept multifaceted uncertainty: the uncertainty of place, time, people, events, but also the uncertainty manifest in the lack of control over emotions or in their ambivalence. Choosing to become immersed in the new, the traveler opens before himself worlds and options unknown before—an exercise of particular importance: as the Dalai Lama supposedly said, one should travel to a place they have never visited before at least once a year. Possibly, we visualize ourselves as *travelers* and our lives as a *journey*, because, as travelers, we cast ourselves into the unknown, embracing the fact the journey will inevitably expose us to uncertainty, teaching us how to *be-in-uncertainty*. Needless to say, I find it hard to accept the commonplace model of life—propelled by the craving for complete control and desire of predictability; I organically resist the idea of life-as-tourism, life subject to economic calculations, motivated by the search for immediate gratification, oriented toward the fulfillment of the conditions of predictability and the enhancement of the levels of perceived luxury—life subjugated to external factors. Everything has its price: the tourist doesn't fully know where he has been—and this is a price I am not ready to pay.

Meanwhile, in planning any journey, when one finds reconciliation with uncertainty and opens up to novelty, it is also important to take into account the possibility of a collision of expectations with the actual reality to be encountered. For a tourist, any discrepancy between the expected and the actual is a sign of failure; for a traveler—each such rift presents a splendid, and welcome, opportunity. Furthermore, each experience of this kind affords the traveler an insight into the sense of his or her projections of the future, and provides a premise to ponder the validity of any of our *self-projections*. A journey is a many-faced unity, a presence of many manifestations, because—as Andrzej Stasiuk wrote—"to travel means to live. Or at least to live doubly, triply, multiple times."[3]

In the course of my many trips to Greece, my perspective of this country has evolved, and, along with this transformation, also my notion of journeying to and in Greece has undergone a metamorphosis. Today, I know that the benefits of slow traveling in Greece definitely exceed those of "checking off the must-sees." A slow journey allows one to *feel* the change, enhances mindfulness, lets oneset the places visited in space, and root them in contexts. As I have mentioned before, for several years now, rather than fly, I *drive* to Greece. Although the trip by car may take longer and be more cumbersome, it affords one a more attentive experience of the road, thereby imparting upon the driver and the passengers a deeper sense of space. A road trip across Greece offers yet another undeniable value: while driving, one organically feels the changing landscape, and thereby develops a much more versatile understanding of the uniqueness of the country than other, faster forms of transportation could allow.[4] For a tourist *flying* to Greece, and then taking a direct shuttle from the airport to the hotel, the primary associations with the country are routinely reduced to the turquoise sea and breathtakingly beautiful coast. Yet, for a traveler traversing Greece, Hellas, far and wide, is primarily mountains. Of course, the Greek seascape, with hundreds of islands, is inseparable from the landscape of the mainland, but it is the towering mountains, whose presence is rarely acknowledged in travel agency folders, that seem to be the country's indispensable, defining—albeit seemingly unspecific—feature. To me, the landscape turning into seascape, and extending from the mountains to the far-off horizon is the emblem of the Greek experience: such terrain formations, naturally hindering movement, simultaneously create conditions conducive to the emergence of many local communities, more or less isolated and independent from one another.

Yet, regardless of whether one's first encounter with Greece takes place in Evzoni, Niki, Mavromati, or Promachonas, or whether it happens in Athens, Heraklion, Kos, Rhodes or elsewhere, the fundamental experience of Greece is the experience of light. The Greek sunlight does not overwhelm; instead, it liberates; it lends distinctiveness to the Greek scape and space; it emphasizes details, it lends rhythm to life. The people of Greece understand the importance of the sun perfectly well. The figure of Helios traveling across the skies in his chariot, the dramatic journey of his son Phaeton, the rites in the Peloponnese, and, above all, the sun cult practiced on the island of Rhodes, are only some of the symbolic testimonies to the importance of the Greek experience of natural light. Sylvain Tesson, describing the immutable perdurance of Homer's epics in his fascinating book *Un été avec Homère*,[5] notes that a Greek, both in the antiquity and now, is content with reality "given in the sunlight," which implies the association of the natural light with the acceptance of human fate. Tesson believes that in the Greek sun things become tangible, present, while people affirm the currency of their own presence without having to resort to any supernatural, or foggy, promises. And I am inclined to add that it is worth noting that the Greek sun exposes the *limits of things*, it renders them in sharper focus, it highlights our location. The world in which we live is thus *given to us*—it is *visible*. And whether or not we adopt the attitude of the necessary attentiveness to immerse ourselves in the world's rich variety—depends solely on our choices. It offers itself to us in light, inviting us to watchfully observe and explore it. It is impossible to remain indifferent to the natural beauty of the world that calls us to itself: the affirmation of reality is inextricably bound with the encouragement of activity. The man of Greece is well aware of this, and can creatively combine the pathos of contemplation with the requirement of action.

"The Greek Journey" is probably more than just a well-known literary theme: one might even argue that the phrase has come to denote a unique literary genre. On the one hand, the narratives of Greek journeys tend to present the country through the lens of its ancient heritage; on the other—they often offer readers personal accounts of the traveler's experiences and reflections arising from their Greek encounters. For me, too, encounters with Hellas provide the fabric for my literary canvas. More importantly, however, I find that each "Greek journey" may potentially reveal the fundamental frameworks of metaphysical—and existential—experiences. In my view, this revelatory potential is of central value to each meaningful encounter, because the motivations, the character, and the purposes of "the Greek journey" may well offer us otherwise unavailable insights into ways, in which humans think about themselves and about nature. In the paragraphs below, I attempt to contrast two such attitudes to travel, reflecting, in essence, two dissimilar modes of living and contemplating reality. To explain my point, however, I must refrain from referencing scholarly or scientific texts; likewise, I abstain from quoting from existing accounts of actual peregrinations in Greece. Instead, I rely upon the atmospheres called into existence by the language of poetry and its metaphors, since poetry, as I firmly believe, oftentimes demonstrates its unique power to accurately expresses what descriptive and discursive languages of scholarship and science fail to grasp.

The philosophical consequences of each of the two dissimilar attitudes to travel become clear in the light of the comparative analysis of two canonical poems: "Ithaka" by Konstantinos Petrou Cavafy[6] and "A Journey" by Zbigniew Herbert. I open the section below by introducing both masterpieces, and then pass on to sublimating from them the fundamental models underlying each of these "Greek journeys"—expanding my analytical insight to include discernible attitudes toward the world, informing each of the poems, respectively.

"Ithaka"
As you set out for Ithaka
hope your road is a long one,
full of adventure, full of discovery.
Laistrygonians, Cyclops,
angry Poseidon—don't be afraid of them:
you'll never find things like that on your way
as long as you keep your thoughts raised high,
as long as a rare excitement
stirs your spirit and your body.
Laistrygonians, Cyclops,
wild Poseidon—you won't encounter them
unless you bring them along inside your soul,
unless your soul sets them up in front of you.

Hope your road is a long one.
May there be many summer mornings when,
with what pleasure, what joy,

you enter harbors you're seeing for the first time;
may you stop at Phoenician trading stations
to buy fine things,
mother of pearl and coral, amber and ebony,
sensual perfume of every kind—
as many sensual perfumes as you can;
and may you visit many Egyptian cities
to learn and go on learning from their scholars.

Keep Ithaka always in your mind.
Arriving there is what you're destined for.
But don't hurry the journey at all.
Better if it lasts for years,
so you're old by the time you reach the island,
wealthy with all you've gained on the way,
not expecting Ithaka to make you rich.

Ithaka gave you the marvelous journey.
Without her you wouldn't have set out.
She has nothing left to give you now.

And if you find her poor, Ithaka won't have fooled you.
Wise as you will have become, so full of experience,
you'll have understood by then what these Ithakas mean.[7]

In contrast, Zbigniew Herbert—one of the best known and most often translated post-war Polish poets—writes thus:

"A Journey"
1
If you set out on a journey let it be long
a wandering that seems to have no aim groping your way blindly
so you learn the roughness of the earth not only with your eyes but
by touch
so you can confront the world with your whole skin

2
Befriend the Greek from Ephesus the Jew from Alexandria
they will lead you through sleeping bazaars
cities of treaties hidden entrances
there on an emerald tablet above an extinguishing athanor
are swaying Basileos Valens Zosima Geber Filalet
(the gold evaporated wisdom remained)
through a half-open veil of Isis
corridors like mirrors framed by darkness

silent initiations and innocent orgies
through deserted tunnels of myths and religions
you will reach the naked gods without symbols
dead that is immortal in their monsters' shadows

3
When you come to know keep your knowledge silent
learn the world anew like an Ionian philosopher
taste water and fire air and earth
because they will remain when everything passes away
and the journey will remain though no longer yours

4
Then your native land will seem small
a cradle a boat tied to a branch with your mother's hair
when you mention its name no one at the campfire
will know which mountain it lies behind
what kind of trees it bears
when really it needs so little tenderness
repeat the funny sounds of its speech before going to sleep
ze—czy—sie
smile at the blind icon before sleep
at burdocks at the brook the brood of chicks
home has gone
there is a cloud over the world

5
Discover the insignificance of speech the royal power of gesture
uselessness of concepts the purity of vowels
with which everything can be expressed sorrow joy rapture anger
but do not hold anger
accept everything

6
What city is this what bay what street river
the rock that grows in the sea does not ask for a name
the earth is like sky
signposts of winds lights high and low
inscriptions crumbled to dust
memories worn by sand rain and grass
names are like music transparent and with no meaning
Kalambaka Orchomenos Kavalla Levadia
the clock stops and from now on hours are black white or blue
they absorb the thought that you lose the lines of your face
what answer can an eroded inscription give to thistles

give back the empty saddle with no regret
give the air back to another

7
So if it is to be a journey let it be long
a true journey from which you do not return
the repetition of the world elementary journey
conservation with elements question without answer
a pact forced after struggle
great reconciliation[8]

Although an interval of several decades separates the publication dates of these two poems, Zbigniew Herbert's "Journey" seems to clearly allude to Cavafy's "Ithaka." The similarity between these works is most evident in their respective structural layers. Both poems serve as personal messages from experienced lyrical subjects, whose wisdom came with age—individuals seasoned by the hardships of the long journey of life. The didactic goal of each of the poems is to provide guidance to those just about to embark upon a journey of their own. However, it is the comparison of the goals of both journeys and the juxtaposition of the forms of their realization that emerge as most fascinating. The journey depicted in "Ithaka" is essentially a journey to return home, to the source; yet, in terms of the tropes employed by the poet, the journey itself is distilled out of the Greek context, thereby becoming a grand narrative with which each of us—every Odysseus—will identify: for each of us, embarking on the journey of life, there are Laistrygonians, Cyclops, and Poseidon, lying in wait.

Cavafy advises that the journey be long, savored slowly, replete with adventures and experiences. It is important to discover that which is new, to reach harbors yet undiscovered. Above all, the journey is about learning lessons from those-who-know; Cavafy's Ithaka is the home to which we all aim to return, and therefore we must never forget Ithaka. Or, perhaps, it would even be more accurate to pose that each of us, like Odysseus, is a pilgrim seeking his or her way home—and that each of us has his or her own Ithaka, which the use of the plural form of the toponym (Ithakas) in the last line clearly underscores. Cavafy emphasizes that "Ithaka gave you the marvelous journey," recognizing that we depart from a certain place only to return to it, and that upon our return, carrying a baggage of experiences, or perhaps even knowledge, we are changed. Moreover, our sojourn outside Ithaka teaches us to understand what Ithaka really is, and why we find her so valuable. The life-journey of the poem is a *return*, and thus a form of *nostalgia*[9]—epitomizing the longing to return to the place from which one originally departed. This desire to return, the sentiment for the past, or even the peculiar kind of pain one feels experiencing the lack—all render the journey's goal as *familiar*, generally recognized, *shared*. Although such a journey brings us toward what is somehow known, ourselves, transformed, we assume that each of our Ithakas has undergone a metamorphosis.

Unlike Cavafy, Zbigniew Herbert addresses another type of journey that is life: "a true journey from which you do not return." For Herbert, the grand journey of life is an "elementary journey" and—like Cavafy—the poet notes that it should be long. In Herbert, however, "long" is not so much about its duration, as it is about the intensity and diversity of experiences. In his poem, the journey resembles wandering, seemingly aimless, erring in the dark. It is important to observe the fundamental difference between the two outlined models: Cavafy looks at the journey through the lens of its goal (home), while Herbert sees it in the perspective of the journeying itself. It is not even about the wanderings of a particular traveler; in a sense, what is crucial to Herbert is the changelessness of the journey that each of us shares—or perhaps even more: the journey as such is key: "the journey will remain though no longer yours," to which, a few lines later, Herbert unambivalently adds that "home has gone / there is a cloud over the world." In a sense, the entire poem is cautionary: it offers advice, or a set of guidelines. The lyrical I recommends that one's heretofore obtained knowledge be silenced, and that, in its place, one should relearn the world (as the Ionian philosophers once recommended). Furthermore, it emphasizes the validity of the realization of the pettiness of speech (in times of the dictatorship of endlessly multiplied words), and, above all, it commands one to *accept everything*. This pre-Socratic theme seems to be of paramount importance: in pre-Socratic philosophy, the acceptance of nature and the elements, of one's own changeability and immortality, of the entire reality that presents itself to us is of central significance: the world is a garden of diversity, growing and blooming, but also, with time, withering and dying. The journey of life is an affirmation of the laws of nature, of the insignificance of man, of wandering and of erring, which is best captured in the concluding phrase of the poem, emphasizing the "great reconciliation." And it is precisely such a journey—the journey in which it is not the *goal*, but *journeying itself* that is important, that is mine. My Greek Journey is not an "abstracted journey" of an Odysseus-Everyman: it is an "elementary journey," focusing on the intensity of sensations, emphasizing their diversity, providing space for the affirmation of the experienced world, and, above all, intensely concentrated on the understanding of the great reconciliation of all things expressed by the word *phusis* (φύσις).

Blue Zones—I Am Not in the Blue

Many years ago, during one of my trips to Greece, I learned about the phenomenon of the so-called "blue zones." Researchers identified several places in the world where people live significantly longer than elsewhere. Among these spaces of longevity were Okinawa (Japan), Sardinia (Italy), Nicoya (Costa Rica), Ikaria (Greece), and Loma Linda (California, USA). The term "blue zones" was first coined by Dan Buettner, a writer, journalist, and restless spirit, who became interested in this phenomenon and, consequently, attracted to it the attention of global audiences.[10] Its popularity is evidenced by the fact that in 2023 Netflix launched an original series titled *Live to 100:*

Secrets of the Blue Zones,[11] in which Dan Buettner himself visits these zones to explore longevity, finding out, as he also emphasizes in his writings, that "the world's longevity all-stars not only live longer, they also tend to live better."[12] Leaving the precise understanding of what "better life" is undefined for a while, it is nonetheless worth noting that the qualitative dimension of "good living" did not escape the researchers' attention. The phrase "Secrets of the Blue Zones" may suggest the existence of some hidden factor responsible for longer and better life, and it was partly the hope of uncovering these mysterious elements that whetted the curiosity of individual people and motivated whole companies, interested in commercializing this knowledge, to explore longevity. As it turns out, however, the "secrets" of long life are anything but mysterious: they lie in the rhythms of life, natural to the inhabitants of the blue zones, in their centuries-old traditions, and in their intuitive approach to well-being.

Unsurprisingly, of all of the Blue Zones listed above, it was Ikaria, an island in the north of the Aegean Sea, that triggered my particular interest. Although scientists have long been trying to explain the uniqueness of the phenomenon of longevity in Aegean Greece, their research tends to preliminarily point to the combination of physical exercise with genetic, dietary, and social factors. Yet, in search for unequivocal answers, it might be worth approaching the issue of the Ikarian's immunity to pathogens from a different theoretical perspective: Ikaria may be looked at as a garden, whose rhythm favors its long-term, balanced development.

Two questions concerning the Ikarian longevity bothered me in particular: the first one concerned the particularity of the rootedness of contemporary Ikarian lifestyle in tradition; the second regarded possible reasons why data indicating the uniqueness of Ikaria differ from data concerning the nearby islands of Samos, Chios, Patmos, or Lipsi, or from those located further away, such as Paros, Kythnos, or Kefalonia. When exploring Ikaria, Dan Buettner was aware that seeking to find his own answers, he was not looking for one, single, clue, but for many smaller elements that could jointly help explain the phenomenon at hand. Contrary to the popularly shared "wishful thinking," no isolated factor—a single substance, plant, or diet—can be identified to account for the islanders' longevity. Yet, at the same time, the accumulation of many factors (including those mentioned by Buettner) does not solve the mystery either. Philosophically speaking, mere multiplicity will not suffice if devoid of a synthesizing unity. Isolated, individual components of life, irrespective of their number, cannot account for its overall uniqueness in a particular place.

Interestingly, the geographical and historical characteristics of Ikaria are not dramatically different from those typical of other islands. Unlike the neighboring island of Samos, the homeland of Pythagoras and Epicurus, the mountainous Ikaria, whose steep cliffs are ruthlessly whipped with the Etesian winds (μελτέμι) in the summer, did not go down in history as home to outstanding personalities. Always perceived as somewhat parochial, since

Greek and Roman times, Ikaria, at best, has been recognized for its thermal springs. In modern times, its population, fearing pirates, fled to the mountainous interior. Yet safety came at a price: work in a harsher, more hostile environment was far more strenuous. Still, Ikarians, ready to pay this price, were left with no other option but to adapt to these new conditions and to somehow carve out a living. The demographics of the island changed after 1948, when several thousand communists were resettled to the island, which factor significantly influenced the increase in the non-egocentric social tendencies, ones oriented toward the commune.

Dan Buettner attempts to catalog the most important factors responsible for the uniqueness of Ikaria. Owing to the lack of mass tourism, Ikarians are less inclined to live a life scheduled according to the demands of the market economy—or money—than the inhabitants of other islands. Their life is paced according to their own biological clock, its rhythms unaffected by norms imposed elsewhere by newcomers from across the sea. Ikarians go to bed late and sleep in until late, taking regular naps throughout the day (regular napping significantly reduces the risk of heart attack, unlike occasional napping, which increases it). Their diet is low in dairy and meat, and rich in olive oil, vegetables (potatoes, beans), fruit, wine, and (to a lesser degree) fish. Of unique significance is the use of local herbs and wild plants (sage, wild mint, oregano, rosemary, milk thistle), including the famous "mountain tea" (*Sideritis*; *Lamiaceae*), the many varieties of which are popular throughout Greece. The tea has documented beneficial effects on the human digestive system; additionally, it lowers blood pressure, and has antispasmodic, antioxidant, antibacterial, and anti-inflammatory properties. Likewise, honey, occurring in the island in many unusual varieties, is widely used by its inhabitants. An indispensable component of their daily diet, it is also used in medicine (both for consumption and for wound-dressing) as an antibacterial agent. Importantly, essential to the Ikarian diet is not *what* is eaten, but also *how* it is eaten, and how particular dishes are prepared. Buettner, for example, mentions healthy combinations of food and drink: "I learned, for instance, that drinking wine along with a plant-based meal increased flavonoid absorption and doubled the healthful effects of drinking wine alone."[13] It is also significant that Ikarians consume home-grown produce, thereby avoiding harmful compounds used in industrial chemical sprays.

The above notwithstanding, apart from the diet (rich in vegetables, fruits, legumes, and olive oil), the key factor affecting longevity is what could be termed as the Greek way of life. The process of life—in terms of particulars—can vary between regions of Greece, but this peculiar lifestyle—a certain philosophy of everyday life—is fundamentally common throughout the country. Naturally, considering the uniqueness of Ikaria, we focus on the rhythm of life shared by the islanders, but many of their practices can also be observed in numerous other locations, perhaps less intensely so in large cities, whose pulse is quite similar to the rhythms characterizing the life of other metropolises around the world. Above all, the way of life in question is an amalgamate of many components, not a homogenous phenomenon. It is for this reason that it is not possible to isolate a single factor, ingest a particular substance, or eliminate a concrete element from our lives to achieve the desired effect of longevity. To the frustration of those whose fascination with the blue zones (at least to some extent) arose from the hope of easily isolating life-extending agents and commercializing them, the life philosophy conducive to longevity proves to be an aggregate of a plethora of components. Sanctified by tradition, attuned to the rhythms of nature, it is immune to shocks, but also resistant to the temptations of easy transplantation to the foreign soil.

In the case of the phenomenon of the Greek culture, several constituent elements can be identified. In Greece, the natural need of safety is catered to owing to the power of shame: an often-overlooked factor related to the diminished need for privacy among the Greeks. People living in their communities function—if one may say so—in plain sight of others. The others' gaze, somewhat inherently, curbs potential antisocial behaviors: a popular Greek joke has it that an old lady in the window is the Greek equivalent of CCTV. At the same time, family and community, warranting the sense of one's physical safety, are also the source of broadly understood support in the event of any adversity—such as illness, job loss, or a conflict—affecting a community member. This sense of community is evident even in small-scale, mundane phenomena, such as sharing conversations over shared meals. In this respect, it becomes clear that it is not only the very substance of the food that affects the well-being of individuals but it is also its form and circumstances. Life is not just a set of repetitive rituals, powered by the right diet and exercise. Greek philosophy teaches us of the necessity of thinking about a person in terms of purposes and causes: the Greeks seem subconsciously aware of this in their daily activities. It is about what the Japanese call *ikigai*—the reason to live, a purpose that makes getting out of bed every morning worthwhile, irrespective of one's advancement in years: even old age should have a vision of purpose, a sense of meaning in life. Western culture has developed a life-system focused on the effort of work, and culminating in consumption. This ever-accelerating whirl *must* spin faster and faster, as stagnation—or uniform motion—would imperil the realization of the goals. In fact, such a system does not allow time for breaks, or for creative stillness; periods of rest—vacations—are tolerated only inasmuch as they allow one to "recharge batteries" to a level necessary for one to be able to return to the treadmill. It is hardly surprising that the life of contemporary cities no longer revolves around former urban landmarks: churches (spaces of reflection) or meeting squares (agoras); today, "the urban" is largely defined by the refined contents of supermarkets and the elegance of shopping malls. Still, there comes a moment when the centrifugal force of the ever-accelerating whirl casts us out of the race, and the frenzied chase abruptly stops. Suddenly decelerated, we fall into the numbing stillness of retirement—and find out that the discourse of retirement favors one particular grammatical form: the past tense. We reminisce about *what was*, we rest after *what was*, we enjoy *what was*, because everything that used to give us "a vision of purpose" happened in the past. Meanwhile, in all the cultures in which zones of longevity have been identified, old age, never "systemically inactive," never loses purpose—and thus the future tense always applies to life. While in Greece, I always have the impression that people, "bustling about," and running their errands at a moderate pace, far away from the madness of the mainstream whirl, and unimperiled by the future of stillness—embrace a philosophy akin to that underlying Robert Neil Butler's apt adage: "None of us know whether we have already had the best years of our lives, or whether the best are yet to come."[14] Incidentally, in his recent book, the researcher formulates nine guidelines for a longer and better life: (1) maintain mental vitality; (2) nurture your relationships; (3) seek essential sleep; (4) set stress aside; (5) connect with your community; (6) live an active life; (7) eat your way to health; (8) practice prevention; (9) stay with these strategies. In conclusion, Butler explains: "I want you to live longer; to accomplish that I want you to do things that rouse the quiet stream of happiness that you know is there. More than anything else, that will enhance your longevity"[15]—and it is worth noting that most of these recommendations are related to social relations and active living. Ultimately, it does not matter what the sought-after purpose of life should be, or what it is that makes one want to get out of bed every morning. What is important is that we do have *some* goal, that there is an attractor that pulls us toward life: it may be a hobby—such as gardening—but also may be regular, or schedule-bound work. When the entirety of life is taken into consideration, and when it is motivated by a vision of purpose in such a holistically conceived structure of human existence, meaning is revealed in the lens of the intentionality of human nature. In the context of the above, one could claim that the most important dimensions of the human being-in-the-world that affect one's longevity and quality of life are one's social contacts, one's perception and experience of time, and one's relationship with the totality of nature (including one's attunement to the challenges posed by physical and meteorological conditions of space), as well as the nature of the adopted life goals.

Community—Wholeness—*Siga-Siga*

A person is a member of a community, but also an active component of the *continuum* of life—not so much life in the individual sense, but life understood as the great procession of generations. Everyone is an important constituent of this whole, and everyone has an inalienable role to play. Everyone *feels* the meaning of their existence and has a deep sense of belonging to the world, community, tradition, and time.

A concert in Sitia, Crete, to which I went several years ago, comes to my mind as a testimony. The group on stage performed folk songs—almost non-stop, not interrupting their performance even to take drinks. In concentration, and with increasing excitement, the audience would listen to their musical narrative, which described the history of Greece—and Crete—from ancient times through World War II to the present day. The artists sang about events of general importance, but intertwined them with personal stories; they recounted breakthroughs of historic magnitude, but interwove them the problems of the present day. Their song was one, unbroken, great story—a poetically musical Shield of Achilles of sorts, representing their entire world, past and present. From time to time, someone from the audience would approach the vocalist to whisper in his ear. Soon thereafter, the artist would weave new threads—topics of importance for the people who made their silent requests—into his performance. In the end, everyone left their chairs, stood up, and started dancing to the rhythm of the music of their life. Ah, if I could get a penny for every evening fete in Greece I witnessed, where countless people would join together in dance, sharing songs familiar to everyone. But I do remember one evening in Milies that was different than others: standing under a tree, watching the joyous spectacle, I suddenly felt acutely embarrassed by my non-Greek-like solitude.

Greeks know their melodies and listen to them often; they love communal singalongs and dancing; they communicate with each other, but also with their ancestors—through culture. This cultural continuity equips them with a unique idiom for mutual communication: one that extends far beyond the schematic language of information transmission. This "extended" language is capable of expressing content that the latter would be powerless to even approximate: emotions, feelings, hope, pride, and many other aspects of individual or shared experience are out of its range. Cemented with it, the sense of community is evident at every turn, manifesting itself as a distinct alternative for the "monadic" cultures, commonly typifying people lacking connection. In Greece, it is obvious that community is neither a group of people nor a collection of solitary monads—not even a nation: community emerges as a result of quotidian activities, shared on many levels. Greece, supposedly, best exemplifies the consequences of the pandemic crisis piling upon other, still unresolved crises in the Western world. In reality, however, adopted by many of the country's residents, the Greek lifestyle proves to be surprisingly resilient to external shocks. Perhaps it is this local community spirit that is key.

No one sings the communal spirit and the spirit of coexistence as exquisitely as does Homer. In book XVIII of the *Iliad*, the poet depicts the famous shield, which, at the behest of Thetis, Hephaestus forged for Achilles. And it is not only the literary refinement of the ekphrastic description that renders the passage striking; equally compelling is its rendition of the world. In the passage, reality is represented as harmoniously organized whole consisting of opposites, of which each has its place and measure: the world is thus an epitome of attunement and unity. Most suggestive is the portrayal of life as a dance, a frolic, a procession of our fates. Even today, as one watches the Greeks dance in taverns or outdoors, this Homeric image comes to mind. The same harmonious world, the same passions shaking and disturbing the order, the same difficult choices people must face, the same inexorable fate above them, and most importantly, the same dance of figures holding hands, the same steady melody and order of mellifluous steps. Whatever this world may be like, it must be accepted as it is in its deep structure. The ancient man, and many Greeks today, appreciate the measured rhythm of the day, content with the modesty of needs achievable by their own efforts, and the repetitiveness of the processes, which stems from the fact that the rhythm of our life is only a function of the atmospheric rhythm of nature.

One can then say that the broadly understood communal space is the proper space for human functioning. Furthermore, it is only the communal space that conditions human life, bestows it with meaning, and determines its health. In this respect, another excerpt from Dan Buettner is worth quoting:

> As Nicholas Christakis, a physician and social scientist at Harvard, found when examining data from a long-term study of the residents of Framingham, Massachusetts, health habits are as contagious as catching a cold. By his calculation, an individual's chances of becoming obese shot up by 57 percent if a friend became obese. Among the Adventists, it was mostly positive social contagions that circulated through networks.[16]

Of course, this phenomenon is not unique to Adventists. As the examples above demonstrate, the Greek communal spirit is precisely this: one of the key "positive social contagions." Living in a particular environment requires that one necessarily adopt the lifestyle of the community sharing it, thereby forcing one to accept the philosophy of life manifested in such a lifestyle. Consequently, one could conclude that the Greek environment forms positive habits, and positive habits lead to flourishing—and, if that is the case, atmospheric health can also be understood as a set of communal, natural, and supra-individual factors shaping a person. By extension, this is how one can understand the claim that it is particular metaphysics that determines human existence—these factors, accumulated, complement each other.

For example, you cannot help but adjust to the daily rhythm of people living in a climate zone necessitating a siesta in the middle of the day: *nolens volens*, no matter how hard one tries, nothing will get done until the siesta is over. However, most matters, irrespective of how simple or how complex, can be settled in the evening, at the tavern, where one meets other people and talks, discussing pressing issues or simply engaging in, apparently trivial, chitchats. Eventually, one adopts the routines of the local community: in sync, one starts to eat when others do, enjoying the same simple food usually available in the area; one takes strolls like others do, hikes in the mountains like others do, participates in communal festivals, and, most importantly, learns to converse. Such atmospheric coexistence shapes one's habits, and the constancy of the latter affects one's health. Quite unconsciously, by immersing in the rhythm of Greek life, one also takes on the subtle set of metaphysical principles organizing the life of the locals. One's health, therefore, is not anchored in our corporeality—but it is also *more* than a physical, psychological, social, and spiritual phenomenon: one's health—primarily and fundamentally—is *atmospheric* and *metaphysical*, as long as we understand metaphysics in terms of the causal foundations of the structure of our being.

Summarizing his chapter on Ikaria, Dan Buettner writes thus:

> After studying Blue Zones for nearly a decade, the big aha for me is how the agents of longevity reinforce one another for the long term. If you want people to adopt a healthy lifestyle, you need to build an ecosystem around them. As soon as you take culture, belonging, purpose, or religion out of the picture, the foundation collapses. There's no silver bullet. The power lies in the mutually reinforcing relationship between lots of little bullets. The secret is silver buckshot.[17]

Above all, however, one cannot ignore the fact that health is grounded in the ecosystem of which Buettner writes: health, therefore, has an *atmospheric* dimension.

Greek life seems to be ticking to a different clock, time flows at an alternative, seemingly slower, pace. Scheduled meeting times, as a rule, are construed as "approximate": scheduled hours expand into loosely framed "slots" on the day's agenda—especially that it is rather unlikely that someone's bus should arrive "on time" anyway, and no one—maybe except tourists—expects to see it at the bus stop at the hour listed in the timetable. The Greek language is replete with idioms encapsulating this unique attitude toward time: *siga-siga*, for instance, means: "slowly," "easy," "no rush," "no hurry"; *avrio* is the word most frequently used to postpone actions or obligations; and "a Greek minute" is a phrase I hear whenever Greeks wish to communicate that they will need an unspecified amount of time to do something. In Greece, the world decelerates, but owing to this it becomes *noticeable*—it becomes present to us in the perspective of our attentiveness. Ourselves, we slow down along with it, and then we get to see more, with greater precision: we notice details, we can savor the minutiae. Yet, it is precisely the slowing-down, the profound tranquility, and the understanding of the peculiarity of Greek meditation that may be the first among the serious barriers outsiders encounter when coming into contact with Greece. Globally, Western culture has embraced a peculiar cult of productivity and consumption: people work long hours, and the intensity of their work affects the intensity of rest—which, in itself, often involves not so much *the break* from work, but *the change of place* where work is performed (more often than not, laptops and smartphones accompany us on vacation). In most Western countries we work a lot, but despite the affluence we generate, we live somehow less: the routine of duties and obligations overshadows the world around us. Meanwhile, in Greece, time, whose flow is perceived as slower, is also atmospheric. A Westerner, used to his or her own accelerated pace of life, who attempts to find his or her bearings in the time space of Greece, may experience a shock, and sometimes—disappointment. Complaining about the "Greek time" will not help; to accomplish anything, one has to adapt to the environment. In Greece, one could say, time manages people, not the other way around. During the siesta—nothing will get done; the rhythm of the Greek person's work is not dictated by the commands of hyperactive managers, but is synchronized with the atmospheric pulse of their life—life, in which work is as important as are holidays, meetings, or celebrations.

The lazy stream of Greek time efficiently dissolves Western egocentrism. The profound peace one attains is all-encompassing, whether we contemplate the sea, the mountains, or an evening at the local tavern. And soon we discover that owing to this ordinary, quotidian, commonplace deceleration of the pace of life, a fundamental axiological recognition comes to us somewhat naturally: we begin to distinguish between that which is really worth our while from that which is of little importance. A vision unattainable from the windows of a vehicle whizzing down the highways of Western agglomerations.

Sitting in a tavern over my morning *elliniko kafe*, I usually have my chair turned toward the street, where the fascinating spectacle of everyday life unfolds. I particularly enjoy spots next to markets. Crowds of people, bustle, conversations, trade—all of which, although seemingly familiar, feels different. Unlike in the markets of the industrialized world, in Greece, there is something majestic to the spectacle—majestic, but at the same time naturalistic; both metaphysical and mundane; simultaneously ceremonial and quotidian. Time slows down, but also becomes liberated; it is both circular and cyclical, yet also vividly unique.

The Stoics advocated a whole set of spiritual exercises aimed at self-enrichment. One of them involved the practice of alertness and attention (προσοχή). The ensuing awakening of self-awareness allows a person to tell things that depends on them from those beyond his or her control. Such attentiveness is also a function of one's becoming sensitized to the present moment, to one's participation in the common, holistic, *being*. And it cannot go unnoticed that the Stoic call remains alive: even today, in the liturgies following the norms of the Byzantine rite, the deacon utters the word: Πρόσχωμεν (let's be attentive). Such a fusion of attentiveness, vigilance, and sensitivity, however, should be built from the ground up, in close relation to specific practices, naturally emerging from particular lifestyles. This was the common practice in Hellenistic philosophical schools, but traces of such an approach are present in Greece today. Meanwhile, in its lieu, the globally understood Western culture offers mindfulness courses and workshops whose contents and methods, more often than not, are severed from, or even entirely stripped of, life philosophy; such contents, artificially transplanted onto the Western soil, must inevitably remain a foreign body.

When we reflect on the defining components of the Greek world, what undoubtedly comes to the forefront are communality, complexity, relationality, humility, deceleration, sense of purpose, meaningfulness of actions, and naturalism. Hence, as it seems, that Greece, indeed, consciously or not, conceives itself as a garden, whose rhythms, conducive to flourishing rather than economic productivity, determine its uniqueness. The garden of Greece pulses in alignment with its life-cycle calendar: it blooms in its own time, and follows its own pace of change. And if, on top of the above, we realize that all the "forefront" elements are based on what is simplest, omnipresent, handy, and that the fruit of such a garden are consumed in simplicity, what immediately comes to mind is one of the fundamental models of human life—a mode to which Greece (first Samos, then Athens) gave birth—the garden of Epicurus. Exploring it deeper, we might arrive at the conclusion that the present-day "garden of Greece" might, actually, be an iteration of the Epicurean garden.

For this purpose, it is worth recalling the already-quoted recommendation of Robert N. Butler's: "I want you to live longer; to accomplish that I want you to do things that rouse the quiet stream of happiness that you know is there."[18] I do not know if Butler realized it, but his declaration has a clear Epicurean undertone: first, it concerns the quiet stream of happiness, which is how Epicurus understood *eudaimonia*; second, the life goal Butler describes is both attainable and close to us, or—at the very least—its achievement does not require of anyone to develop any specialized competencies, nor does it call for supernatural interventions.

Ethical Geography

Analyzing the ethics of gardening in ancient Greece and Rome, Meghan T. Ray noted: "The farm and garden were settings that evoked a rich and complex set of cultural meanings about human nature."[19] Indeed, garden imagery allows us to understand the multifaceted and transgressive nature of human existence, but also the relationship of man to nature.[20] The first such representations appears as early as in the Homeric descriptions of the garden of King Alcinous and of Laertes's trees (in the *Odyssey*), and in the famous depiction of the shield of Achilles (in the *Illiad*). Hesiod, in turn, points to a set of values associated with life in-and-with-nature, and emphasizes the importance of working the land. Meghan T. Ray adds: "The poet has created an 'ethical geography' where physical distance from the farm and the work of the soil equals moral distance from the good and the just."[21] For Hesiod, the main source of evil and suffering is *hubris*. Addressing his brother Perses, Hesiod tells him directly to heed the law (*dike*), and not to follow the path of *hubris*.[22] *Hubris*, therefore, is *impiety*, construed as lack of reverence for the god—but also alienation from nature and its just order. Humility, in turn, is a choice not to transgress the thresholds of righteousness,[23] i.e., to respect the boundaries appropriate to our human status. Therefore, we can conclude that humility with respect to nature, gods, and law is associated with the toil of work,[24] with piety,[25] and critical wisdom.[26] A person aware of the evil of their time is primarily interested in how to minimize it.

In this context, a passage from Xenophon's *Economics* seems particularly relevant: "the earth, being a deity, also teaches justice to those who are capable of learning; it is to those who show it the most regard that it grants in exchange the most goods" (ἡ γῆ θεὸς οὖσα τοὺς δυναμένους καταμανθάνειν καὶ δικαιοσύνην διδάσκει: τοὺς γὰρ ἄριστα θεραπεύοντας αὐτὴν πλεῖστα ἀγαθὰ ἀντιποιεῖ).[27] Meghan T. Ray, commenting this excerpt, notes that:

> Xenophon develops several ideas concerning cultivating the soil as part of estate management. Book 5 is a list of all the pleasures and benefits that come to the farmer either through his labor or its fruits. This includes health, fitness, gifts for the gods, wealth for men, and lessons in integrity [...].[28]

It is not difficult to interpret the kind of justice that Xenophon addresses. Firstly, the earth is insomuch just in itself, as it *teaches* justice: its role is to demonstrate certain principles that need to be adhered to. Secondly, the earth treats everyone engaging in work related to it equally—it favors no one and excludes no one. Thirdly, it cannot be deceived; one must reckon with its rules. Anyone who cultivates land is forced to learn the laws of nature and can only exercise their own freedom within these confines. Fourthly, the earth is for everyone, but one must heed the earth, and, accordingly, humbly estimate one's own capabilities—as Rick Rubin aptly wrote: "Listening is suspending disbelief."[29] Fifthly, the amount of goods increases as one's respect for the earth grows. And sixthly, agriculture can be conceived in terms of a just exchange between man (farmer) and the earth. As Leah Kronenberg notes in her book *Allegories of Farming from Greece and Rome: Philosophical Satire in Xenophon, Varro, and Virgil*,[30] the common denominator of the reflections of these thinkers is the fact that agriculture is not only a positive symbol, traditionally associated with such an exchange, but also an ironic—and allegorical—excuse to contrast the active model of life with its contemplative counterpart. In my opinion, however, interacting with the earth—working the land or tending the garden—necessitates, in a sense, a unique combination of both ways of life. Just as—ontologically—gardening and agriculture combine nature and culture, in the anthropological dimension, they are overlapping elements of active and contemplative life. In any case, while this is not always stated *expressis verbis*, conversations with many Greeks gave me the impression that in their view these models cannot be treated as mutually exclusive. Another passage from Xenophon sheds even more light upon the subject: "the earth does not deceive but makes truly clear what it is capable of and what it is not capable of."[31] In this sense, the earth/nature is not only an epistemological touchstone of the validity of one's judgments, but also one revealing human limitations on many levels: a pretext for human self-knowledge.

Extrapolating Xenophon's statements onto a more general plane, one could state that his observations concerning the earth retain their validity in relation to all of nature: the farmer, the gardener, the fisherman, the sailor interact with elements that "do not deceive." One could then say that *nature as such* teaches us justice, humility, activity, responsibility, community, hope, planning, patience, diversity, and respect.[32] It seems that this set of values—derived from the patronage of nature—continues to actualize itself in traditional models of Greek life today. Furthermore, one could even claim that the ancient teachings have not gone in vain: forming the core of the Greek tradition, they are so firmly embedded in the culture of the nation that alternative models of life seem to be little more than just grotesquely extravagant. In my view, Greeks may disregard established law, they may defy rules imposed by others, they may ostentatiously manifest their freedom and independence (for instance, by parking their cars or mopeds wherever they feel like, with no regard to road signs), they may treat regulations they deem ineffective and severed from life with a pinch of irony—but they can never disrespect nature, urged to care for it in a profound, metaphysical way. I have yet to meet a Greek person who would not be proud of his or her country and its natural beauty.

Toward the Garden of Epicurus: *Eudaimonia* and *Autodidactos*

In the context of the argumentation presented above, the Garden of Epicurus—both as a particular location of a philosophical school, but also as a distinctive microcosm—emerges, in these reflections, as a "model garden." As follows, it comes as no surprise that Epicureanism would not lend itself to being reduced to an amalgamate of theoretical speculations alone: Epicureanism (and Hellenistic philosophy *sensu largo*) is an *engaged* philosophy; the type of insight it fosters is purposefully oriented to cater to the health of the soul. For Epicurus, associated first with Samos and then with Athens, the central theoretical challenge is the problem of human life, and not just the ontological description of the world's structure. From an ethical perspective, Epicurean philosophy epitomizes the reversal of the *polis*/individual scheme: in its light, it is the private individual, not the *polis*, that is granted the status of subjectivity—one can think of people as private *beings*, not just as *citizens of the polis*. Epicurus's project was largely revolutionary, anti-Platonic, and polemical with respect to other philosophical schools. This revolutionary nature of his insight is also expressed in the location of the school. It is not the city center, nor the gymnasia of Plato and Aristotle, but a building with a garden—a *vegetable* garden at that—in the outskirts of Athens that houses the school. Giving free rein to our imagination, we can envisage this place: away from the city's hustle and bustle, somewhat idyllic, close to the silence of the cultivated fields, where one interacts with nature, enjoys the balsamic caress of the wind, undisturbed by logorrhea. "Philosophers from the garden," as the Epicureans were called, is a nomination of paramount significance: one reflecting their unique way of thinking about man. The school was more of a community of friends who worked and philosophized together, striving toward self-sufficiency. Besides, the garden also had educational and therapeutic purposes, it was meant to liberate those frequenting it from the sufferings of body and spirit, and thus, catering to pleasure, to foster self-development. In his *Epistle XXI*, Seneca cites an inscription, allegedly located above the entrance to the garden:

> Go to his garden and read the motto carved there: "Stranger, here you will do well to tarry; our highest good is pleasure." The caretaker of that abode, a kindly host, will be ready for you; he will welcome you with barley-meal and serve you water also in abundance, with these words: "Have you not been well entertained?" "This garden [...] does not whet your appetite; it quenches it. Nor does it make you more thirsty with every drink; it slakes the thirst by a natural cure, a cure that demands no fee. This is the 'pleasure' in which I have grown old."[33]

When I first came to Greece (and each time since), I realized that this text was a silent motto for many people in Greek villages. Hospitality combined with simple food and simple needs, which, rather than escalate over time, undergo gradual reduction. The garden "slakes the thirst" with the "natural cure," that "demands no fee": nature itself is the peak of subtlety and the ultimate fulfillment of fundamental needs.

Epicurus and his disciples preached that *eudaimonia*[34] (usually translated as "happiness") is attainable in life, and essentially consists in the absence of physical suffering and freedom from inner turmoil. To achieve this state, a person primarily needs *themselves* (αὐτάρκεια), but to achieve its fullness, one's life must be complemented by the community of friends. From this perspective, external and circumstantial factors, such as birth, wealth, or state citizenship, should be considered unnecessary for achieving eudaimonia. The garden renders people equal because—as is always the case in any garden—everyone is subject to the same law. In his *Letter to Menoeceus*, Epicurus expresses his belief that life is not only about reconciling oneself to the necessity of nature's *logos* (in the Stoic spirit) but about practical wisdom, a kind of prudence, which is responsible for every positive choice and every avoidance; invariably, the goal of these reconciliatory efforts remains the health of the body and peace of the soul. These goals are attainable by way of self-restraint, allowing one to be content—and satisfied—with the simplest things. Stobaeus illustrates this claim with the following passage from Epicurus: "I revel in the pleasure of my humble body, employing water and bread, and I spit upon the pleasures of extravagance, not for their own sake, but because of the difficulties which follow from them,"[35] and Cicero confirms it: "[...] natural wealth is easily won, because nature is satisfied with little."[36]

It is worth paying attention to these continuous assurances of Epicurus about the simplicity and ease of a happy and peaceful life. Of course, the key issue is to build such an inner attitude that would agree with these goals—an attitude of satisfaction with a simple life. One could say that the "ease of happiness" goes hand in hand with the difficulty in understanding that this, which is so easily attained, is indeed happiness. The benchmark of happy life is Nature (φύσις), which regulates the laws governing the world: "Thanks be to blessed Nature because she has made what is necessary easy to supply, and what is not easy unnecessary."[37] To further emphasize this minimalist dimension of happiness, another testimony must be mentioned: "Nothing satisfies the man who is not satisfied with a little."[38] The passage, advocating minimalism, simultaneously warns one of the dangers of the endless multiplication of needs: someone who cannot be content with "a little" will ceaselessly strive to accumulate more and more. Therefore, developing the skill of self-restraint conditions the possibility of self-fulfillment: the agency in the area of our happiness lies within us: "the greatest wealth is self-sufficiency."[39]

The self-sufficiency of which Clement of Alexandria writes, however, may be understood in several ways. Certainly, the key point of the claim is that the conditioning of happiness lies within us; the potential for the attainment of happiness is independent of external circumstances. At the same time, self-sufficiency is associated with reasonable self-limitation and hinges upon the belief that what is most important is easy to obtain. Finally, self-sufficiency understood in this fashion has educational and therapeutic implications. Let us dwell on that a little longer.

The philosophers of the Garden would teach that our happiness is in our hands because we are entirely responsible for our lives. At the same time, we cannot see ourselves as isolated monads—mainly because we sprout out of the same garden. For the Epicureans, friendship was a fundamental human bond, a value in itself, one unbased on benefit—even if it is mutual. Friendship thus conceived, free from the usual utilitarian undertones, gains an independent and fundamental merit: it *becomes* happiness, it *becomes* pleasure—not temporary, but for a lifetime.

Even this brief overview of the elements central to the Epicurean lifestyle demonstrates how many of them remained in sync with the natural rhythm of life shared by people in modern Greece. This is not to claim that Epicurean philosophy has somehow been "grafted" into the Greek mentality, but rather that it was—at least in part—an expression of our atmospheric attitude toward the world, and, in any case, that it contained elements that have proven incredibly durable in the face of the historical change and variability of temporary stances. What is especially important to me, in these ancient fragments, I recognize the world I always experience during my Greek travels. Underneath the layer of civilizational superstructures pulses the same rhythm of life, the same beliefs about the relationship of man with nature, the same potential of easily accessible happiness, and—above all—the same philosophy *of the garden* and the same philosophy *emerging from within it*.

As Diogenes Laertius reports,[40] the Epicureans led an uncommonly simple and modest life; they would be content with a cup of wine, but most often drank water. Elsewhere we find information that, according to Epicurus, a sage (an ideal of personal life) enjoys rural life,[41] and in a certain sense—prefers life in hiding, or "to live hidden" (λάθε βιώσας). It is worth emphasizing that Epicurus—contrary to the Cyrenaics—mainly recognized the importance of the "static" pleasure, the so-called καταστηματική ἡδονή, not only the pleasure-in-motion (κατὰ κίνησιν ἡδοναί). The former is spiritual in character and becomes manifest in the absence of spiritual sufferings, anxieties (hatred, envy, contempt), and in the suspension of disbelief in achieving happiness. In such a perspective, ἀγωγή (education) becomes a mode of life and a principle of its organization (διαγωγή). I do admit that I find this model of self-sufficient life in the garden of nature, away from the bustle of the city, the quintessence of Greece, and I still find it astonishing how easy it was for me to adapt to the Greek rhythm of the day from my very first visit. This may have happened because this model is unforced, established by centuries-old tradition, and re-lived by countless generations. Travel is a form of intent listening to the rhythm of the place visited; such an attunement is tantamount to the acceptance of this rhythm and, at the same time, a mode of submitting to the rules of the place.

This revisit of the Epicurean garden philosophy as a philosophy of rural life allows us to reach back to other, similar, descriptions of rural philosophy. Almost mechanically, Socrates's extraordinary walk, described in Plato's dialogue *Phaedrus*, comes to mind. The action of this dialogue takes place outside the walls of Athens, which is exceptional not only for Plato, but also when considering the practice of philosophy by other ancient thinkers, always associated with gymnasiums and urban schools. At one point of his conversation with Phaedrus, Socrates asks: "do you think one could understand the nature of soul satisfactorily without knowledge of the nature of the whole?"[42] Phaedrus, referring to Hippocrates, responds that without adopting such an approach, one cannot even speak of the body. What is particular must be related to the whole, and the whole makes sense only in the perspective of its components. Moreover, each individual thing must be considered both in terms of its impact on other things and in the perspective of being the object of influences from other objects. The whole and the part are mutually interconnected, and methodologically complement each other—not unlike "the one" and "many," which is the subject of Plato's dialogue *Parmenides*.[43] This is why the quoted passage from *Phaedrus* is also important for Gadamer, who highlights the connection between healing—and therefore health—and the understanding of the whole.[44] In his commentary on the Greek model of thinking about health-as-a-whole, Gadamer writes:

> Socrates forbids the purely isolated examination of symptoms and precisely thereby demands real scientific knowledge. But in this connection Socrates goes beyond what modern medical science recognizes as its methodological basis. The nature of the whole that is at issue here is not merely the unified whole of the single organism. We possess abundant evidence from Greek medicine to show us how the weather and the

> seasons, how temperature, water and general sustenance, in short how all possible climatic and environmental factors were seen to make up the concrete ontological constitution of what it is that the physician helps to restore, namely health. But the context in which this passage stands also permits a further conclusion to be drawn. The nature of the whole includes and involves the entire life situation of the patient, and even of the physician. Medicine is compared with the true art of rhetoric which allows the right kinds of discourse to exercise an effect on the soul in the right kinds of way.[45]

The key issue of this passage seems to be not only thinking in terms of the whole-of-the-spirit and the whole-of-the-body, but of *wholeness* as such. We cannot analyze the elements of our life in isolation, or in the context of their nearest colligations. It is the whole that is the broadest environment, which should not be overlooked and which determines the perspective in which we think about life. Living the natural rhythm of life in a small Greek village, consciously or unconsciously, we live in the perspective of the wholeness of nature. Only then is this life rooted and atmospherically present.

Let us return, for a moment, to the dialogue *Phaedrus* (276b–277c), in which the image of the farmer-sower appears as a model for proper rhetorical art. According to Plato, as in actual gardens, where a prudent sower should have an expertise in agricultural art—including the profound knowledge of the soil, in which the seed is to be planted—so it is in the "gardens of writing," in which the "gardener" should know the nature of souls. The passage 277b–c contains a summary of this aspect of rhetorical activity: the central idea of it is that when speaking the truth about something, one must first recognize the nature of the whole, and then be able to divide it up into its components, down to the elements no longer divisible. This applies not only to the message sender/farmer, but also to the object of their influence (soil/soul/nature).

> First, someone has to know the truth of every matter he's speaking or writing about, which is to say that he has to be capable of defining a whole as it is in itself and then know how to divide it up class by class until he reaches something indivisible. He also has to be able to distinguish souls in the same sort of way, discover the kind of speech which naturally fits each kind of soul, and organize and arrange his speeches accordingly—offering a complex soul a complex speech which covers the whole range of modes, and a simple soul a simple speech. Until he can do all this he will be incapable of tackling speeches in as much of a professional manner as their nature allows, either for teaching or for persuasion.[46]

Doubtless, in the quoted passage gardening and agriculture indeed serve as a model for thinking about what proper rhetoric is. But, of course, the analogy opens many more venues. It would be hard not to notice that the methodology outlined above applies to *all* philosophical (dialectical) reflection and—most importantly, in the context of our insights—to life as such.[47] It is work in the field or in the garden that serves as a model of correct thinking about the world—and about the influence we exert upon it. Of paramount importance to this model is the premise of the necessity of one's engagement in the reflection about the whole, then—about its relation to its component parts, and finally—about the relationship among and between the individual elements, contributing to the whole. In the garden of the world, one should construe reality as a network of mutual connections between elements in the perspective of the whole of being. In such a garden, correct consciousness is a function of one's ability to think holistically, but also relationally. It is a mistake to think of the world in terms of complex relations of entities without grasping the whole, but also the holistic perspective is inappropriate if the specificity of individual relations between entities is dismissed. In such a garden, health cannot be conceived of as a function of an isolated element (e.g., the body): it is a function of the simultaneity of the whole and of the relationality of its constituent elements.

In the context of *Phaedrus*, Anne Cotton draws her readers' attention to an important educational—and formative—aspect of gardening:

> Plato's image of the wise and foolish gardeners presents us with a memorable and striking image of education. [...] In doing so, it helps us develop the ability to be self-conscious and self-critical commentators on our own progress as learners. To do this is not to reject the value of external advice, guidance, and stimulation, or the role of discussion and collaboration. It is simply to recognize our potential, as individuals, to become independent and self-sufficient learners; plants that can take responsibility for their own development even without the routine care of a gardener. [...] By suggesting, in this image, that we can become gardeners of our souls, Plato is affirming our potential to do the same.[48]

Importantly, the emphasis on independence and self-sufficiency does not exclude the validity of external stimulation and learning from others. We do not have to choose between self-education and external education: what seems recommended is a certain form of learning which combines both—albeit with an emphasis on self-development, self-criticism, reliance on one's own experience, or, in other words, acknowledgment of one's own wisdom, stemming from our own, unmediated encounters with life.

In almost every conversation with Greek seniors, I experience a unique mixture of pride, wisdom, and kindness that my interlocutors emanate. Their faces, lined and wrinkled, tell the story of a lifetime of hardships—but, at the same time, radiate serenity, or even a vivacious cheerfulness, resulting from these experiences: having faced countless challenges, they never gave up, never withdrew from the fight. Their pride, however, is far from *hubris*: it is a non-invasive testimony of daily experience, of their successful efforts to find solutions, and from their critical assessment of their capabilities. Their life wisdom is largely based on self-education and the legacy of tradition, which they not only cultivate but also creatively assimilate to everyday circumstances. The educational challenge, to which Cotton alludes, is worth exploring in the context of the Greek garden, a space of flourishing.

In a certain sense, the dilemma of choice between self-education (αὐτοδίδακτος) and external teaching (διδαχή) once again directs us back to ancient Greece—back to Epicurus and his garden. At first glance, it seems obvious that we should contrast these two types of education: we tend to think that "experience is the best teacher" and that one "only learns by doing," which suggests that we *either* learn by ourselves, *or* derive our learning from external sources—and, because the proverbs evidently prefer the former, we are often inclined to believe that the principle of *tertium non datur* applies. However, the reality of the learning process does not have to be perceived unilaterally, and the assumption of the exclusive alternative might be, to say the least, ill-motivated. Even a cursory observation of the origins of our knowledge will make us admit that our cognitive resources are embedded both in individual experiences and in learning from others—especially if we consider the impact of family and community education, as well as formal schooling. The above notwithstanding, the ancient tradition of self-knowledge undoubtedly refers us to the Delphic Oracle, which advises: "Know thyself" (Heraclitus, in fragment B101, writes: "I searched for myself"), and also to Hesiod, who, in his *Works and Days* writes thus:

> The man who thinks of everything by himself, considering what will be better, later and in the end—this man is the best of all. That man is fine too, the one who is persuaded by someone who speaks well. But whoever neither thinks by himself nor pays heed to what someone else says and lays it to his heart—that man is good for nothing.[49]

Hesiod's observation suggests that it is possible to distinguish three types of people:

1 people, who understand everything themselves (v. 293);
2 people, who listen to good advice (v. 295);
3 people, who neither think for themselves nor listen to the advice of others (v. 296–297).

The passage leads one to the conclusion that someone who understands everything on their own is "the best," does not need anyone else, and therefore does not need to listen to anyone. His or her self-sufficiency, resulting from self-knowledge, appears to be of a quite extreme variety. Therefore, it is worth contrasting Hesiod's view with the division alluding to Epicurus's worldview, which we find in Seneca:

> "Epicurus remarks that certain men have worked their way to the truth without any one's assistance, carving out their own passage. And he gives special praise to these, for their impulse has come from within, and they have forged to the front by themselves" (Quosdam ait Epicurus ad veritatem sine ullius adiutorio exisse, fecisse sibi ipsos viam. Hos maxime laudat, quibus ex se impetus fuit, qui se ipsi protulerunt).
>
> (LII, 3)

> "Again, he says, there are others who need outside help, who will not proceed unless someone leads the way, but who will follow faithfully" (Quosdam indigere ope aliena, non ituros, si nemo praecesserit, sed bene secuturos.).
>
> (LII, 3)

> "You will find still another class of man—and a class not to be despised—who can be forced and driven into righteousness, who do not need a guide as much as they require someone to encourage and, as it were, to force them along. This is the third variety" (Praeter haec adhuc invenies genus aliud hominum ne ipsum quidem fastidiendum eorum, qui cogi ad rectum conpellique possunt, quibus non duce tantum opus sit, sed adiutore et, ut ita dicam, coactore. Hic tertius color est).
>
> (LII, 4)[50]

While the description of the first group resembles that offered by Hesiod, the characterizations of the second and the third groups are different. For Hesiod, the second group consists of those who listen to good advice and are thereby directed, whereas for Epicurus, the second group consists of those who *need* external support to see the right path. The greatest discrepancy occurs in the definition of the third group: for Hesiod, these are useless people, for they neither think for themselves nor listen to others' advice; Epicurus, on the other hand, does not disdain the third group, but simply considers its members as people in need of a guide and driver at the same time. Attempting to reconcile both these typologies, one could gradationally single out *four* categories of people, in which division Hesiod's third group would be located at the very bottom, and just above it—the third group from Epicurus's typology. However, an issue of equal importance is the suggested opposition between *autodidactos* and external guidance, which may lead to the exclusion of any form of their complementarity. Can we then assume that Epicurus—like Phemius, who proclaims: "Self-taught am I"[51]—considers himself an autodidact learning only from Nature? But then again, it is well known that Epicurus drew his inspiration from Democritus and that he contested many views of other thinkers, which proves his knowledge of different doctrines and positions. We may have to agree with Michael Erler, who writes that "Epicurus' claim to originality does not conflict with openness to other viewpoints, and that strict adherence to Epicurus' doctrinal system precludes neither flexibility nor individual emphases among his followers."[52] In my opinion, members of the first group of people in the combined typologies of Hesiod and Epicurus are not those who reject existing knowledge and external (including personal) sources of information *a priori*, but those who are the source of inspiration and creation for themselves (in the spirit of Nietzsche), and who, ultimately, must also analyze and critically examine every problem on their own. Such individuals are simultaneously creative and self-critical. Therefore, self-education and listening to others are not mutually exclusive, whether these "others" are "other people" or Nature herself. Ultimately, the key point is to be self-sufficient, self-aware, self-critical, and creative, capable of understanding the world, and thereby also capable of achieving happiness understood as the absence of physical suffering and spiritual pain. Such a position puts emphasis on interiorized self-education based on self-knowledge and on individual discovering of atmospheric health. As such, it is a response to the solipsistic, monological, and egocentric models of contemporary education and development.

Coda: Nature Loves to Hide

And so—this is how I perceive the complex process of wisdom-based adaptation to Nature, practiced by the inhabitants of Greece. In my mind's eye, I always have the image of a mountain village on the island of Kos, where a Greek friend took me one evening. All the tourists had already left, the sun had long set, and a largish group of people were sitting at a massive table in one of the taverns in the central square, sipping *ouzo* or *tsipouro*. As usual, they discussed everything and nothing; some of them would raise their voices to drive their points home, others would listen, reflectively silent. An elderly man sitting next to me caught my attention. Dignified yet smiling, almost frivolous, he would speak in a commanding voice, only to fall into a meaningful silence, looking at everyone with keen interest. I was intrigued. My presence did not escape his notice. We started talking; he asked me a lot of questions, listened attentively to the answers, and when he learned that I knew ancient Greek, I saw a flicker of a happy emotion in his face. At some point, I asked him if he would like a drink, as I was about to order something for myself. "Maybe a beer?" I asked—and he laughed heartily. His friend, sitting next to him, also chuckled, pointing out that probably never in his life had my companion had anything but *tsipouro*. This, however, was not the biggest surprise: that night, I learned many more astonishing facts about the elderly gentleman. He had spent almost his entire life tending sheep in the mountains and, as he stated, had not been to the coast or at sea for many years: he felt no need to go there—it was not his world. He began to tell me about his life, about the importance of family, about the beauty of the mountains, about the cruelty of war, about people, about his own, homemade, *tsipouro*, which was allegedly better than the tavern's, and, spinning his yarns, he would finally assert that he was generally satisfied with his life. And that made me think of the centuries of our civilization, of science and philosophy, of great literary works, of school education, of my own education, of various skill improvement programs, of human depressions, of being lost, of fear, of all sorts of therapies—of everything that this old man had not experienced and *did not need to experience* in order to sit among people close to him every evening, chat with them, or tell a stranger from another country that his life had been good.

His wisdom stemmed from his own experience, from the choices that determined his own path in life, but a part of it was also what one could call "the wisdom of the routines," resulting from the patterns evolving throughout centuries-old Greek tradition of coping with the challenges of existence. This wisdom, founded upon his fascination with the pulsing life around and his ability to listen to nature, allowed him to transpose all global connections to a specific place, making his small corner of the world the cosmos containing all the truly important elements the universe offers. I still see the smiling face of that old man, I see his deep wrinkles and curious, vivid eyes. After we said our goodbyes, he walked away, and I watched him calmly blend with the shadows of the night. And I thought that it was perhaps about him that Hesiod and Epicurus wrote.

It did not take me long to come up with the ideal representation of Greek space, which would embody the complex structure of Greek life rhythms. It seems to me that this is best exemplified by the agora in a Greek mountain village, where tables and chairs, set under a few plane trees offering generous shade, invite people to gather for morning coffee and return in the evening for a leisurely dinner. At the foot of one of the plane trees, a spring shimmers, trickling with cold, refreshing, tasty water. The agora serves as a refuge from the heat, noise, turmoil, haste, anxiety, and fear. It is here that people form a community, converse, debate, resolve shared problems, and address issues of importance to individuals and to the group. The square changes in accordance with both daily and annual rhythms, yet it consistently remains a central hub, focusing the life of the entire community. Certainly, the agora in Athens, hosting figures like Socrates, Plato, Gorgias, and others, earned a special place in the Western imagination, but the truth is that such agoras can be found everywhere. They are unlikely to be replaced by shopping malls, which, utilizing the concept of community to intensify trade, impose a new rhythm: the artificial rhythm of consumer routines. The agora, pulsing with life, operates across, and blends, the communal and the natural, much like the garden described by Mara Miller in the already quoted passage:

> Organized and systematic, gardens operate at the border between the explicit and tacit, between communal or "universal" and individual, between objective and subjective, between the conscious and the unconscious or prereflectively conscious, between deliberate and taken-for-granted. They occupy, or perhaps create, a vast of transitional territory between the amorphous, unrepeatable, incommunicable chaos of irreducibly idiosyncratic experience and the readily communicated and comfortable norms of socially generated experience.[53]

In my view, the Greek communities—consisting of extended-humans—scattered across the mainland and throughout the islands alike, are precisely these gardens. Referring to Marc Augé's vocabulary,[54] these gardens-as-extended-communities are specific and local places within the great place that is Greece. These local places-gardens operate precisely at the intersection of various values, connecting what usually remains separate, and they possess a meaning that transcends apparent contradictions. At the same time, they create a unique garden space where health manifests itself in its atmospheric dimension.

Leaving Greece, I am often imbued with a profound sense of having internalized its essence—I carry Greece back home, transplanting her "local places" into my own place. The certainty of return—whatever the formula—lingers with me. Departing from Greece, I gain space, necessary to reassess Greece's singularity, to appreciate its virtues while acknowledging its shortcomings. At the same time, such returns home are often hard: each time, I must acclimate to different rhythms of life, altered experiences of time, and new challenges. And yet, as it seems, such shifts in perspective allow one to better discern global issues in relation to local realities. Observing the world, I can see it through Homer's far-seeing eyes, sitting beneath a majestic plane tree. What do I see? The same. I see that we have disrupted nature's equilibrium, that our greed and short-sightedness, leading us astray, made us cross a line that should never have been crossed. I see that summer heat intensifies from year to year, that violent storms and floods surprise us, and that people are forced to abandon their homes as a result. We are all Achilleses: proud and valiant, yet seriously flawed, burdened with anger, petulance, violence, and excessive emotionality. Like Achilles, we are brutal on a global and environmental scale, yet we yearn to be immortalized in songs. Before our eyes a war, triggered by the disruption of the global order, is playing out. We have learned nothing, always unfulfilled, always seeking to transcend our own boundaries. We cannot—unlike the Epicureans—be content with the simple abundance of what we already possess, we are unable to enjoy small things. We flee our gardens, thinking they are "too tight" for us, that they will limit our development and growth, thus effectively supplanting our own flourishing. We must devour the vast world to rekindle our longing for Ithaka. Restless, uneasy, unreconciled with our inner selves, we sense our world vibrating on edge. We are simultaneously proud and fearful, insatiable and arrogant, brutal and sentimental. Homer reminds us that such a war must eventually be brought to an end, that balance must be restored, emotions must be calmed, and each must find their rightful place and return there. The war of the *Iliad* must evolve into the *Odyssey* of return. We must return to the gardens.

In fragment B123, Heraclitus writes: "nature loves to hide" (φύσις κρύπτεσθαι φιλεῖ). Hesiod, predating Heraclitus, makes a similar observation: "For the gods keep the means of life concealed from human beings" (Κρύψαντες γὰρ ἔχουσι θεοὶ βίον ἀνθρώποισιν).[55] Thus, our return to the gardens is not blind fumbling in the dark, nor is it aimless wandering or mere experimentation with available options. Our recognition of the fullness of nature in its diversity, our attunement to the natural rhythms of flourishing, our acknowledgment of the rightful place of a holistic perspective that transcends artificial dualisms, our consent to participate in communal presence—are but stages of our journey, yet each of them gives us purpose and direction. Let us then return to the gardens, where we can rediscover the profound value of atmospheric health and our potential to attain well-being, our natural state. If nature delights in concealment, then so too should we cherish what is hidden, what lies beneath the surface, no matter where—whether under the shade of a plane tree in a mountain village tavern, around us, or within us.

Sometimes I feel that my perception of Greece might be tainted with naïvete or love, that I am painting an idealized, and therefore untrue, image for myself. I realize that the people I meet in Greece might be different than I see them, and that beyond my gaze the Greek world may be altogether unlike my projections. Yes, that is quite possible. And yet, despite my belief in caution and critical distance, I want to keep that world as it is in my eyes. After all, gardens have always been planted at the intersection of the tangible and the imagined, and the same is true for the Greek places-gardens. They, too, call into existence a unique space, one bordering on the limits of reality.

Notes

1 Lawrence Durrell, *Prospero's Cell: A Guide to the Landscape and Manners of the Island of Corfu* (London: Faber and Faber, 2000), 1.
2 Paul Theroux, *The Happy Isles of Oceania: Paddling the Pacific* (New York: Putnam, 1992), chap. 1.
3 Andrzej Stasiuk, *Fado* (Wołowiec: Wydawnictwo Czarne, 2006), 39. Translated by Paweł Jędrzejko.
4 For a philosophical typology of journeys, see, for instance, Dariusz Kubok, "Philosophy of / as a Journey," *Forum Philosophicum* 29, no. 1 (2024): 9–27, https://doi.org/10.35765/forphil.2024.2901.01. See also: Paweł Jędrzejko, "Poza murami Sandomierza—czyli refleksja nad językiem, kulturą i wzrokiem turysty. (Miniwykład z ćwiczeniami)" [Outside of the Walls of Sandomierz: A Reflection on Language, Culture, and the Tourist Gaze] in *Archeoturystyka: nowy produkt turystyczny*, ed. Tomisław Giergiel (Sandomierz: LOT Partnerstwo Ziemi Sandomierskiej, 2008), 37–45.
5 Sylvain Tesson, *Un été avec Homère* (Paris: Équateurs, 2018).
6 Greek: Κωνσταντίνος Πέτρου Καβάφης, known as Constantine P. Cavafy or C. P. Cavafy.
7 Constantine P. Cavafy, "Ithaka," trans. Edmund Keeley, in *C. P. Cavafy, Collected Poems*, trans. Edmund Keeley and Philip Sherrard (Princeton, NJ: Princeton University Press, 1975).
8 Zbigniew Herbert, "A Journey," trans. John Carpenter and Bogdana Carpenter, *Salmagundi*, no. 185/186 (Winter – Spring 2015): 373–375.
9 "Nostalgia" is a compound of two Greek lexemes: νόστος (homecoming), and ἄλγος (sorrow or despair).
10 The term "blue zones" was first coined by Dan Buettner, author of such books as: Dan Buettner, *The Blue Zones: Lessons for Living Longer from the People Who've Lived the Longest* (Washington, D.C.: National Geographic, 2008); *The*

Blue Zones Solution: Eating and Living Like the World's Healthiest People (Washington, D.C.: National Geographic, 2015); *The Blue Zones of Happiness: Lessons from the World's Happiest People* (Washington, D.C.: National Geographic, 2017); *Thrive: Finding Happiness the Blue Zones Way* (Washington, D.C.: National Geographic, 2010); *The Blue Zones Kitchen: 100 Recipes to Live to 100* (Washington, D.C.: National Geographic, 2019); *The Blue Zones Challenge: A 4-Week Plan for a Longer, Better Life* (Washington, D.C.: National Geographic, 2021).
11 *Live to 100: Secrets of the Blue Zones*, directed by Angus Wall, featuring Dan Buettner, Netflix, 2023, based on the books by Dan Buettner.
12 Dan Buettner, *The Blue Zones, Second Edition: 9 Lessons for Living Longer from the People Who've Lived the Longest* (Washington, D.C.: National Geographic Books, 2012), 237.
13 Dan Buettner, *The Blue Zones, Second Edition: 9 Lessons for Living Longer from the People Who've Lived the Longest*..., 219.
14 Robert Neil Butler, *Why Survive? Being Old in America* (New York: Harper and Row, 1975), 141.
15 Robert Neil Butler, *The Longevity Prescription: The 8 Proven Keys to a Long, Healthy Life* (New York: Avery/Penguin Group, 2010), 264.
16 Dan Buettner, *The Blue Zones, Second Edition: 9 Lessons for Living Longer from the People Who've Lived the Longest*..., 229–230.
17 Dan Buettner, *The Blue Zones, Second Edition: 9 Lessons for Living Longer from the People Who've Lived the Longest*..., 232.
18 Robert N. Butler, *The Longevity Prescription: The 8 Proven Keys to a Long, Healthy Life*..., 264.
19 Meghan T. Ray, "Cultivating the Soul: The Ethics of Gardening in Ancient Greece and Rome," in *Gardening Philosophy for Everyone: Cultivating Wisdom*, ed. Dan O'Brien (Chichester, West Sussex: John Wiley & Sons Ltd., 2010), 30–31.
20 "Farming and gardening were often described positively in literature either in contrast to less honorable professions or as essential components of a well-ordered society. [...] In addition to providing a living, the benefits attributed to this way of life accrued to its practitioners in two primary arenas. First, in civic life, the diligent farmer builds up a surplus so that he can share with his neighbors and with the state. [...] In contrast to the civil incentives for farming, the rewards of divine approbation are received by the individual himself, since by pleasing the gods he averts divine anger and he has a better chance of keeping his earnings. This juxtaposition of divine will and the life of the soil gave philosophical weight to otherwise pragmatic enthusiasms for farming." Meghan T. Ray, "Cultivating the Soul: The Ethics of Gardening in Ancient Greece and Rome"..., 27.
21 Meghan T. Ray, "Cultivating the Soul: The Ethics of Gardening in Ancient Greece and Rome"..., 29.
22 Hesiod, *Op*. 213.
23 Hesiod, *Op*. 226.
24 Hesiod, *Op*. 306–316.
25 Hesiod, *Op*. 138–139.
26 Hesiod, *Op*. 293–297.
27 Xenophon, *Econ*. 5.12.
28 Meghan T. Ray, "Cultivating the Soul: The Ethics of Gardening in Ancient Greece and Rome"..., 30–31.
29 Rick Rubin, *The Creative Act: A Way of Being* (New York: Penguin, 2023), 74. Available at: https://ia601205.us.archive.org/11/items/the-creative-act-a-way-of-being-by-rick-rubin/The%20Creative%20Act_%20A%20Way%20of%20Being%20by%20Rick%20Rubin.pdf (access: June 4, 2024).

30 Leah Kronenberg, *Allegories of Farming from Greece and Rome: Philosophical Satire in Xenophon, Varro, and Virgil* (Cambridge University Press, 2015); for a more detailed examination of Xenophon's *Oeconomicus*, see Gabriel Danzig's article, "Why Socrates was not a Farmer: Xenophon's *Oeconomicus* as a Philosophical Dialogue," *Greece and Rome* 50, no. 1 (2003): 57–76. Additionally, James A. Stevens discusses the themes of friendship and profit in Xenophon's work in the chapter "Friendship and Profit in Xenophon's *Oeconomicus*," in *The Socratic Movement*, edited by Paul A. Vander Waerdt (Ithaca, NY: Cornell University Press, 1994), 209–237. For further insights into classical conceptions of reciprocity and the economics of friendship, refer to Tazuko Angela van Berkel's *The Economics of Friendship: Conceptions of Reciprocity in Classical Greece, Mnemosyne Supplements* (Leiden; Boston: Brill, 2020).
31 Xenophon, *Econ*. 20.13.
32 Xenophon, *Mem*. 2.1.28.
33 Lucius Annaeus Seneca, *Ad Lucilium Epistulae Morales*, Volume I, trans. Richard M. Gummere (London: William Heinemann; New York: G. P. Putnam's Sons, 1920), XXI.
34 It is worth noting that the contemporary concept of "happiness" somewhat diverges from the ancient understanding dating back to Aristotle. For him, εὐδαιμονία was neither a matter of life success nor a positive emotional state, but rather the development of individual excellence and the realization of values accessible to humans.
35 Johannes Stobaeus, *Anthology*, III, 17, 33 (Usener, fr. 181), quoted in https://www.attalus.org/translate/epicurus.html.
36 Marcus Tullius Cicero, *De finibus bonorum et malorum*, II, 28, 91 (Usener, fr. 468), quoted in https://www.attalus.org/translate/epicurus3.html#us4f.
37 Johannes Stobaeus, *Anthology*, III, 17, 22 (Usener, fr. 469), quoted in https://www.attalus.org/translate/epicurus3.html#us4f.
38 Claudius Aelianus, *Varia historia*, IV.13 (Usener, fr. 473), quoted in https://www.attalus.org/translate/epicurus3.html#us4f.
39 Clement of Alexandria, *Stromata*, VI.2 (Usener, fr. 476). See also: Porphyry, *Ad Marcellam*, 28.
40 Diogenes Laertius, *Lives of Eminent Philosophers*, X.11.
41 Diogenes Laertius, *Lives of Eminent Philosophers*, X.120a.
42 Plato, *Phaedrus*..., 270c.
43 See, for instance, Merle G. Walker, "The One and Many in Plato's *Parmenides*," *The Philosophical Review* 47, no. 5 (1938): 488–516, https://doi.org/10.2307/2181200.
44 "Plato speaks here about the true art of rhetoric and draws a parallel with the art of healing. For in both cases it is a question of understanding nature, either the nature of the soul or the nature of the body, at least if we are hoping to act on grounds of genuine knowledge rather than merely on the basis of routine and everyday experience. Just as we must know which remedies and what sustenance should be administered to the body if it is to become healthy and vigorous once again, so too one must know which laws and ordinances and which kinds of discourse should be introduced to the soul, if it is to acquire the proper convictions and attain its authentic being or virtue (*arete*)." Hans-Georg Gadamer, *The Enigma of Health*..., 47.
45 Hans-Georg Gadamer, *The Enigma of Health*..., pp. 45–46.
46 Plato, *Phaedrus*..., 277 b–c.
47 "At the very heart of the image is a connection between education and natural growth. The wise grower outdoes the foolish one by choosing conditions suitable for growth (the time of year, the soil, the growing period): his role is to help his

seeds achieve their full potential as natural organisms. The wise educator does the same, by choosing a soul that provides suitable 'soil' for the seed of discourse to flourish." Anne Cotton, "Gardener of Souls: Philosophical Education in Plato's *Phaedrus*," in *Gardening: Philosophy for Everyone*, ed. Dan O'Brien (Sussex: Wiley Blackwell, 2010), 233–234. See also Kenneth Dorter, "Imagery in Plato's Phaedrus," *Journal of the History of Philosophy* 9, no. 3 (1971): 279–288; Cynthia Freeland, "Imagery in the Phaedrus: Seeing, Growing, Nourishing," *Symbolae Osloenses* 84, no. 1 (2010): 62–72; Dennis J. Schmidt, "The Garden of Letters. Reading Plato's *Phaedrus* on Reading," in *International Yearbook for Hermeneutics* 12, ed. Günter Figal (Tübingen, Germany: Mohr Siebeck, 2013), 61–76.

48 Anne Cotton, "Gardener of Souls: Philosophical Education in Plato's *Phaedrus*"..., 243–244.

49 Hesiod, *Theogony. Works and Days. Testimonia*, ed. and trans. Glenn W. Most, Loeb Classical Library (Cambridge, MA: Harvard University Press, 2006), *Works and Days*, 293–297.

50 All translations of the original quotes come from the following edition: Lucius Annaeus Seneca, *Ad Lucilium Epistulae Morales*, Volume I, trans. Richard M. Gummere (London: William Heinemann; New York: G. P. Putnam's Sons, 1920).

51 Homer, *The Odyssey*, XXII, 347.

52 Michael Erler, "Autodidact and Student: On the Relationship of Authority and Autonomy in Epicurus and the Epicurean Tradition," in *Epicurus and the Epicurean Tradition*, ed. Jeffrey Fish and Kirk R. Sanders (Cambridge: Cambridge University Press, 2011), 26.

53 Mara Miller, *The Garden as an Art*..., 4.

54 Marc Augé, *Non-Places: Introduction to an Anthropology of Supermodernity*, trans. John Howe (London; New York: Verso, 1995).

55 Hesiod, *Works and Days*, in *Theogony. Works and Days. Testimonia*, ed. and trans. Glenn W. Most, Loeb Classical Library (Cambridge, MA: Harvard University Press, 2006), 42.

5 In Search of the Garden of the Isles

Greece, indeed, is infectious. Whoever has experienced the multilayered complexity of Greece—whether in the populous cities or slow fishing villages, whether in the mountains of the mainland or in the islands of the Aegean and Ionian Seas—forms a rather strong opinion about the country and its inhabitants: indifference is not an option. Sharing Dariusz Kubok's fascination with this amazingly multifaceted country, and driven by the desire to impart his contemplative observations on the mechanisms of well-being to others, Marcin Fabjański developed an idea of organizing a philosophical workshop in Greece—a workshop specially tailored for a small group of very tired individuals in highly responsible positions. Originally, the project was to emulate the experiences that Marcin had previously shared with his students of the Apennine School of Living Philosophy, which he had founded in Trevi Nel Lazio—a place of breathtaking natural beauty—several years before.

> The area of Parco Naturale dei Monti Simbruini, where The School is located, is a perfect place for anyone who would like to explore Italian landscape in a philosophical way. The place makes it easy to feel that under the surface of what we see, houses, streets or forest paths still pulsate [with] thoughts, emotions, and visions of exceptional people, who used to life there: philosophers, monks, ascetics. The place [offers] an ideal environment for enhancing [one's] insight into the sages' way of thinking by means of investigation, meditation, observance of nature, as well as other philosophical exercises. One can sit in the cave in area [close to the one in which] Saint Benedict meditated for three years, or observe whirlpools on the streams similar to those which Leonardo Da Vinci observed.[1]

The intuition that contemplation, leading one to the self-attunement to the rhythms of nature, and thereby to flourishing, would be facilitated by the proximity of pristine national parks and nature reserves, goes hand in hand with Marcin's distrust toward "armchair philosophy" and institutions monetizing "pop-mindfulness." Explaining his fundamental motivations to offer students a viable—and responsible—alternative, Fabjański writes thus:

DOI: 10.4324/9781003328469-8

> Where to study [the] art of living? Philosophy has closed itself today at universities and reaches out, at best, to coffee places, where it assumes [the] form of [a] dry discussion. It has lost its therapeutic sting. Nothing [thus far] has successfully replaced [...] philosophy as understood in antiquity, i.e. as the medicine for the soul. Psychotherapy stays away from metaphysics and fundamental questions about death and existence, religions [proselytize] their revealed truths, and meditation is offered most often as a mechanical technique, which is supposed to make you more relaxed and effective.[2]

As opposed to courses offered within venerable university walls, or taught by commercially oriented organizations, the Apennine School of Living Philosophy was envisioned as a center in which life itself would be the object of examination—

> [...] a place of reflection, intensive contact with nature, and psychophysical regeneration. The practice in our School is modelled around exercises proposed by ancient philosophical schools (both Western and Eastern). It also includes micro-expeditions into wilderness and trips to places crucial for [the] shaping of the Western identity, such as the Monastery of San Benedetto in Subiaco [...].[3]

As these short descriptions demonstrate, Fabjański's method, which he aptly dubs "self-off," combines non-verbal immersion with a variety of forms of intellectual insight, which allow the practitioner to learn to "switch off" their ego and embrace the rhythms of the reality, of which he or she is, fundamentally, an inextricable part. It therefore comes as no surprise that one of the essential areas of Marcin's philosophical interest involves the contemplation of the classical elements—earth, water, fire, and air (sometimes expanded to include ether/space, or, in ancient and mediaeval philosophies, also the "fifth essence," the fabric of the universe). As is well known, these elements play a significant role in the process of the attunement to the rhythms of nature in various meditation practices across different traditions, but also in the classical and early modern understanding of health-as-balance. In the context of philosophical exercises, when used metaphorically, the classical elements serve the purpose of giving the meditative experience a focus, or are used to enhance specific qualities within the person engaged in the practice. But when the presence of the elements is also *physically* felt, when the insistence of wind, water, fire, earth, and space is impossible to dismiss, the outcome of the meditative practice may be significantly reinforced. This was the premise of Fabjański's Greek workshop, which was supposed to be based on a "micro-expedition" into the wilderness of the sea. However, before I recount my experience of this event, let us refresh our memories concerning the meaning of the classical elements in the contemplative practice.

Commonly, earth is associated with "grounding" or "stability"; the practitioners are often asked to visualize themselves as trees whose roots extend deep into the soil, focusing on the gravity pulling their bodies downward. Sitting or lying down, the meditating person concentrates on his or her contact with the earth, which results in the development of a more profound sensation of being "rooted" in the ground, and thereby more balanced. Water, in turn, representing flow, fluidity, adaptability, and purification, is often visualized as an ocean, a river, or a lake; the practitioners are asked to remember their experiences of free-floating in calm water or being carried by the slow stream, which gives them a sensation of profound bodily relaxation and allows them to "wash away" disturbing emotions, affecting both the body and the mind. Fire, epitomizing change, energy, and passion, becomes the focal point for the cultivation of one's strength, determination, and one's potential to transform. It is often visualized as a flame or sunray, heating particular body parts (especially the solar plexus), and such a visualization enhances one's focus and mental clarity (a good example of a fire-oriented meditative technique is candle-gazing, a form of yogic *Trāṭaka*). Air, naturally associated with breath, is also the symbol of intellect and communication. Focusing on the breath, noticing its natural rhythm, and practicing yogic breathing (*pranayama*) helps the meditating person, imagining themselves as weightless or light, to connect with the air element, and thereby to embrace the sensation of freedom. Finally, ether, or space, representing openness and expansion, is the object of meditations oriented toward the recognition of the expansive nature of consciousness and of the interconnectedness of all things. Fostering a sense of openness and unity, the space-oriented meditations often involve the visualization of the vast expanse of the sky or the limitlessness of the cosmos, enhancing the sense of the connectedness of one's own inner space with the universe, and producing the sensation of oneness, resulting in the attainment of the peace of mind. The meditative practices focusing on individual elements are often integrated into practices (known, for example, in certain forms of yoga) oriented toward the equilibrium of the five elements within the practitioner's body and, consequently, mind, which—ultimately—helps the meditating individual arrive at a more harmonious state of being.

With the above in mind, it is easy to understand why Marcin's project, integrating earth, water, fire, air, and space to demonstrate our fundamental unity with the universe and our dependence on its rhythms, would reverberate with me—an ocean-going captain, whose experiences, spanning over three decades, unquestionably confirm the healing power of a sea voyage. Likewise, it is comprehensible why the philosopher himself was willing to try his own concept out by embarking upon three cruises: a floating workshop oriented toward the experience of the elements in the Aeolian Islands, in which the participants would be the passengers of a sailing yacht, and two sailing trips to "test the philosophical waters" along with the viability of a different model—a model, in which the workshop participants are also the crew: one in the Stockholm Archipelago, and one, immediately preceding the workshop proper, in the Aegean Sea.

The Greek Amazement

On July 3, 2021, "The Greek Amazement," organized as the first workshop within the frame of the Seven Seas Academy project, became a fact. In our Facebook post hosted on the Seven Seas Academy profile, we described its assumptions thus:

> "The Greek Amazement" is a philosophical workshop aboard a yacht (the *Misty*, Sun Odyssey 519) with Marcin Fabjański (instructor/counsellor) and Paweł Jędrzejko (skipper)], designed for leaders: a reset, a reconnection with vitality, and a boost to creativity. [...] This time, the trip is not only about Greek beaches, wine, food, and monuments. This time—it is about the Greek amazement. Through mindfulness exercises, engagement with the most inspiring thoughts of ancient Greeks, and in-depth conversation in a friendly atmosphere, we will return to a time when the world was young. We will remind ourselves how to marvel at the beauty and logic of reality to refresh our thinking. We will sensitively attune our attention to the reality around us through meditation. Besides [your interactions with the instructors] [...] you will also engage in a conversation with ancient sages such as Socrates, Epicurus, and Epictetus through discussions, readings, and exercises. In search of inspiration from the living philosophy, we will sail for a week along the routes associated with these thinkers, finishing our trip in Athens, [the quintessential philosophical hub]. We will sail, swim, meditate in quiet coves, and observe human behavior in port towns. We will enjoy the simple joys of life, in a way accessible only to philosophers.[4]

Within a very short time, our list of participants was complete. Coming from different walks of life and representing different age groups, most of our "philosophical crew" were recruited from among the members of the Assistant of Conscious Living group, set up as a methodological extension of the Apennine School of Living Philosophy. All of them, however, were the seekers of what Fabjański described in his interview with Przemysław Poznański as "daimonistic happiness, derived not from the intensification of stimuli, but from attuning oneself to a certain natural harmony," born in the "transcendence of the self, empathetic resonance, rather than the mere experience of personal satisfaction."[5] Such philosophical happiness, as Marcin asserts elsewhere in the interview, does not require strong stimuli and, independent of a variety of external and internally motivated factors, can be energized by small things. The *sine qua non* condition of the arrival at such a state, however, is a properly executed, meditative departure from the *episteme* assuming the centrality of one's ego: a departure free from self-deception.

> [...] It happens that people consider themselves altruists, but this is often tinged with egoism, ultimately leading to burnout. This sometimes happens with people working in foundations, who burn out very quickly. They have so many reasons for genuine satisfaction, but do not know how to build contentment on the fundament of the development of virtues, how to draw it from the strengthening of their characters. [...] In the Stoic school—and in other schools, which may express the same concept in different words—we speak of a kind of cosmic intelligence that responsible for wisdom, happiness, and flourishing. *Oikeiôsis* [complete attunement with the universe] is our connection to this cosmic resource. It is not a matter of our individuality or the expression of our ego, but the transcendence of certain barriers between individuality and the rhythm of nature in general. [...] I am not speaking of God. And indeed, in none of the schools I refer to does cosmic intelligence imply personal immortality. Rather, I am talking about the possibility of participating in collective intelligence, and thus about a possibility of transcending individuality.[6]

Sharing the same goal, and to a greater or lesser extent, philosophically prepared to embrace the intellectual and physical aspects of the "Greek Amazement," the workshop participants, of whom only two did have prior sailing experience, arrived in Athens in joyful anticipation of what the new adventure was about to bring—and after a day of exploring the spectacular Athenian sights and a night of bonding and merriment, they boarded the *Misty*, docked at a busy pier in the Agios Kosmas marina. And, to their surprise, the first physical experience of the elements was not that of the water, but that of the earth.

Sea-legs

Every skipper has seen it countless times, and jokes spinning off the mockery dressed up as "offshore sympathy" are probably among the most popular among the members of the sailing community. The "greenhorns," being the butt-end of these jokes, however, rarely appreciate the humor: their first encounter with the sea makes them realize just how much out of time they are with the rhythm of the waves. The simplest activities, such as walking, shaving, cooking, urination, or defecation—pose a challenge. Ashore, none of us wastes a second to think about the principles of gravity or the physiological mechanisms responsible for our balance when we put one foot in front of the other; at sea, the "ground" pitches, rocks, and rolls. To avoid extensive bruising, one needs to quickly learn to walk anew, factoring in the yacht's heel, the swell, the wind force and direction, and the possible errors of the person at the helm. Ashore, one's problems in keeping balance might be a result of the labyrinth not working properly or, for instance (as it is in my case), of pain originating in spinal degeneration; at sea—the labyrinth is fed data impossible to process quickly. Yet, with time, everyone on board learns to be mindful of the swell, of the waves made by passing ships, of the tiderips; even the least experienced of the crew become somewhat attuned to the rhythms of the sea: the sooner they do, the fewer bruises they collect, the fewer plates or coffee mugs left unattended on the messroom table they break, the fewer meals they prepare with dedication worthy of utmost admiration will end up in the bilge, the fewer burns they will have to dress. Getting ones "sea-legs" is, in fact, tantamount to developing a renewed vision of the very purpose of legs in general, involving the amazement at the very simple realization of how very little we appreciate the miracle of our extremities while ashore.

Importantly, such an experience can only be gained "outside of language," powerless despite its evocative might. It is precisely the immediacy of the experience—swept under the magnificently embroidered carpet of the paradigms of the Age of Reason—that usually proves to be instrumental as granting access to a much broader reflection, which, moderated by Marcin, would allow our crewmembers to personally discover the philosophical dimension of individual attunement to the rhythms of nature: the attunement that—until the seasickness and bruises made the necessity to revise all that may have been taken for granted painfully obvious—could have otherwise been dismissed as "pure rhetoric," or "ideology." Although some of them would later expand their horizons sailing in the Stockholm Skärgård, the Ålands Skärgård, the Ionian Sea, the Canary Islands, the archipelago of Madeira, in the Caribbean, the Bermuda, the Azores, or the French Polynesia—learning how to "read the elements" and recognizing the paradoxical hostility of the lee shore—it was during the first days of the "Greek Amazement" that our workshop participants profoundly recognized the difference between themselves and the Greek fishermen, whose daily catch we would occasionally purchase, suddenly realizing that each octopus, each dorada, each shrimp they enjoyed in the local restaurants or on board the *Misty* ended up on their plates only and exclusively because the sea, an indispensable component of the Greek way of life, is a part of the Greek's fishermen's extended, nature-attuned, expertly walking, and very well "anchored" self.

Sensing the Elements

It is common knowledge that the organisms of people and animals suffering from disabilities somehow compensate for the dysfunctional body parts. What is not a part of the common knowledge is the extent to which the arrogance of the "able-bodied" blinds them to limitations of their *hic-et-nunc* functionalities. A simple experiment allowed our crew to become amazed at the discovery that in their daily life they prioritized one sense over all the others: the sense of sight. As opposed to the participants of the workshop in the Aeolian Islands, our "philosophical sailors" were expected to work as the crew of the yacht, which—apart from performing certain functions during the docking maneuvers or anchoring, involved handling the sails and steering the *Misty*. The latter proved to be a challenge: used to the firm grip of the tires of their cars, whose heading would be controlled by easily operated power steering systems, our crewmembers found it hard to keep the yacht on course: a moray eel in our wake would probably require a serious spine treatment, should it choose to run the risk of following us long enough. Interestingly, when we asked the people at the helm to cover their eyes with a bandana, the situation changed. Once each of them would get over the initial fear (naturally, preparing them for the experiment, we would assure everyone that irrespective of what they would do steering blind, the vessel was entirely under control), they would let go of their attachment to eyesight, recognizing the importance of the senses of touch, hearing, balance, and even smell and taste. To their surprise, the information that yacht gave them sufficed to keep her on course: any alteration of the heading would result in the helmsperson *hearing* the change in the sounds that the yacht's hull, its rigging, and its sails—interacting with the elements—would produce; the difference in the melody of the vessel would go hand in hand with the change in the song of the wind, whose angle of attack is both *felt* on the face and *heard* in the ears, depending on the course. Finally, the sun operating on the helmsperson's body would be *felt* as changing position, its rays warming different parts of the body along with the yacht altering her course. Relative to the heading, all sensory information would palpably change, and it did not take long for each of the volunteering helmspersons to realize that. Attuned to the "desired" melody of the ship, mindful of the direction of the wind (carrying the aroma of the vegetation ashore and the salty spray from the bows), discerning the preferred rhythm of the waves, the experimenters, although blindfolded, proved to be able to use *all* of their remaining senses to keep the *Misty* on course, thus testifying to their adaptability, readiness to change, recognition of freedom and, ultimately, of their primal connectedness to the universe, which another philosophical session—led by Marcin—made very clear to everyone amazed with their new experience.

The Night Sea Journey

One of the questions most frequently asked by the first-timers on board concerns the skipper's intentions of sailing at night. As a rule, suspended between their trust for a stranger (whose license is the only document testifying to their experience) and the simple instinct of survival, the "greenhorns" project their fears of the nighttime darkness from the land to the sea. Failing to understand the essence of the old saying that "at sea things get clearer when it gets darker," they are torn between excitement and anxiety. Indeed, it is often easier to determine the courses and speeds of the ships in the vicinity when one sees their navigational lights; indeed, the lighthouses, beacons, lateral and cardinal marks—when lit—are unmistakable landmarks by which one may safely navigate. True, it is impossible to see the shoreline and true—objects, like semisubmerged containers blown off the decks of the cargo ships, loose superbuoys, or unlit boats, pose a significant danger, but usually the night sky provides enough contrast to allow diligent observation. But it is already the first experience of the night sea journey—an experience inevitable in the case of transoceanic yacht transfers or other long-haul cruises—that gives the first-timer a memorable lesson of attunement to the rhythms of nature: eyes learn to penetrate darkness in search of light; ears listen for any change in the song of the rigging and for the distant rumbling of ships' motors, noses develop a particular sensitivity to the odor of the fumes of the passing vessels or the aroma of greenery; the whole body adapts to the regularity of the swell. When the sea is stormy, time expands; watches seem endless, and each cup of hot tea brought to the cockpit by the crewmember taking a break from his or her navigational duties marks a moment of joy—a necessary recharge of one's batteries before the first strokes of the sunrise paint the skies pink. Each little pleasure, no matter how small, makes one appreciate being alive, of which the waves rolling over the deck and the howling wind never allow us to forget, enforcing a most precise, utterly nonverbal, utterly focused, attunement to the raging elements from which we are *never* separate.

But when the seas are calm and the canopy of the sky, unobstructed by clouds, reveals itself in its full splendor, time contracts. Even those off watch, even the most concerned of the crew, will crawl out of their cabins, as if magnetically attracted to the vastness of the universe waxing wide above the perfect circle of the horizon. Free of any light pollution, glittering splendidly with the plethora of constellations rarely admired by those on land, the night sky—providing guidance to skilled navigators—offers a humbling experience of the ether: the endlessness of space in contrast to the insignificance of an isolated human self. And then, suddenly, the realization of the amazing fractal geometry of the universe, in which the same planetary patterns at the level of solar systems recur at the level of atomic structures, awakens one's sense of belonging: every atom that belongs to me as well belongs to the cosmos.

Toward *Oikeiôsis*: On Some Fringe Benefits of the Search

Observing the behaviors of the group of "newbies" seeking self-knowledge proved to be an eye-opening experience, not only because watching them I was able to reconnect with a much younger version of myself, but also because the stringencies of my responsibility for their safety reminded me to appreciate the wisdom and care of all of the excellent commanders under whose (sometimes callous) wings I once studied the ins and outs of the seafaring trade. All of our "attunement seekers" somehow benefited from the "Greek Amazement." Some—by finding out that, having overcome their fears of the "inhuman" sea, they prefer the exposure to the immediate experience of the elements to philosophical contemplation, sensing that their attunement to the rhythms of nature comes more naturally by way of painstaking concentration rather than in the course of meditative relaxation. Others—by discovering a more profound connection with the observations of the ancients (and a more substantial sense of philosophy-as-*therapeia* in general) owing to the lived experience of the transcendence of one's ego in the encounter with the power of the elements. Still others—by realizing the fundamental sense of the following reflection of Michel de Montaigne:

> Pyrrho the philosopher being one day in a boat in a very great tempest, shewed to those he saw the most affrighted about him, and encouraged them, by the example of a hog that was there, nothing at all concerned at the storm. Shall we then dare to say that this advantage of reason, of which we so much boast, and upon the account of which we think ourselves masters and emperors over the rest of all creation, was given us for a torment? To what end serves the knowledge of things if it renders us more unmanly? If we thereby lose the tranquillity and repose we should enjoy without it? And if it put us into a worse condition than Pyrrho's hog? Shall we employ the understanding that was conferred upon us for our greatest good to our own ruin; setting ourselves against the design of nature and the universal order of things, which intend that every one should make use of the faculties, members, and means he has to his own best advantage?[7]

In the story of Pyrrho's pet hog, the issue at stake is that of one's ability to tell these elements of reality which we can control from those we cannot and, consequently, one's readiness to let go of anxieties and projections concerning matters he or she cannot influence. This seems to be the problem that comes to the forefront of our attention when we experience unfamiliar contexts, but becomes invisible when we return to functioning within the frames of familiar routines. To generalize an experience of one of our "philosophical" crew-members into a more legible example, let me offer an illustration of this claim.

A lifelong CEO of a major company, responsible for hundreds of employees and for multi-million-dollar transactions, despite their vast knowledge and undeniable experience in business management, is still a "greenhorn" aboard a ship. When the vessel is beset by a storm, the CEO learns that, in the context, their knowledge is of no value, or worse: it may prove a disturbance to everyone, including themselves. Therefore, he or she *must let go*—both of their projections and of their professional urge to control reality around them. Hard though "letting go" might be at first, trusting those already familiar with the language of the sea is the only option they have. True: a humongous wave breaking over one's head might be terrifying, but an experienced sailor, even if he or she is afraid, "understands" the wave and the ship better than the CEO, and it is the sailor's experience that affords everyone a fighting chance to weather the storm. The "newbie's" knowledge, which, in professional contexts, allows him or her to exercise control and agency, but proves useless in the storm, only aggravates their sense of helplessness—which can only be alleviated if and only if they learn (or choose) to trust the mate and other, more seasoned, shipmates. Seeing the officer casually lighting a cigarette, which miraculously refuses to get wet despite the rain and the spray, changes the "greenhorn's" perspective on his or her own position in the "natural order of things." As a result, rather than trying to exercise their "CEO self" and (unprofessionally) get in the way, they arrive at the understanding that there exist situations in life which others are far better equipped to resolve—and, by extension, that there may be some beyond *anyone's* control. Thus, acknowledging the necessity of leaving the decision-making to others, the CEO allows him or herself not to shoulder an impossible responsibility, thus opening up to the possibility of regaining the sense of tranquility and balance. Similarly, when the experienced watch mate becomes anxious, they will look to the captain of the ship, whose experience should have equipped them with knowledge superior to that shared by officers. Yet, when the ship's master finds himself or herself in a tight spot—there is no one left to look to. The captain is the only person who can never afford to "be at sea" with the ship and the crew while at sea: he or she is obliged to make decisions until the crisis abates, or until, incapacitated, he or she no longer can make any decisions. The wisdom flowing from the story of Pyrrho's "stoic" hog thus gains substance, embodied in the experience of all the CEOs, who have found themselves in the context of a predicament of which the resolution would require relinquishing control to someone else.

Finally, in the aftermath of our philosophical journey, two important realizations dawned upon Marcin and myself, the "conceptual fathers" of the "Greek Amazement." The first concerns the criteria by which each of us assess the success of our project; the second—the physical, pedagogical, and methodological limitations of the organization of philosophical workshops aboard sailing vessels.

On the first count, we both agree that although the *complete* attunement with the universe might well be unattainable, the exposure to the experiences of *momentary* or *temporary* synchronization with the rhythms of nature may safely be described as impactful enough to motivate a person to change their life and seek "daimonistic happiness" through repeated acts of conscious participation in what Fabjański dubbed "collective" intelligence. I believe that in this sense, even the choices of those, whose divorce with philosophical exercises led Marcin to briefly question his own, superb, pedagogical skills, testify to the success of the project. Departing from the philosophical insight that they had gained in the course of land-locked workshops in which they had participated prior to the "Greek Amazement," these seekers made an informed decision to follow their own mode of engaging with the pulse of the universe. In their own accounts, the "Greek Amazement" made them realize that direct exposure to the elements, afforded by sailing, allows them to become attuned to the rhythms of life "more naturally" than does meditative insight. At the same time all others, although sea sailing has since become an indispensable element of their annual vacation plans, continue their philosophical explorations under Marcin's guidance, searching for *eudaimonia* both at sea and on land—and more open-minded individuals join them every year. Yet, although his emotionally driven theory of the project being a "pedagogical failure" has thus been refuted, I do share many of his concerns about the organization of similar workshops in the future.

Firstly, when workshop participants simultaneously function as students and as crewmembers, the effort related to the work they perform on deck or in the galley inevitably affects their concentration during the meditation sessions, as well as their efficiency in engaging with demanding philosophical texts. When the sailing time exceeds six hours, the very excitement with the journey, combined with the stress of the diligent observation of the sea, monitoring the weather, watchful control of the behavior of other vessels, the intensity of the physical exercise that operating the sailing yacht entails, and the exposure to the scorching Greek sun and to the strong meltemi, results in fatigue whose level defies the point of any meditative practice aimed to translate the experiences of the day onto the plane of philosophical reflection. On such days, routinely, the workshop participants would fall asleep during the first relaxation-oriented exercise, rendering a substantial number of the goals envisioned in the syllabus impossible to attain. Although the physical experience of sailing facilitates the crewmembers' attunement to the rhythms of the sea, the intellectual recapitulation of that experience poses a challenge.

Secondly, the combination of two strategies of guiding the students toward the mindful realization of the fundamental unity of the universe—one, requiring meditative tranquility of focus, the other demanding focus on and in energy-consuming action—backfires when their deployment is simultaneous rather than consecutive. Because each strategy necessitates a polarly different state of mind (adrenaline rush sharpening the focus on action-in-tune-with-the-elements versus relaxation being the prerequisite of the focus on insight), the students find it nearly impossible to switch from one to the other when their watch ends and their meditations start. Although, theoretically, such a switch would be easier in a calm, the unpredictability of weather—despite excellent forecasts—makes the assumption of the viability of systematic realization of preconceived sessions at sea impractical at best.

Thirdly, when students and instructors are all actively involved in sailing, the "chain of command" favors the sailing instructor: unlike it is in the context of the land-locked workshops, the position of the instructor responsible for philosophical guidance becomes "secondary" to that of the skipper. This imbalance may lead to the conflict of interest—for instance, in situations in which the weather forecast prompts the skipper to seek and find a well-sheltered mooring, and the meditation instructor needs time to carry out the session planned for a given day—a session, which, needless to say, the students expect to be delivered. In such and similar contexts, the "safety first" principle will always prevail, which disrupts the regularity of the philosophical practice, questioning the fundamental goals of the workshop itself.

The three observations lead to four major conclusions:

a when students and instructors are crew of the yacht, their involvement in active sailing supports and anchors the intellectual experience of meditation in the direct, unmediated, felt experience of the elements, of agency in operating the vessel, and of the (reciprocal) responsibility for the well-being of one's shipmates;
b the above notwithstanding, philosophical workshops at sea are efficient only if sessions oriented toward meditative reflection and active sailing follow a consecutive order, whether *en bloc* (a week of sailing followed by a philosophical workshop) or day-by-day (a day of sailing followed by a day of meditation); additionally, such a solution minimizes, or altogether eliminates, the possibility of the conflict of interest in the face of the "safety first" principle;
c the "philosophical" syllabus should be prepared in such a fashion as to render the "physical" experience of the elements relevant to personal development at the contemplative plane;
d in light of the above—and bearing in mind that getting one's sea-legs may take more than a few days, and that the standard charter begins on Saturday and ends on Friday of the following week—an efficient workshop at sea, affording the students both the felt experience of the elements and the meditative experience of insight, should last no less than weeks (plus travel time).

The implementation of the principles sketched out above might allow one to eliminate (or minimize) the problems identified in the course of the "Greek Amazement," thus elevating the educational experience of the workshop to a new level of excellence. A thus organized workshop affords the students a chance to become anchored in the felt reality, sensitizing them to the pulses of life that they subcutaneously feel, even if their daily routines, prejudices, or ossified convictions render them imperceptible beyond layer upon layer of discourse. Equally importantly, the concurrence of the physicality of sensation and the meditative insight into the nature of our experience open the students' eyes to the fact that their land-locked anxieties, frequently groundless, fundamentally differ from the fear they experience while facing the wrath of the "inhuman" elements: fascinating though they may be, they require attunement, but disallow control. In the context of the sea, attunement may mean much more than a step toward daimonic happiness. It may be—and most often is—*the sine-qua-non* condition of our survival.

Oikeiôsis and the Will to Be

When we are young, we tend to believe in our own immortality. Especially when raised in families nostalgically cherishing chivalric traditions, we also develop a naïve belief in our own invincibility, often imagining ourselves as potentially equal to the revered, mythical, or historical heroes. Yet, the first near-death experience, be it a car accident or a false diagnosis of cancer, questions the validity of such self-projections rather substantially, perhaps even more so than the first passings of family members or close friends. We mourn the departed ones even though we know for a fact that atoms never die and that the closed circuit of matter and energy logically questions the substantiality of death. And even though we are perfectly capable of employing scientific knowledge and combine it with philosophical reflection to rationalize death, the affective reality of the experience of the loss prevails. We experience mourning organically: even if death itself is not "real"—the pain of the brutal loss can physically kill us. With each experience of the loss of someone loved, we are less likely to arrogantly declare that we are not afraid of death, realizing that we do not really care for atoms: we care for people. And, at the most instinctive level, we cling on to life for as long as our own or someone else's suffering is bearable; it is beyond that threshold that we are likely to welcome death.

Although the beginnings of my fascination with ocean sailing date back to mid-80s of the previous century, my first relevant near-death experience—the truly groundbreaking experience of the inhumanity of the elements—that shattered my youthful self-projections and utterly re-shaped my way of thinking about my own place in the universe came to pass between September 6 and 16, 2002, when the *Oskard*, a racing sailboat carrying ten people on board, en route from Lisbon, Portugal, to Szczecin, Poland, lost its mainstays and, soon thereafter, its engine propulsion, in the Bay of Biscay. The following report from the proceedings of the Maritime Chamber at the District Court in Szczecin, Poland, quoted *in extenso* below, meticulously describes the accident and the eventual loss of the vessel. Needless to say, the description, striving for objectivity, is unemotional: evidence-based, it aims solely at determining the responsibility of "P.J.," the captain of the vessel, the only person on board, whose being "at sea" with the vessel, the crew, and the elements *while at sea* would be unpardonable.

In the document below, a careful, compassionate reader will certainly recognize several voices of the members of the panel: the voice of a professional judge, the voices of professional commercial masters, the voice of a yachtmaster, the voices of the commanders of three ships, the voices of the yacht's crewmembers, and finally—the only voice unequivocally defined: the voice of the Delegate of the Polish Minister of Infrastructure (an institution no longer in existence). Each of these voices, returning to the same events and facts in circles, contributes to what resembles a cubist narrative; one in which a four-dimensional phenomenon is spread out into a sequence of 2D images, each painted with a slightly different tint, each stimulating new reflections. As one reads the simple, unadorned, stylistically plain sentences of the formal document below, one discovers that the emerging vision of what truly transpired in the Bay of Biscay in September 2002 is, in fact, laden with emotions: emotions of those, who survived, but also emotions of those who, in their (more or less compassionate) attempts to understand, invoked their own memories of crises they themselves had had to resolve. And, as is probably clear already, the initials of the commander of the vessel used throughout the report in its publicly available version are mine—and so was the *crisis*, that eventually led to the "change of my course."

The *Oskard*—2002—Sinking—The Bay of Biscay

The Ruling of the Maritime Chamber at the District Court in Szczecin. Ref. No. WMS 46/2002

On September 6, 2002, the *Oskard* (registration number: PZ–2025; type: Cetus; rig: sloop; built 1987; home port: Kołobrzeg; LOA: 13.65 m; width: 3.91 m; gross tonnage: 11 tons; maximum draft: 2.20 m; sail area: 79.8 m^2; equipped with a Volvo Penta MD17 D diesel engine, power 26.5 kW, with a clockwise propeller; owner/operator: GTKŻ Gwarek Zabrze Sailing Club) was boarded by its crew in Lisbon. The intention was to transfer the yacht to Szczecin, Poland. The crew consisted of:

Captain P.J.—yacht skipper PYA (licensed in 1990, as captain since 1989);

First Mate K.G.—yacht helmsman PYA (Polish Yachting Association License);

Second Mate P.C.—yacht helmsman PYA;

Third Mate M.B.—yacht helmsman PYA;

Fourth Mate A.W.—yacht helmsman PYA;

Bosun J.E.—without a Polish sailing license (professional seaman, graduate of Main Marine Academy, holder of a diploma in ship construction design, with experience in maritime rescue and engine operations);

Crew member M.S.—yacht sailor PYA;

Crew members K.E., Z.S., and M.W.—without sailing licenses.

Captain P.J. organized the crew into four watches, each consisting of a mate and a deckhand. All mates were holders of valid sailing licenses. Each officer on watch had the relief watch at his disposal, standing by. In Lisbon, the inspection of the vessel and its equipment was carried out. [As a result, an] additional bilge pump and a small backup battery (12 V) were purchased. The engine was inspected; the fuel filter was replaced; oil was changed; the Morse control cable was exchanged, and the throttle was serviced. The newly purchased electric bilge pump was installed. The fuel was topped up—100 liters of diesel was bunkered in the main tank and additional 50 liters in reserve tanks.

The yacht was equipped with a set of the following sails: jib, mainsail, storm jib, heavy mainsail, genoa, spinnaker. 3 GPS devices were carried on board—one stationary, two portable. The other devices carried on board were: a handheld anemometer, a sextant, a longwave radio receiver, a VHF radiostation, and a radar reflector. Safety equipment included: 10 life belts, 2 ringbuoys, and a 10-person liferaft. The yacht was not equipped with the EPIRB 406 MHz emergency beacon, the legality of which was confirmed by the proper authorization document issued by the Maritime Office in Gdynia.

On September 8, 2002, the *Oskard* began its journey from Lisbon to Leixões (Portugal), where 60 liters of fuel were bunkered. After setting sail, alarm drills were carried out: flooding drill, abandon ship drill, fire drill, as well as the crew training in the use of lifesaving equipment. On September 11, 2002, the *Oskard* departed from Leixões. A compass heading of 350° was set, leading along the coast of Portugal and then Spain, up to Cape Finisterre. After rounding the cape, the compass heading was changed to 010°, aiming for the Cork harbor in Ireland.

Synoptic charts and weather forecasts for the next 2, 3, and 4 days, were on board, and current weather information was regularly received on VHF channel 11 from the Cabo Finisterre Traffic Control Station, with which constant communication was maintained. The forecasts predicted winds SE to E, force 4–6° Beaufort, sea state 2–5. Up to Cape Finisterre, sailing was smooth with winds of 1–2°B and sea state 1. The *Oskard* passed Cape Finisterre on September 13, 2002, ca. 0200. From the early afternoon, the force of the easterly wind gradually increased, reaching 5–6°B at night, with the sea state 3–4. Sailing proceeded without issues. When the wind reached the force of 5°B, the mainsail was shortened (11 reefs); when the wind force increased to 6°B, the mainsail was dropped and the journey continued under a moderately high-cut jib. The yacht sailed smoothly, in a slight heel, achieving speeds of 7–8 knots.

On September 14, 2002, the yacht continued its journey in easterly winds, force 6°B, the sea state 4–5, and excellent visibility. The *Oskard* maintained the heading of 010°, carrying the jib, and averaging 6–7 knots of speed, which, occasionally, increased to 11 knots. At 1200 hours, the watch was taken over by the fourth mate, A.W. (22 years old, university student, holding a yacht helmsman's certificate obtained in 2002, his sailing experience dating back to 1997) and his deckhand K.E. Having taken the helm, the officer ordered the left jib sheet to be trimmed, as the sail was slightly luffing when the 010° heading was attempted to be maintained. At 1210, when the *Oskard* was at the position 46°31.2′ N and 009° 29′ W, a loud bang was heard from the bow section of the yacht; the carried sail was torn and the jib sheets loosened. The watch mate ordered "all hands on deck." The captain was the first to arrive; immediately, he headed to the foredeck to drop the shredded sail. The officer changed the course to windward, to compass heading 090°.

On inspection, it turned out that part of the yacht's foredeck was torn open. Through the opening measuring 30×30 cm to 40×40 cm, [the location of the] severing of the forestay brace from the stem was visible. The connector—a steel rod approximately 20 mm in diameter connecting the bow stem to the forestay deck collar plate—had broken. The fracture surface was smooth, without irregularities or distortions, perpendicular to the axis of the rod.

[...] The engine was started, the sail was dropped, and steps were taken to secure the mast from falling. The anchor chain was run between the double forestay tensioners' mountings installed on the brackets welded to the forestay collar—the metal plate, which, after the connector rod had broken, was torn out of the fiberglass deck. At one end, the chain was secured with shackles to the bow hawsehole, and at the other it wrapped around the pawl wheel of the anchor windlass and secured with ropes. After all the halyards (except one) were moved to bow and tightened with the use of the winches, the chain on the anchor windlass was tensioned and secured. Despite these measures, the mast remained unstable, although its movements were significantly reduced.

The opening in the deck was plugged with the torn sail, yet water would still penetrate the seal. The water entering the hull was continuously pumped out with the Attwood bilge pump. At the same time, K.G., the first mate, was ordered by the captain to mark the yacht's position and to plot course to the nearest harbor of refuge. That port was La Coruña, located approximately 180 nautical miles southwards.

Having covered about 20 Nm, considering the weather conditions—wind E 6°B, sea state 4, the significant weakening of the rigging, and the possibility of further breakage of elements stabilizing the mast or the mast breaking, which would also lead to the loss of the yacht's communication capabilities (as the VHF antenna was mounted at the masthead)—the captain, P.J., decided to call for assistance and evacuate part of the crew—except for himself and two volunteers: the first mate, K.G. and bosun, J.E.

The captain explained the situation to his officers and crew, ordering the preparations for evacuation. He ordered his crew to dress warmly, to put on the life belts, to secure personal documents, money, and valuables. The second mate was ordered to secure food and drinking water. The first mate secured the passports, which he handed over to the care of fourth mate. The bosun secured the small backup battery.

Then, at 1300, a PAN PAN signal and a message requesting assistance and explaining the desired method of help were transmitted on

VHF channel 16. A RORO ship of an undetermined name responded to the call and approached the yacht within the range of 2 Nm. The seven-member crew, led by second mate, P.C., was prepared to abandon the yacht. Captain P.J.'s request to launch a lifeboat to take on the yacht crew was refused by the captain of the foreign ship, who justified his refusal by the high sea state and by the fact that the *Oskard* did not call MAYDAY. [After several minutes of radio exchange, the] ship left without providing assistance. Further PAN PAN calls remained unanswered.

The captain ordered the removal of life belts and return to duties. He re-organized the crew into emergency watches. On deck, watches were led alternately by the captain himself and the first mate, with younger officers and the bosun assuming the deckhand roles. The unstable mast continued to exert severe strain on the weakened rigging; the wind force was increasing. There was a risk of further damage. To minimize the shocks to the mast, the most experienced helmsmen were entrusted with the steering duty: the captain, the first mate, and the bosun. The captain ordered that the helmsmen steer in a fashion that would ensure the minimization of the wave impacts on the bow. Sailing proceeded smoothly, the engine operated at half ahead (1500 RPMs), and the yacht's compass heading was 175°.

At 0000, on September 15, 2002, the first mate took the watch. Before 0000, a brief drop in the RPMs has been observed. After about 10–15 minutes, i.e., around 0215, another drop occurred, and then RPMs spontaneously increased back to about 1500, only to drop again—after which the engine stalled. The *Oskard* was at the position 45°44.5′ N and 009°13.5' W, sailing in good visibility, with wind E 6°B, and sea state 4. The captain and the bosun attempted to determine the cause of the engine stalling with the intention of restarting it.

At 0245, a storm jib of about 4 square meters was set. The operation of this sail rendered the yacht's heeling smoother, and the strong jerks [to the rigging that the instability of the mast caused] could be minimized. Nonetheless, [deprived of propulsion] the yacht lost manouverability. Carried by the wind and the current, it began to drift southwards at 1–2 knots (compass heading 090°–150°, wind E 6°B). The bosun, J.E., took actions to start the engine. He checked the fuel level, cleaned the fuel filter, checked if the fuel pump was working, if the system was not air-locked, and if fuel was reaching the cylinders (injector lines). He inspected the oil levels, and made sure that the engine shutdown lever was in the correct position. Attempts to start the engine continued until the morning, but were all unsuccessful, causing significant battery discharge. The captain did not allow the use of house batteries [powering the VHF radio, the GPS, the bilge pump and the lights] for further attempts to start the engine. The reasons for the engine's stalling were not determined.

As a result of the engine failure, the yacht's batteries could not be charged any more and the *Oskard* lost her manouverability. Considering the above—along with the damage to the forestays, the impossibility to set sails larger than a storm jib, the increasing wind force and swell, the uncertainty about the physical condition of other elements of the rigging [possible breaking of the unnaturally working shrouds], as well as the distance to the port of La Coruña (about 140 Nm)—the captain, P.J., resolving that these combined factors posed imminent threat to the life and health of the crew, decided to call for rescue.

At 0815, the MAYDAY signal was sent on VHF channel 16, along with the position of the vessel and the explanation of the nature of the distress. Around 0900, the call was answered by the ship *Selendang Sari*, en route to Gibraltar. The commercial ship's master communicated with the Finisterre rescue station, which commanded the commencement of the rescue operation. The master of the *Selendang Sari* and captain P.J. discussed the terms and methods of assistance via VHF, and arrived at an agreement. The [initial] plan included lifting the yacht with its crew onboard [using the ship's crane and belts]. When it turned out that the yacht's mass exceeded the permissible lifting capacity of the crane (10 tons), and [also due to the weather conditions] that it was not possible to take the yacht onboard, captain P.J. requested that the *Selendang Sari* should launch a lifeboat, to which eight of the crewmembers of the *Oskard* would be transferred. (The captain himself and the first mate intended to stay on board to manually pump the bilge, maintain radio communication, and steer the yacht during the towing by the SAR tugboat [summoned by the master of the *Selendang Sari* and dispatched by the Finisterre Rescue Station]). However, the master of the *Selendang Sari* refused to launch the lifeboat justifying his decision by harsh hydrometeorological conditions (wind ESE 6–7°B, sea state 4) and the risk to which his own crew could be exposed should he decide in favor. Instead, he proposed that he would approach the yacht from its leeward (starboard) side, pass heaving lines, and then move the yacht to his own ship's leeward (starboard) side, where his crew would stabilize the *Oskard* and transfer her crew onboard using a storm ladder and safety harnesses.

Around 0940, the *Selendang Sari* (flag: Malaysia, single-screw chemical tanker, gross tonnage 29,956, length 182.92 m, width 32.26 m, no cargo) was observed. Moving by inertia, it positioned itself so that the yacht slowly drifted with its starboard side to its windward (port) side. The yacht's crew prepared to abandon ship; everyone put on warm clothing, safety harnesses, and life belts. Valuables and personal documents were secured. Fenders were placed on the yacht's starboard side.

Around 1100, the ship approached to a distance of about 50 m. As opposed to the initial plan [previously agreed upon by both commanders], according to which lines would be fired from the ship's pyrotechnical line throwers to maintain safe distance [between vessels], the lines were thrown by hand, upwind, and were only taken when the yacht was alongside the *Selendang Sari*. After the lines were picked up, they were secured to cleats. As a result of the yacht's movement on the high waves, the cleats were torn out of the deck, pulling the yacht's bow toward the side of the *Selendang Sari*.

The *Oskard* hit the ship's side with its bow pulpit, resulting in its obliteration and ensuing damage to the storm railings on the sides. The yacht, working on the wave, hit the ship's port side several times with its own starboard side, rubbed against it with [what remained of] its railing, and then drifted toward the ship's stern.

When the yacht was already behind the stern and drifting westwards, the master of the *Selendang Sari* briefly worked the machine in reverse and stopped, which caused the ship's propeller's stream to attact the yacht closer to its starboard (leeward side). Fenders on the yacht were moved to her port side. Heaving lines were thrown again, and then mooring lines were secured to emergency devices: anchor windlass at the bow (as the ropes were too thick to pass through the hawsehole, and the cleats had been torn out) [and the portside stern cleat]. The yacht was stabilized alongside the ship. A storm ladder and safety lines were lowered from the chemical tanker (the free side was 14–15 m high).

Members of the yacht's crew transferred to the *Selendang Sari*. During the evacuation, the captain steered the yacht to enable the crew to safely transition to the ship's deck. Despite this, the yacht, [tied off to the ship,] hit the *Selendang Sari*'s side several times with its own port side, bow, and mast—and cracking sounds were heard (the fenders did not break; they worked without failure). No leaks in the hull were apparent. Given the risk of losing the mast and rigging, Captain P.J. decided to abandon ship as the last member of the crew, salvaging some of the yacht's documents with him. The evacuation of the crew was completed on September 15, 2002, at 1230 at 45°11.1'N and 008°42.3'W. At that time of the operation, visibility was good, wind ESE 6–7°B, sea state 4–5.

The yacht was abandoned with all its equipment, personal belongings of the crew, nautical charts and navigation manuals, the navigation computer, and other items, still on board. The electric bilge pump with a float switch (purchased and installed in Lisbon) was left working. The yacht, unmanned, hit the ship's side several times with its mast, which resulted in the breaking of the upper portside spreader and the bending of the lower spreader. Fenders on the port side did not break.

On board the *Selendang Sari*, the yacht's crew was well taken care of. The ship's master informed the captain P.J. that he had called a tug from the Finisterre rescue station, which had already set out to sea to tow the yacht to the port of La Coruña—and that he canceled the helicopter previously ordered. The yacht, whose mass precluded lifting it on board, and with wind gaining force, was eventually cut off from the ship. With the mast slightly tilted to the starboard side, it remained adrift under the small, four-square-meter storm jib.

Around 2230, off the entrance to the La Coruña port, the crew of the *Oskard* were transferred to a pilot boat, which transported them to the yacht harbor. A minibus and ambulance were waiting at the quay. None of the crew members suffered any injuries. Around 2300, the *Oskard* crew were transferred to a hotel.

On September 16, 2002, in the early hours of the morning, Captain P.J. informed the yacht's owner and the Polish Embassy about the incident. He also made contact with the Finisterre rescue station, where he obtained the information that the yacht had been found and was being towed to port. The estimated time of arrival at the port was September 17, 2002, at 1200, but around 1900 on September 16, 2002, the Finisterre rescue station informed captain P.J. about the sinking of the yacht with a broken mast in a 9°B gale, and about the fact that the *Oskard's* liferaft was salvaged onboard the tug. Neither the position of the yacht's sinking, nor details about the rescue operation were not provided.

The value of the yacht *Oskard* was determined to be 150,000 PLN (Ergo Hestia Insurance policy).

The Position of the Delegate of the Minister of Infrastructure

The Delegate of the Minister of Infrastructure pointed out that the yacht *Oskard* was in good technical condition and manned by a qualified crew. Emergency drills were carried out during the voyage. On September 14, 2002, the weather was favorable: wind 6°B, sea state 4, very good visibility, speed 7–8 knots. The yacht only carried only a jib. [...] That day at 1215, while trimming the left jib sheet, the forestay collar broke, causing the shrouds to loosen and the carried jib to tear. The engine was started and steps were taken to secure the mast from falling with the use of the anchor chain. The resulting hole in the deck (30×30 cm) was sealed with the torn sail, yet water continued to enter the hull. Captain P.J. assessed the situation as dangerous. He decided to evacuate those who were not essential for navigation and transmitted a PAN-PAN signal around 1300. One ship responded, but did not undertake to take the crew aboard due to the sea state. Considering the damage to the yacht, weather forecasts, wind direction and strength, as well as wave action—the captain directed the yacht to the port of La Coruña. All actions related to the planned evacuation were carried out. [...] On September 15 at 0215 the engine stalled. Despite attempts, it could not be restarted. The journey was carried on under an emergency small sail, to diminish the load to the mast. [...] The weather deteriorated, the speed was minimal, and it was difficult to steer. In this situation, the captain transmitted a distress signal on VHF channel 16. The call was answered by the ship *Selendang Sari*, which in communications with the Finisterre rescue station, undertook the rescue but did not agree to lower their lifeboat due to the sea state. Both captains agreed that the yacht would drift near the windward side of the ship, lines would be shot, and the yacht would be led to the leeward side of the ship. There, the yacht would be sheltered, and the crew would transfer to the ship using a storm ladder and safety harnesses. Captain P.J. considered this a feasible and, under the circumstances, a sensible maneuver. For unknown reasons, the lines, unfortunately, were not fired from the ship's line throwers, and the captain of the chemical tanker decided to bring the yacht to his ship's windward side. This highly dangerous maneuver was successfully carried out, i.e., the entire crew safely abandoned the yacht, when it was already transferred to the ship's leeward side. While the yacht was alongside the ship's side, it hit the ship's side with its hull, which undoubtedly affected the overall weakening of the yacht's structure. The chemical tanker's captain contacted the rescue station to call a rescue tug to tow the yacht to the port of La Coruña. The rescued yacht's crew, in good condition, was transferred to a pilot boat off the port of La Coruña. In the port, Captain P.J. made phone calls to the embassy and the owner. He also inquired about the fate of the yacht. On September 16, 2002, around 1900, the Finisterre rescue station informed the yacht captain about the sinking of the vessel with a broken mast in a 9°B gale and about the salvaging of the life raft onboard the tug.

The Delegate requested that that the loss of the yacht should be attributed to the following causes:

1 the breaking of the forestay collar from the stem connector, and the hole torn in the deck covering, measuring 30×30 cm. It is presumed that intercrystalline corrosion of the rod securing the collar plate occurred.
2 The stalling of the auxiliary engine and the impossibility to restart it. The cause of the engine failure was not determined.
3 Further damage—sustained during the rescue operation—to the rigging; broken portside upper spreader and bent left lower spreader, the damage to mooring devices, and the weakening—possibly cracking—of the hull.

The Delegate emphasized that due to the lack of documentation from the unsuccessful towing of the yacht, it can only be presumed that the sinking occurred due to hull leaks and the overall poor condition of the yacht after the accident. According to the Delegate, the bad weather, i.e., wind strength of 9°B and high waves, also contributed to the sinking. The crew's actions during the accident were assessed positively by the Delegate. He pointed out that the choice of the form of the rescue action mainly depended on the captain of the rescuing ship. This action, although dangerous, was successful in terms of saving people. The yacht, however, was lost.

Considering the documentation and the course of the proceedings, the Delegate concluded that Captain P.J. acted efficiently within his capabilities and his actions did not contribute to the loss of the yacht.

The Maritime Chamber's Assessment

The Maritime Chamber took the following factors into advisement: The technical condition of the yacht; the preparedness of the yacht for the voyage; the crew.

The yacht *Oskard* was built in 1987. The yacht had been operating in the Mediterranean Sea since 1995, where it served as a leisure vessel. In June 2002, the yacht was inspected—and its class renewed—in Fiumicino. The Polish Register of Shipping, which carried out the inspection, issued its report on the inspection of the yacht's hull and its equipment and the report on the inspection of the underwater part of the yacht's hull. As a result, PRS issued a classification certificate for the yacht, and the Maritime Office in Gdynia issued a safety card no. 613/GM/2002, valid until June 22, 2003, at the request of the yacht's owner. The Gdynia Maritime Office's Navigation Safety Inspectorate exempted the yacht *Oskard* from the duty of being equipped with a 406 MHz emergency EPIRB for the 2002 sailing season. The yacht's engine was last inspected in June 2002 in Fiumicino. The inspection was carried out by a licensed VOLVO service inspector from Zabrze, Poland. Among the documents of the yacht were a list of spare parts, purchase receipts for parts, an oil change book, and an inspection book. These documents were not salvaged from the yacht during the evacuation of the crew on September 15, 2002. The last voyage of the yacht *Oskard* began on September 11, 2002, from Lisbon. The yacht was to be transferred to the Baltic Sea. Before the voyage, a navigation computer, bilge pump, small backup battery (12V), two sails including a storm jib, and a throttle control line were purchased. The yacht and its equipment were inspected. The equipment and safety devices were removed from the yacht and inspected on land. Bulbs in life jackets were checked and replaced. This equipment was then placed back on the yacht and stored amidships, in an accessible location. On deck, there were two ring buoys equipped with throw lines and a light buoy. The yacht carried a ten-person life raft with a valid certificate. Pyrotechnic devices and their certificates were also checked. After setting sail, training in rescue techniques was conducted on the yacht, including the timing of the fitting of the life belts. Drill alarms were properly conducted (water, abandon ship, fire). The captain emphasized the necessity of attaching of safety harnesses to solid deck structures in case of heavy wave action. Barometric maps were received in ports with a nine-day lead time, and in the first phase of sailing, information from Cabo Finisterre Traffic Control about current weather and forecasts was received. The yacht's crew consisted of ten people. The captain held a PYA offshore skipper's license, four officers held PYA yacht helmsman's licenses. Of the remaining crew, only one person had a PYA yacht sailor's license. The bosun's position was filled by an American citizen—a professional seaman. He was a graduate of a maritime academy, with extensive maritime experience. He and the yacht captain had twice crossed the Atlantic together.

The accident on September 14, 2002—the breaking of the rod (connector) securing the forestay collar plate and the stalling of the engine on September 15, 2002.

As explained by the yacht captain and confirmed by the corroborating testimony from crewmembers, the first stage of sailing proceeded without difficulties. By the end of September 13, the wind began to strengthen. At midnight, the wind reached a force of 6°B, sea state 4. In response, the mainsail was dropped. Sailing under a jib, the yacht sailed smoothly, with a slight heel, reaching speeds up to 8 knots. On September 14, 2002, at 1200, the rod—the connector securing the forestay collar plate—broke. The standing rigging (lines, connectors, fittings) had been inspected during the rigging class review in June 2002. The yacht captain described that the break of the connector rod, approximately 20 mm in diameter, was perfectly even, with a smooth surface, without grains, irregularities, or distortions. The interested party presented two metallurgical opinions, whose authors hold master's degrees in engineering, and are graduates of the Wrocław University of Technology's Mechanical Engineering Faculty with a specialty in metallurgy. These expert also hold yachtmaster's licenses. Their opinions are consistent. They present hypothetical causes of the breakage of the connector-rod securing the forestay collar plate. Based on the [description of the] fracture presented by the yacht captain, there were no rest lines characteristic of fatigue fractures. Although tensile stress may be the cause of fatigue-related fracture, in the case of the *Oskard* there is no basis to assume that the cause of the failure was material fatigue.

The fracture has been determined to be the effect of intergranular corrosion, which is a type of electrochemical corrosion occurring at the grain boundaries in stainless steels, especially chrome-nickel steels. This corrosion occurs when an electrolyte (such as seawater) is present on the surface, and impurities that precipitated during the solidification of the steel in the metallurgical process are present at the grain boundaries. These impurities are always anodic and create an electrochemical polarity with the cathodic surface of the metal.

The factor accelerating the aforementioned phenomenon is the stress in the steel due to the constant load on the forestay exerted both by the preload tension through turnbuckles, and by the action of the sails as well as the vibrations of the entire rigging while she vessel is working in waves. The phenomenon described above is known as *stress corrosion*. As a result of this stress, a temperature difference arises between the center of the component and its surroundings, leading to an acceleration of corrosive chemical reactions.

The final factor, possibly decisive, is the brittle hardening of the rod. That is, the steel, due to both constant and variable tensile forces, elongates to its yield point, after which point the material breaks. The progression of steel elongation to the yield point occurs each year with the trimming of the rigging.

After the damage to the forestay occurred, the yacht's crew began work to stabilize the mast. The captain reported that, although the mast was not fully stabilized, its movements were significantly reduced. Consequently, the yacht continued its journey under power. The commander decided against setting any sails: according to the captain, setting sails would increase the stress on the emergency mountings and endanger other elements of the standing rigging, which, being of the same age as the forestays and shrouds, were at risk of breaking due to the improper operation of the mast. This could ultimately lead to the breaking, or loss, of the mast. The above notwithstanding, the captain opined that the only sail that would not endanger the mast's [stability] was the storm jib (approximately 4 square meters). This sail could be set on the babystay and, in this situation, its load would only affect the part of the mast, whose rigging was still intact (the babystay and column shrouds). From that moment onwards, emergency watches of 6 hours were introduced, led alternatingly by the captain and by the first mate. Experienced helmsmen were charged with the steering of the vessel. The captain stated that the navigation under engine proceeded efficiently and relatively safely.

Attention was paid to proper steering—specifically, steering away from the crests of the waves to eliminate the swell's immediate impact on the bow. The hole in the deck from the torn fitting was secured with a torn sail. The amount of water that did get inside the yacht was continuously pumped out by a newly installed bilge pump. At this stage of the voyage, the captain made the first attempt to evacuate most of the crew members. The captain and two volunteers were to remain on the yacht to safely navigate it until a tow vessel could take over. The ship that responded to the assistance call refused to lower a lifeboat due to severe hydrometeorological conditions. Nonetheless, the captain decided against sending distress signals: the yacht was not sinking, the engine was operating correctly, and it was still possible to steer the vessel in a way that minimized the effects of the waves.

However, on September 15th at 0215 hours, the engine failed. Before the engine finally stopped, the officer on watch recorded a drop in RPM, stabilization of RPM, and then after 10–15 minutes, another drop in RPM leading to the engine stalling. This moment was not preceded by choking or other signs of improper operation. The bosun, the crew member most experienced in the area of maritime engines, was charged with the task of fixing the motor. Numerous diagnostic tasks were performed. Subsequent attempts to restart the engine were unsuccessful, and these efforts were halted due to the depletion of the batteries. During the yacht's inspection for class renewal in June 2002, the engine was thoroughly checked. Also, the engine had previously been run in windless weather conditions—also to recharge the batteries—and in these contexts its RPMs and performance were well within the norm. The commodore of the Club owning the yacht stated that VOLVO PENTA MD 17 D engines were ill-famous for becoming air-locked.

The crews are informed of this issue, but they are not provided any instructions for its resolution, as the repair can only be performed by a specialized service. The only recommendation is to keep the fuel tank at least half full, which practice was adhered to by the yacht's commander. In this situation, it was decided to set the storm jib (approximately 4 square meters) on the babystay. After the engine failure, the yacht lost steerage and experienced uncontrolled heeling, behaving like a cork on the waves. The storm jib's operation caused the yacht's movement to become smoother, and it began to drift more stably. The wind pushed the yacht westwards, while the current moved it eastwards, resulting in a consistent southerly course. The yacht achieved the speed reaching up to 2 knots. With the storm jib set, the yacht climbed the waves more gently, although not as comfortably as it did with the engine running. The yacht's course stabilized, heeling decreased, and the hull's violent motion lessened. It was possible to maintain a course with a deviation of 10–20 degrees to either side. With the storm jib set, the yacht [partly regained] steerage. Yet, the lack of success in re-starting the engine, coupled with the earlier rigging failure and the worsening of the hydrometeorological conditions, prompted the captain to send a distress signal. As before, the captain intended to stay on board with two crew members and manage the yacht until the summoned tugboat could commence the rescue operation. The assisting vessel [the chemical tanker, *Selendang Sari*] did not consent to lower a lifeboat, and its crane equipment was not capable of taking the yacht aboard.

The yacht's captain intended to approach the *Selendang Sari* from the leeward side, but the latter ship's commander, citing poor maneuverability, recommended that the yacht approach from the windward side, where, at a safe distance, heaving lines would be shot in its direction, and then the yacht would be guided to the leeward side. However, this plan was not executed by the rescuing party. As a result, the yacht suffered from a strong impact, with its bow pulpit hitting the ship's port side (which also damaged the railing), and from the tearing out of the mooring devices (the lines secured on the ship's deck worked vertically and pulled out the cleats). After the initial impact, the yacht, moved to the leeward side—despite being equipped with fenders—sustained multiple strong impacts against the ship's starboard side, as well as damage to the mast's spreaders. In this situation, the captain decided to evacuate the entire crew onto the ship. After cutting the mooring line, the yacht began to drift.

The Cause of the Sinking of the Yacht Oskard

After the accident, the captain [P.J.], following proper procedures, did not receive any information about the circumstances surrounding the sinking of the yacht. He only received a notification on September 16, 2002, that the yacht had sunk, and that, at the time, the wind was blowing at 9° Beaufort, and that it was being towed; the tugboat had been summoned by the master of the *Selendang Sari*—charged with the rescue operation—while the yacht's crew was still on board, just before evacuation.

Analysis of barometric charts provided by the yacht's captain confirmed that on September 16, the wind was blowing from the east at the force of 9° Beaufort. The sea state was 7–8. Given the yacht's last known position (at the moment of crew evacuation at 1230 hours on September 5) and considering the hydrometeorological conditions (wind direction and force, sea state, current of 90–150 degrees, 1–2 knots), as well as the yacht's drift speed, the Maritime Chamber determined that the yacht sank approximately 60 nautical miles west of position 45°11.1′ N, 008°42.3′ W.

The cause of the yacht's sinking was the loss of buoyancy due to the flooding of the yacht's interior with seawater through hull leaks. This primarily concerns the large tear in the deck that occurred after the rod connecting the forestay collar to the stem of the vessel broke. Although this tear was secured with an old sail, and a working bilge pump was left on, it cannot be ruled out that during the yacht's drift, this measures became ineffective, and that the pump stopped working.

Furthermore, the crew reported multiple hull impacts against the ship conducting the evacuation, and although they did not report hull leaks at that time, such leaks cannot be ruled out. It should also be noted that the yacht's drift occurred in severe gale conditions. The captain was also informed that [when the tugboat found the vessel] the yacht's mast was [already] broken.

The sinking of the yacht was preceded by two major failures—the breakage of the rod connecting the forestay collar plate to the hull and the engine failure.

Following the first failure, the yacht's commander decided to continue sailing under power. No attempts were made to set sail. According to the captain's testimony, making his decision, he took into consideration the condition of the rigging after the damage sustained, the lack of mast stability, and the need to maintain communications (the VHF antenna on the masttop). It was only after the engine failure that he did decide to raise the storm jib, which proved effective in stabilizing the course and minimizing heeling.

The Maritime Chamber believes that setting emergency sails on the course to La Coruña (beam reach)—such as the storm jib and reefed mainsail—would have allowed the yacht to sail with a slight and constant heel to the starboard side, reducing the load on the forestays, thus minimizing mast stress and improving its stability. Under these conditions, the yacht could have achieved a speed of up to 5 knots.

Arguments raised by the captain, justifying his decision quoting the inviability of setting emergency sails, affecting the mast's instability, have not been found convincing, considering that no attempt was made to set emergency sails after the forestay connector broke, and the yacht's behavior after setting the storm jib was not tested. Setting emergency sails would have allowed the crew to test the yacht's actual behavior, rigging performance, and mast stability. With constant monitoring of these elements and careful maneuvering, the 120–140 nautical mile distance to a safe harbor should not have posed significant problems. Should such a possibility have been explored, the option of starting the engine may have been retained. However, the commodore of the club owning the *Oskard* reported that discussions among sailors continue to this day about whether emergency sails could have been set under these circumstances. Still, the assessment remains theoretical, since no such attempt was made.

In summary, the Maritime Chamber determined that the cause of the sinking of the s/y *Oskard*, abandoned by its crew and towed by a rescue ship in the Bay of Biscay, approximately 60 nautical miles west of position 45°11.1′ N, 008°42.3′ W, on September 16, 2002, in wind conditions of 9° Beaufort from the East and sea state 7–8, was the loss of buoyancy due to the flooding of the yacht's interior with seawater through hull leaks resulting from the deck tear after the rod connecting the forestay collar to the hull broke on September 14, the impacts with the ship running the operation of the crew's evacuation on September 15 under wind conditions of 6–7° Beaufort from the East, sea state 5, and the breaking of the mast.

The crew's actions following the rigging damage raises questions, because the captain did not attempt to set emergency sails, but his decision to evacuate the crew after the engine failure on September 15 was appropriate.

Maritime Chamber Ruling of May 9, 2003

Having considered, on April 25, 2003, the case of the sinking of the yacht *Oskard* on September 16, 2002, The Maritime Chamber at the District Court in Szczecin ruled as follows:

1 The cause of the sinking of s/y *Oskard*—abandoned by the crew and towed by a rescue ship, in the Bay of Biscay—in the area about 60 Nm West of position 45°11.1′N, 008°42.3′W, on September 16, 2002, under conditions of wind E 9°B and sea state 7–8 was the loss of buoyancy due to the flooding of the yacht's interior through hull leaks caused by the tearing of the part of the deck after the rod

connecting the forestay plate to the stem broke on September 14, impacts of the yacht against the sides of the ship conducting the evacuation of the crew from the yacht on September 15 under wind E 5–6°B, sea state 5, and breaking of the mast.

2 The crew's actions after the rigging damage raise concerns in that the captain did not attempt to set emergency sails, and the decision to evacuate the crew from the yacht, after the engine failure on September 15, was justified.

3 A fee of 835.57 PLN is charged to the owner of the yacht *Oskar*, Górniczy Turystyczny Klub Żeglarski "Gwarek" Sailing Club in Zabrze, on behalf of the State Treasury (Maritime Chamber in Szczecin), to cover the costs of the proceedings.[8]

***

We survived. Not a single person was physically hurt. I was cleared of all charges. Invitations to speak on safety at sea for full houses in yacht clubs and sailing schools would flood my mailbox. I, unwittingly, became "the hero of the day." Yet, for years, I could not understand why I was not able to shake off the nauseating Lord-Jim-feeling, despite the very clear difference between his decisions and mine. What the report fails to show is the extent to which much of my pre-crisis "courage" was a product of a vague combination of self-projections, based largely on literary models, and retrospections of relatively easy experiences, to which the ocean had exposed me before. In fact, although I had skippered yachts before September 2002, although I had sailed with numerous captains before I chose to become one myself, and although I had always found the *bravado* in others suspicious while myself, in my own self-perception, I had always approached the sea with ultimate respect—the events described above gave me a painful epiphany. I realized that for years I had harbored a naïve belief that *accidents only happen to others* because *they are not careful enough*—and hence, because *I am careful*, nothing could possibly happen to *me*, *my crew*, or *my ship*. But when it *did* happen, my eyes opened wide. It dawned on me that, indeed, I had witnessed numerous situations of crisis before: situations, in which my commanders—experienced, careful, respectful of the power of the elements—had no other option but to face challenges whose gravity I would not realize, because their faces, gestures, and words would make me believe that whatever we faced was "under control." Under *their* control. When "my" accident happened, I was the one responsible for the atmosphere of tranquility, I was the one to chase away the fears in nine pairs of eyes, looking up to me. Mortally terrified, I would smile; utterly uncertain, I would issue orders to others to keep them working on particular tasks rather than panic. Furious at the unresponsiveness of the RORO commander, I would calmly explain to everyone that our situation would not allow us to call MAYDAY, because we were not in any immediate danger of the loss of life or the vessel. And yes, the sheer fact that

I could, but did not, call MAYDAY—worked: hour after endless hour, it worked. And perhaps it is only the absolute necessity to keep composed on the surface that allowed me not to utterly fall into pieces inside, but, at that moment, I sincerely hated the Ocean; I hated myself for my youthful ambition of becoming a captain; I hated every book and movie that lied to me about the glory of the life at sea; I abhorred the splendor of the night sky that mocked my helplessness among the infinite tons and tons of water around me; I hated the Sun. Every dolphin, sea turtle, or luminescent jellyfish that would otherwise spark an explosion of the *joie de vivre* in every fiber of my body struck me as a cold reminder of how much the *sea is not my element.*

Bitterly, I would remember my own ill-camouflaged foolhardiness in my reckless declarations of disbelief in the agency of transcendence in conversations with my friends on land. There, in the middle of the angry Bay of Biscay, I found myself fervently praying to every god and goddess that humankind had ever called into existence. I prayed to the Ocean to stop flogging my ship; I begged the wind to abate; I even spoke to the rigging, beseeching it not to break; I prayed to the stars. I prayed, I cursed, I cried, I laughed, and I prayed again, promising the Ocean that I would never arrogantly encroach upon its vast, unconquerable, territory *ever again.* Yet, the massive waves and the howling wind, utterly deaf to my prayers, continued to batter my tiny ship, a plaything amidst inhuman matter, with nine scared human beings, whose lives depended on me, on board. I was haunted by a desperate terror of a scenario in which someone dies and I survive to face their family; ashamed of my own selfishness, and at the same time, ashamed of my hypocrisy, I begged the ocean to kill me first to spare me the agony of facing the consequences of my possible survival.

The absurdity of the question "why me?" presented itself to me with full clarity: Why me? Why anybody? Why did *my* rigging break? Why would *my* engine die on me? Why would *my* deck sport an unsealable hole the size of Massachusetts? Why? Oh yes. Me, because I had been arrogant enough to go to sea in a nutshell of a boat and pulled others into this. I, because it was I to have believed in my invincibility. I, because I had been able to successfully fool myself into the conviction that I had always been "a creature of the sea." And then, amidst the ebbs and flows of hope, each cracking sound of the mast whiplashing on the weakened rigging would jolt me into attention, dismissing the pointless self-flagellation, rendering the guilt-ride void, and energizing my will to be. Never before had I experienced such a horrendous concentration of presence in the here-and-now: a presence that was no longer *my-self.* In the course of those endless hours of the struggle to be, *my*-self became *a* self that extended to include me and every wave, every gust of wind, every element of the fiberglass, aluminum, and steel on board, every human being, and every star. It was not *my ship* that was being ruthlessly flogged by the wind, it was *us*; it was not the *Oskard* that pitched on the waves, it was *us*. During my lonely night watches—as I ordered all the non-essential crew and the navigator, responsible for plotting our position, to hide down below, afraid that should any of the cables still supporting the mast cracked under pressure, the vehemence of the breakage would decapitate anyone in its range—the "I" has become utterly attuned to the heaving of the sea, to the wind, to pulse of the ship, to every sound, every vibration. Intertwined with everything around me, I would lose track of time, I would forget how much my sleep-deprived body hurt, I would become a part and particle of fluid, constantly inconstant matter, a body extended to all bodies, an embodied consciousness that had nothing to do with divinity.

And yet, absurdly hopeful, I clung on to life, caring, like never before, about the precision of each movement of the steering wheel: slide off the wave, climb the next wave, slide off the wave, climb the next wave; make sure you do not run bow first into the wave, or the next jolt can bring the mast crashing down on your head, or kill the people inside, climb the next wave, slide off the wave, and in the trough beware of the crest so that it does not come tumbling down all over your leaky deck, and hey, there is an antenna on top of the mast, don't lose it, don't lose it… Fatigue would render my steering less and less precise, the frequency of the dangerous jolts would increase; if I put a bit of the mainsail on, the power of the wind would hold the mast in a more stable position, but if I—or anyone else—lose control over the helm even for a second, and ram into another wave, the jolt, whose intensity would multiply, could break the shrouds. Should I do it? Should I wait? Schemings and mantras, mantras and schemings, concentration and action, fatigue and will to live, all amongst the endless rocking, rolling, and pitching of the yacht, playing a discordant symphony, a nautical Schomberg.

And then someone would give me a cup of hot coffee, and I would start scheming again; if the mast breaks, I cut the rigging off and have it sink so that it does not puncture the hull; if I hear the cable swishing, I duck behind the wheel; if the wheel gets crushed by the falling mast, I break the lid of the aft peak to pull out the emergency tiller; I keep people out of danger, I keep people out of danger. Yes. It is time to call MAYDAY.

I remember clearly that when my replacement took the watch, I went down below and called MAYDAY, as silently as I could, not to wake anyone up, not to stir panic. But they did hear. Everyone opened their eyes, staring at me, attentively, waiting to see if the call would be answered. I was aware that they might have to wait long, so I ordered the watch on duty to prepare breakfast and hot tea for the whole crew, in lieu of the "everything under control," or "nothing to worry about" clichés, that nobody would believe anyway. It worked, the crew went about business as usual, and when the *Selendang Sari* answered, relief and joy exploded. With a cup in my hand, having a casual smoke in the cockpit, I explained to the crew, still crammed down below, the nature of our situation and the essence of my decision. The weather conditions were good: clear skies, bright sun, and we were close to our port of refuge. But without the engine, we would take too long under the small storm jib to get to A Coruña before the weather conditions worsened. I stressed that my steps were *preemptive*, and that I wanted the crew to be

Figure 5.1 The *Oskard* under the emergency jib (Paweł Jędrzejko's private collections).

safe well before the weather forecasts should come true, and that—bearing in mind that weathering any gale is always safer on board of a ship than in a storm-tossed liferaft, and that our VHF antenna, on the top of the mast that could fall, is still operational—it only makes sense to ask for help *ahead* of the time of struggle, and not in the *middle* of it. And then—things on board went back to "normal." People would return to their casual chitchats, cracking occasional jokes, commenting on the beauty of the ocean, and making plans just to kill time before the *Seledang Sari*'s arrival. When the ship finally loomed on the horizon, we recapped our safety drills and made sure that everyone understood the procedure and knew what to do; every detail of the action had been discussed and explained. Morale was high: the sight of the tanker gave everyone hope, and with every passing minute the orange shape, promising the end to our tribulations, would grow more and more real.

But hope ebbs and flows. When the *Selendang Sari* was close enough to start the procedures of the transfer of my crew, her captain informed me of the sudden change of plans. "The weather is too harsh, I cannot risk two rescue actions rather than one, which would be the case, if I lowered my lifeboat to pick up your crew." Right, ok. I know the law. He has the right to refuse to follow such a procedure if he deems it dangerous to his own crew.

Figure 5.2 The *Selendang Sari* approaching (Paweł Jędrzejko's private collections).

"Why don't you open your liferaft, transfer your crew to the raft, and let her drift to my side?"

"I cannot do it, sir, for two reasons: first, I cannot control the course of the liferaft and if it hits your enormous side, people die; second I may need that liferaft it I am to stay on board and wait for the tugboat from Cabo Finisterre."

"Ok then. Let me pick up your yacht with the crew using my crane. Is it more than 10 tons?"

"Affirmative, it is more than ten tons. Plus, how do you imagine fixing the boat to the belts in this swell? Why don't we stick to the plan: you come to my windward, shelter me from the wind and the swell, and launch your heaving lines from the shooters."

"Negative, captain, no can do: I am under ballast, and I am a single-screw vessel. If I lose you, I would take hours to catch you again. I shall come to your leeward and you will drift onto me."

What the hell... *Hours*? I was not born yesterday: the captain of the *Selendang Sari* is clearly in a hurry; he knows his obligations to save lives, but he also knows that saving the yacht is not mandatory. And he already made it clear that he will do *nothing* above and beyond his duty. Can I blame him? I can't. But I do. I am furious, but beggars can't be choosers.

"No, sir, this is not a viable plan. If I hit your side, it is as good as if I hit a solid rock. That is death."

"No, captain, we will pass the heaving lines to your yacht from a safe distance. Have your crew get down below; we will shoot the lines when you are a cable away. Then we will lead you on the leash behind our stern and to our leeward, where you will be sheltered; we will then drop the ladder and bring your people on board."

"Affirmative, captain, affirmative."

I had my crew, fully prepared to transfer to the tanker, lie on the floorboards down below, and myself I tied off all of our fenders on the *Oskard*'s starboard side. I blocked the steering wheel so as to ensure the stable drift and rushed below deck. We were approximately a cable away from the monstrously big *Selendang Sari*.

We waited. And waited. And nothing happened. I grabbed the radio, and, desperate, demanded an update on the situation.

"Not to worry, captain: you are close enough for us to throw the lines by hand."

And then I saw death. Nearing and inevitable, in the form of an enormous orange wall in my cabin's window, rolling up and rolling down on a heaving swell. I ran on deck, without thinking, without knowing what I was doing. I grabbed our boathook on a long wooden pole and stood in the bows, as if I could hope to push my 11 tons away from hundreds of tons of steel and compensate for the power of several meters of the wave. I rested the blunt end of the pole on my lifebelt and waited, every second a hundred lifetimes. Finally, the wave threw my whole universe against the orange impenetrability. First hit; the bow pulpit was gone with a bang, but I, miraculously, was still standing, pressing my pathetic impromptu lance against the monster. With the next wave, the boathook bent and cracked, its splinters passing my face in slow motion, by millimeters, and then the next wave pushed us and the whole world away from the ship. Then, a thick mooring line hit my head; I grabbed it and intuitively tried to tie it off on my remaining bow cleat, which was way too small to accommodate the ship's line. I do not recall how, but I somehow managed to run it under the cleat and wrap the bitter end around the anchor winch; an extra pair of hands helped me. One of the crewmen, M.W., broke my orders, and came out to the deck. Slowly, the yacht went down into the trough of the next wave; I saw the line tighten, I heard it moan, and pull the cleat out of my deck with a massive crack. The heavy cleat flew slowly toward my face, so slowly that I was able to move away, and in the process observe loose remains of the sicaflex on its mounting plate. The cleat bounced of the ship's side and fell into the sea, when the yacht hit the side again. And then, right in front of my face, I saw an enormous propeller. It was all over. We were behind the stern of the ship, and the seamen from the *Selendang Sari* dropped another, much thinner, line, to our deck.

I was motionless. In the state of utmost stupor.

"No, change of plans, let them go. Let them go. I do not want the fucking line..."

"Paweł. *Grab that line*." My first mate, god only knows when he got on deck, retained the clarity of mind. "Grab the fucking line. We are there. It's over."

These words jolted me out of my stupor. We tied the line off at the windlass, and waited for the seamen to lead us to the leeward side of the ship. And then, pulled by what I thought was the massive ship's inertia, I was able to steer our little *Oskard* along the enormous side of the tanker, whose crew have already dropped the ladder and a line with a safety loop. One by one, we evacuated all of the crewmembers, when suddenly one of them broke down: fear of heights, combined with the enormous stress of the last days, rendered her immobile. I felt helpless and weak, unable to convince her to stand up and fight this one last time—and again, someone else saved the day: it was my bosun's calm words that reactivated the woman's will to live: "Whether you like it or not, you *will* be saved. Only, without your active cooperation, without you climbing that ladder, you will be forcibly pulled up, and that will hurt. A lot." The woman—later my wife—stopped crying. Mustering up her energies, she overcame her fear of heights, and minutes later, she was on deck of the *Selendang Sari.*

After a while, I was alone on deck of the *Oskard*, getting ready to cut the line with which we were still attached to the tanker. Relieved that may crew-members are all safe, I forgot how ardently I wished for gods to exist only a few hours later, how fervently I prayed for survival, how much I wanted to live; my ego kicked in, with force. I could only think of the shame that would burn into my face forever if I abandoned *my ship*, which was still afloat. Suddenly I heard a familiar voice in the loudspeaker of my VHF down below. It was my first mate, calm, and methodical, explaining to me with angelic patience that the *Oskard* had sustained so much damage, that staying on board to wait for the tugboat would be tantamount to death. I let go of the

Figure 5.3 The *Oskard* moored to the side of the *Selendang Sari* (Paweł Jędrzejko's private collections).

helm and peeked down below to see the damage for myself, but before I could dive in, another voice, that of the captain of the *Selendang Sari*, called me to order: "Captain, you will climb my side now of your own will, or you will do it with a barrel to your head. Either way, you are not staying behind." I woke up again. I let go of the steering wheel, grabbed the yacht's logbook, and as I did it, I had a quick look at the beams, all cracked when the *Oskard* hit the side of the tanker. I secured the companionway, blocked the steering wheel, and climbed the ladder.

Minutes later, assured by the ship's first mate that my crew were all safe, I ended up on the bridge of the *Selendang Sari*. My wife-to-be and some of the crew were already there, waiting to hug me. The captain of the tanker, a small, smiling man with vivid eyes, too, gave me a big hug, followed by a powerful handshake, and a massive tumbler of whisky, which I emptied in one shot. And then, standing on the bridge, we all watched our *Oskard* being cut adrift, a tortured white ghost with broken spreaders, tilted mast, and the handkerchief of the white sail flapping sadly in the strong wind, tossed among the waves. It was time to let go.

What happened next was one absurdist, Monty Pythonesque, colorful blur. Immediately after the formalities on the bridge were finalized, I was taken to the messroom, where all of my crew, already inebriated, were having the time of their lives as guests of honor during a karaoke birthday party of one of the *Selendang Sari*'s crewmembers. Clearly, no words of encouragement from me were needed: everyone celebrated life. We hugged, had a few

Figure 5.4 The *Oskard* cut loose from the *Selendang Sari* (Paweł Jędrzejko's private collections).

drinks, and then I asked to be led to my quarters, where, hugging up to my future (now ex) wife, I sobbed like a baby.

***

As one can imagine, upon our arrival in La Coruña, I asked all of the members of the *Oskard* crew to communicate with their families and friends *before* the news of the sinking of the yacht would reach them through the official and social media. And, as one can expect, the accounts that the families and friends received were much less unemotional than the extensively quoted report of the Maritime Chamber. Predictably, when the information of the accident reached the sailing milieus home—the Internet exploded. Theories of what *could be done* and what *has not been done* energized discussion groups, and in effect, the whole Polish sailing community split into my "ardent supporters" and "righteous critics." When our friends and shipmates from the GTKŻ Gwarek Sailing Club finally brought us back home,[9] I first faced the overjoyed and grateful parents and spouses of all of the *Oskard* crew. Immediately thereafter, I clashed with ultraenergetic verbal acts of hate, concurrent with equally powerful declarations of solidarity among the sailing *aficionados*, active on the social media. The projections of the "moron, who made so many mistakes" would coincide with the praises to the "captain who saved his crew"; the sheer number of posts astounded me. Luckily, before I could get pulled into any debate, my former captains and excellent teachers—Tomasz Szwed, Andrzej Drapella, Jacek Mitka, and especially Wacław Petryński—all of whom I contacted while still in Spain, reacted immediately, warning me against any engagement in unprofessional interactions prior to the announcement of the ruling of the Maritime Chamber. The principle "do not feed haters or trolls" proved more timely than ever: the accident cost no lives, and the know-it-alls *were not there* when we all struggled to survive. And thus, immune to challenges, I waited patiently for the Maritime Chamber and the Delegate of the Minister of Infrastructure to issue their verdict. And after weeks and weeks of waiting and countless hours of investigation, the ruling was announced, jump-starting my risible career as "expert" in survival at sea. The same people, who would vilify me a few weeks before would now reach out with lame apologies, inviting me to speak in their clubs, sailing schools, or conferences dedicated to safety at sea. Although initially my ego protested, upon reflection, I let go: teaching others what I have learned on so many levels is, after all, of paramount importance.

Nonetheless, I would suffer from nightmares for a very long time. But when I finally regained my composition and came back to my ordinary life, I returned a changed person. I came to see problems that would have given me insomnia just a few months earlier as trifles: the word "deadline" would lose its ominous ring; a dent on the chassis of my car would not even raise my eyebrows, and saying "no" to my superiors at work became easier than ever.

Even more importantly, perhaps, having come to terms with the gravity of the experience, I followed the advice of some of my former commanders, and returned to sea almost immediately, ready to face waters in many ways much more challenging than the open ocean—the Gardens of the Isles, where watchfulness with respect to the sea and the wind must be extended to meticulous navigation among rocks or reefs. The above notwithstanding, however, is the accident-related transformation of my understanding of the philosophy underlying Melville's criticism of Ralph Waldo Emerson's serene Transcendentalism that allowed me to intellectually apprehend the essential difference between *oikeiôsis* and the obliteration of self, which has colored my perceptions of the universe ever since.

Surviving in "the All"

It is likely not a coincidence that the sea becomes an object of apotheosis for writers.[10] At the beginning of *Moby Dick*, a particular, "magnetic" quality of the sea is distinctly emphasized:

> Circumambulate the city of a dreamy Sabbath afternoon. Go from Corlears Hook to Coenties Slip, and from thence, by Whitehall, northward. What do you see?—Posted like silent sentinels all around the town, stand thousands upon thousands of mortal men fixed in ocean reveries. Some leaning against the spiles; some seated upon the pier-heads; some looking over the bulwarks of ships from China; some high aloft in the rigging, as if striving to get a still better seaward peep. But these are all landsmen; of week days pent up in lath and plaster—tied to counters, nailed to benches, clinched to desks. How then is this? Are the green fields gone? What do they here? But look! here come more crowds, pacing straight for the water, and seemingly bound for a dive. Strange! Nothing will content them but the extremest limit of the land; loitering under the shady lee of yonder warehouses will not suffice. No. They must get just as nigh the water as they possibly can without falling in. And there they stand—miles of them—leagues. Inlanders all, they come from lanes and alleys, streets and avenues—north, east, south, and west. Yet here they all unite. Tell me, does the magnetic virtue of the needles of the compasses of all those ships attract them thither?[11]

Why is it that the sea possesses this property, and not, for example, the prairie or the dense forest? Perhaps it is because the sea is more difficult to "arrest" in the categories of direct perception. The experience of the sea seems to most vividly bring one close to the elusive, ineffable, experience of being. Even a cursory analysis of some of the passages in Melville's novels, where the author philosophically describes the ocean, indicates this.

These sections merit closer attention. The sea, although, like other elements, resisting human will, unlike earth, air, or fire, leaves no doubt about whether or not it can be controlled. Every day, local reality deceives the "inlanders," whom the narrator mentions, allowing them to believe that they have tamed fire (which now flickers in their fireplaces), that the air is at their service (turning windmill blades), that the earth (which they cultivate as the Bible instructs) has been ultimately subdued. However, the sea remains an element stronger than man—it is perpetual fire, an eternal hurricane, and a ceaseless earthquake, and although sometimes it behaves tamely (like fire in a fireplace) or allows its riches to be extracted (like the earth)—it never lets one forget that it is an "inhuman" element. Melville writes:

> [...] we know the sea to be an everlasting *terra incognita*, so that Columbus sailed over numberless unknown worlds to discover his one superficial western one; though, by vast odds, the most terrific of all mortal disasters have immemorially and indiscriminately be fallen tens and hundreds of thousands of those who have gone upon the waters; though but a moment's consideration will teach, that however baby man may brag of his science and skill, and however much, in a flattering future, that science and skill may augment; yet for ever and for ever, to the crack of doom, the sea will insult and murder him, and pulverize the stateliest, stiffest frigate he can make; nevertheless, by the continual repetition of these very impressions, man has lost that sense of the full awfulness of the sea which aboriginally belongs to it. The first boat we read of, floated on an ocean, that with Portuguese vengeance had whelmed a whole world without leaving so much as a widow. That same ocean rolls now; that same ocean destroyed the wrecked ships of last year. Yea, foolish mortals, Noah's flood is not yet subsided; two thirds of the fair world it yet covers.[12]

Menacing and powerful, the sea poses a threat to humanity not only because it is not our natural habitat: the sea is also Earth, yet an unknown, perhaps unknowable, and unmastered Earth. After all, sea charts define the boundaries of what has already been delineated. The sea is that part of the Earth whose nature is not and may never be fully penetrated—and contrary to our cartographic illusions, it covers the greater part of the globe.

To reach the New World, as Melville points out, Columbus sailed over many worlds, of which he was entirely unaware, and upon his arrival in America, he only saw a fragment of reality protruding from the sea. He might have thought that what was revealed to his eyes was all that there was to see, but his discovery of the "New Earth" was only "superficial." What he saw could indeed be *part* of the "truth about the world," yet, when it is extended to the entirety of existence, any interpretation of one's immediate perception will, at best, be a (risky) approximation. The majority of "data" is still hidden from us within the ocean of fluidity. It is this fluidity that characterizes two-thirds of the globe, and it is the same fluidity that has swallowed the lives of many a daring individual. Perhaps because they failed to perceive the constant within the fluid.

> Childe Harold not unfrequently perches himself upon the mast-head of some luckless disappointed whale-ship, and in moody phrase ejaculates:—
>
> *"Roll on, thou deep and dark blue ocean, roll!*
> *Ten thousand blubber-hunters sweep over thee in vain."*
>
> Very often do the captains of such ships take those absent-minded young philosophers to task, upbraiding them with not feeling sufficient "interest" in the voyage; half-hinting that they are so hopelessly lost to all honorable ambition, as that in their secret souls they would rather not see whales than otherwise. But all in vain; those young Platonists have a notion that their vision is imperfect; they are short-sighted; what use, then, to strain the visual nerve? They have left their opera-glasses at home.[13]

Fluidity engenders respect for constancy and unequivocally demands it. The reality of the sea, even if it allows only a glimpse of the superficial, does not tolerate a philosophy of doubting the existence of what is plainly seen. Although Columbus might not have known the "truth about the world," he survived without charts at his disposal because he did not dismiss visible reefs as mirages, and exercised caution when navigating among them. But Melville understands the naïvete of youth; he addresses it warmly. In the cited passage, the inexperienced man is merely the subject of a benign jest: the mature writer is aware that, viewing the world through the lens of predetermined beliefs, this maritime "Childe Harold" will not only fail to find answers to his pressing questions, but may also seriously risk his own safety. In the vast ocean, the price to pay for "short-sightedness" could be high: unwilling to "see the whale"—convinced of the solidity of his own reasons, the young man (understandably preferring to avoid the confrontation with the giant) will not notice it until a tragic collision with the animal ultimately occurs. Melville knows from experience that it is sometimes hard for a young person to believe what they see: wanting to bend reality to a preferred vision of the world, the inexperienced "Platonist" will not see it as it truly is. However, if in the face of direct experience the young idealist refuses to revise his stance and fails to recognize the danger as real, his allegiance to his philosophy could cost him his life.

Ocean sailing, which has not coincidentally become one of the most popular metaphors for life in the Western culture, is clearly subordinated to the peculiarity of the element. The fluidity of the ocean requires mindfulness; the condition of survival is constant vigilance; neglect can lead to disaster. The ocean demands a continually heightened readiness to "suspend" what is not always punishable in "terrestrial" existence: (discursive) relativism, disbelief in the factualness of being, and doubting in its indifference to human fantasies. However, even among those who understand that vigilance saves lives, moments of inattentiveness are unavoidable. Melville writes:

> Sailors can not be too wakeful and cautious, when keeping their night look-outs; though, as I well know, they too often suffer themselves to become negligent and nod. And this is not so wonderful, after all; for though every seaman has heard of those accidents at sea; and many of them, perhaps, have been in ships that have suffered from them; yet, when you find yourself sailing along on the ocean at night, without having seen a sail for weeks and weeks, it is hard for you to realize that they are near. Then, if they are near, it seems almost incredible that on the broad, boundless sea, which washes Greenland at one end of the world, and the Falkland Islands at the other, that any one vessel upon such a vast highway, should come into a close contact with another. But the likelihood of great calamities occurring, seldom obtrudes upon the minds of ignorant men, such as sailors generally are; for the things which wise people know, anticipate and guard against, the ignorant can only become acquainted with, by meeting them face to face. And even when experience has taught them, the lesson only serves for that day; inasmuch as the foolish in prosperity are infidels to the possibility of adversity; they see the sun in heaven, and believe it to be far too bright ever to set. [...] And even, as suddenly as the bravest and fleetest ships, while careering in pride of canvass over the sea, have been struck, as by lightning, and quenched out of sight; even so, do some lordly men, with all their plans and prospects gallantly trimmed to the fair, rushing breeze of life, and with no thought of death and disaster, suddenly encounter a shock unforeseen, and go down, foundering, into death.[14]

The seafarer, more frequently than the land-dweller, finds himself in situations of direct life-threatening danger, which could likely be categorized as "liminal." More often than the land-dweller, he or she experiences the existence between life and death, in a realm where there is no place for any discourse. Melville subjects his characters to such experiences, and these experiences *fundamentally* influence their intellectual evolution. The sea teaches that what exists simply exists, regardless of how illogical it might seem from the perspective of relativistic philosophies. At best, the penalty for "blindness" to the manifest reality of being is the mockery of the fellow crewmembers:

> "Why, thou monkey," said a harpooneer to one of these lads, [...] "thou hast not raised a whale yet. Whales are scarce as hen's teeth whenever thou art up here." Perhaps they were; or perhaps there might have been shoals of them in the far horizon; but lulled into such an

> opium-like listlessness of vacant, unconscious reverie is this absent-minded youth by the blending cadence of waves with thoughts, that at last he loses his identity; takes the mystic ocean at his feet for the visible image of that deep, blue, bottomless soul, pervading mankind and nature; and every strange, half-seen, gliding, beautiful thing that eludes him; every dimly-discovered, uprising fin of some undiscernible form, seems to him the embodiment of those elusive thoughts that only people the soul by continually flitting through it. In this enchanted mood, thy spirit ebbs away to whence it came; becomes diffused through time and space; like Cranmer's sprinkled Pantheistic ashes, forming at last a part of every shore the round globe over.[15]

Clearly contesting Emersonian concepts, Melville points out that the pantheistic Oversoul is merely an intellectual construct, a mirage that shatters against material reality. The "surface" of things is indeed not the whole truth, but to ascribe any characteristics other than indefiniteness to what is *not* the surface is to confuse the verbal structure and the-being-that-exists. Such an error exposes a person to the danger of being sucked into Descartian "vortices":

> There is no life in thee, now, except that rocking life imparted by a gently rolling ship; by her, borrowed from the sea; by the sea, from the inscrutable tides of God. But while this sleep, this dream is on ye, move your foot or hand an inch; slip your hold at all; and your identity comes back in horror. Over Descartian vortices you hover. And perhaps, at midday, in the fairest weather, with one half-throttled shriek you drop through that transparent air into the summer sea, no more to rise for ever. Heed it well, ye Pantheists![16]

The sea teaches that transcendence—if it is understood as one's ultimate union with the Oversoul, or as the transgression of the boundaries of the local world into the transparent world-beyond-constructs—is, in itself, merely a verbal construct. The "vortices" over which the philosopher-dreamer hovers represent the mechanical principle of the universe: a magnetic fluid keeps all matter in whirling motion, set into life by the initial divine impulse, and all phenomena result from the continuous collisions of ever-moving particles. On one hand, vortices are *matter*, but on the other, they are elements of a perfectly logical world—the flawless creation of an infallible and omniscient God. However, in Melville's particular interpretation—as Howard P. Vincent points out—the concept of "Descartian vortices" is more metaphorical than cosmological: in Melville, "vortices" are "brutal, material facts."[17] They kill those who cannot distinguish the conceptual structure—the appearance of truth—from the immediate, non-verbal truth about the world. Due to the fault of Ahab's, whose tragic error lay in trusting the theoretical construct of "the thing in itself" and in his failure to recognize in the whale a material being, the sailors of the *Pequod* perish in such a Melvillian vortex; it is not "the thing in itself" that kills them, but water:

> For an instant, the tranced boat's crew stood still; then turned. "The ship? Great God, where is the ship?" Soon they through dim, bewildering mediums saw her sidelong fading phantom, as in the gaseous Fata Morgana; only the uppermost masts out of water; while fixed by infatuation, or fidelity, or fate, to their once lofty perches, the pagan harpooneers still maintained their sinking lookouts on the sea. And now, concentric circles seized the lone boat itself, and all its crew, and each floating oar, and every lance-pole, and spinning, animate and inanimate, all round and round in one vortex, carried the smallest chip of the *Pequod* out of sight. [...] Now small fowls flew screaming over the yet yawning gulf; a sullen white surf beat against its steep sides; then all collapsed, and the great shroud of the sea rolled on as it rolled five thousand years ago.[18]

However, the vortex, as Melville presents, further problematizes the issue of surface and depth: just as the sea is an "unknown land," and the earth—despite all "superficial" appearances of stability—is in fact "fluid" in its essence, the vortex suspends the simple phenomenological distinction between what is visible and what lies beneath. The vortex is both surface and depth at the same time: thereby, it is neither purely surface nor purely depth. By dividing the world binarily into "cardboard masks" and what lies beyond them, humans are thus doomed to a cognitive failure. This, however, is not to say that the brutality of the material facts contradicts the validity of contemplation as a method of their exploration. Melville clearly admits: "Yes, as every one knows, meditation and water are wedded for ever."[19] Yet, to Melville, the experienced sailor, the Emersonian vision of the order of the universe is flawed insomuch as it projects the experience of transcendence as a form of the annulment of the self in the transparency of the divine, and thus construes it as a desirable state. Unlike Emerson, the Author of *Moby Dick* "can neither believe, nor be comfortable in his unbelief":[20] questioning the existence of any pantheistic divinity, he appears to point out to the fallacy of Emersonian premise. To Melville, there is no room for reflexive "I" outside of language: the direct, unmediated experience of being, in his vision, is the nauseating experience of the obliteration of self in the "indifferentiation of essence."

> Why did the old Persians hold the sea holy? Why did the Greeks give it a separate deity, and own brother of Jove? Surely all this is not without meaning. And still deeper the meaning of that story of Narcissus, who because he could not grasp the tormenting, mild image he saw in the fountain, plunged into it and was drowned. But that same image, we ourselves see in all rivers and oceans. It is the image of the ungraspable phantom of life; and this is the key to it all.[21]

And thus, Melville's existentialist position seems to advocate exercising one's contemplative powers to arrive at the state of *oikeiôsis* not in order to place ones' self "under erasure," but to seamlessly attune it to reality, thereby warranting its survival in the battlefield of warring Ontic Empires.

The Ontic Empire versus Sauna

Unlike the "young Pantheist," fascinated by the transcendental experience of being nothing, and seeing all, who might painfully discover the palpability of the "inhuman essence," an experienced philosopher, abhorring the absurdity of the human condition, but also aware of the fact that elementary particles, of which the universe is constructed never die, might make a *choice* to embrace the possibility of ultimate self-erasure. Such explorations in "seeking death without dying," as Andrzej Chojecki—a poet, writer, literary translator, excellent Scandinavist, yachtsman, university professor, husband, father, and a dear, dear friend—formulated it,[22] are often attractive to those unable to believe, yet uncomfortable with their unbelief.

Many years ago, while we were engaged in a light dinner conversation revolving around our beloved archipelagos in the Baltic Sea, the Scandinavian seascapes, their symbolism in popular songs and folklore, and deconstructivist criticism, Andrzej, smilingly, communicated that, driven by his fascination with reality unmediated by categories, he wished to explore absence by way of self-erasure. I only nodded, excited at the prospect of yet another contemplative excursion into the theoretical wilderness of non-existence, which was a recurrent *topos* of Andrzej's academic texts and his poetic insights alike. Lightheartedly, we would play with the idea of suicide as an act of the "ultimate deconstruction," and, amidst general merriment, pose questions concerning the very premises of our professions, entangled in discourses, yet never immune to the dominance of what Chojecki dubbed "ontic empires." Over many months these debates continued, often in letters scribbled on photocopies of texts that we would share—our own, or other people's. One such exchange was spurred by my impromptu comments on Andrzej's fascinating book, *OE*, which he had published several years before. Upon the receipt of my letter, Andrzej responded thus:

> Hi Pablito!
>
> 2×2B=?
>
> I am hugging you as mightily as if I were holding a sheet in 30 knots of wind.
>
> PS: If you would like to add anything to what you wrote about *OE*, that would be wonderful. I eagerly await it, and if nothing comes of it, I will take care of the prototext.
>
> Andrzej[23]

The "prototext" mentioned in the postscript to the note, in fact, served as a "pre-text" prompting me to resume the thereto related conversation with its author, especially that Chojecki's *OE* was essentially an epistle in itself: a *belle lettre* about the Ontic Empire. The following exchange is a condensed, glossed history of a correspondence that began with a letter, with which Andrzej opened his *OE*:

Gdańsk, 15 Feb., 1987

Dear Friend,

I wrote this, but I must confess, I wasn't entirely convinced I was doing the right thing. I wanted to express what I truly believe, though that belief concerns the futility of expression. So why create yet another entity that merely declares the absence of the need to create entities? Perhaps, however, one must try at least once, perhaps at least once everyone must attempt to save the world. But then, if this need to "save the world" becomes a habit—and I come to be a professional savior… the prospect is as unpleasant as it is ridiculous.

Thus, the drama of Penelope: what you create at night, you destroy by day, and vice versa. Conflicting thoughts, feelings, and intentions compel one to seek meaning in one's actions, only to see their banality, their commonness, their obviousness, their superfluity. Penelope, weaving and unweaving, defended her own right to her dreams, longings, and love against (the passing) life that intruded on her only too insistently. She is a human Parca/Maya/Norn, striving to weave her own life, resisting Others, who usually try to do it for us. Penelope is somewhat the reverse of Don Juan. And so, my writing is a dilemma: to write or not to write. Choosing to write is choosing "to be."

If I am to write, know this:

I am weary of genres, conventions, plots, types. Everything is clichéd, easily decipherable, and even if complex, playing this game demands too much effort and time, while life is hasty—time fleeting—and there are many other, more easily assimilated, forms at hand.

After all, everything is text: philosophy, poetry, non-fiction, popular science, academic writing, language, life, the Other—I look for something for myself everywhere, and I keep discovering that all these domains are merely the matter of the languages in which one speaks about me, about Others, about reality.

Literature itself is, essentially, a letter—a beautiful letter written to the Other, to oneself, to God. It is a plea for the acceptance of the author's existence. The more beautiful and original the letter I write, the greater my chances. Beautiful letters—*belles lettres*—beautiful writing. Much like poetry was once the ornament of speech—beautiful speaking.

"To write or not to write," obviously, echoes the dilemma of "to be or not to be"—but they are not the same dilemmas. One can choose "to be" and not to write. One can choose existence as a granted gift, a gift not to be squandered. No one has more starkly exposed the frailty of the humanities, of writing, of knowledge than Goethe, who, speaking in Mephistopheles's voice, chastised the student overly eager for knowledge and forgetting his youth:

Grau, theurer Freund, ist alle Theorie,
Und grün des Lebens goldner Baum.[24]

The green tree of life is the gift I call *creatio Dei*. But man persists in his belief that through his actions he can create many other, more magnificent, entities. The more he creates, the further he strays from *creatio Dei*, from God. The more intensive his creation, the more disorderly, chaotic, and varied it becomes. For John Scotus Eriugena, the situation was clear: unity was God, multiplicity was Not-God.

Seneca contrasted knowledge with goodness when, in his moral letters to Lucilius, he wrote thus: "Frankness and simplicity beseem true goodness. Even if there were many years left to you, you would have had to spend them frugally in order to have enough for the necessary things; but as it is, when your time is so scant, what madness it is to learn superfluous things!"[25]

I mentioned to you the warnings of the Bible and [of the German Romantics] ("Die Elemente hassen das Gebild der Menschenhand"[26]), [but in this regard] one can also cite the Quran:

But never will they long for [the ultimate dwelling],
because [they are aware] of what their hands have wrought in this world.[27]

And more:

[...] woe to those
Who write the Book with their own hands,
and then say:
"This is from Allah"[28]

With such warnings at hand, and bearing the current situation in mind, one must truly reflect—not on humanity (whose fate seems already decided) but on oneself. Today, man can do anything he wishes to do: one can deny everything, approve everything, question everything, turn everything into irony, destroy everything, create almost everything (or so he believes)—everything is within his reach. But perhaps it would be better if he stopped at acknowledging his capabilities and, at the very least, if he chose to limit his—joyful and neurotic—ontic activity, *creatio humana* (assuming that he has not abandoned it already).

Thus, everything is *being*, but the geneses of *beings* vary. I dismiss the old division of the world into the realm of matter and the sphere of ideas. The belief that only matter counts and the rest is illusion is ingrained in our worldview. I have kicked that material rock and felt *myself*, *my pain*. I still know nothing about the rock—I only know that the rock, in its clash with me, is my pain. Clearly, there exists another conviction, which has come to be the shared experience of many generations, and which surfaces in the words of T.S. Eliot:

Man's life is a cheat and disappointment;
All things are unreal...

And, somehow, at the same time, it is no longer universally believed that man is the measure of all things, that man can save the world. Here, too, Eliot's words seem fitting:

Only
The fool, fixed in his folly, may think
He can turn the wheel on which he turns.

Earlier, Lombroso used a similar metaphor:

Noi confitti al nostro orgoglio
Come route in ferrei perni,
Ci stanciamo in giro eterni,
Sempre erranti e sempre qui!

—which one could [philologically] render thus:

Nailed to our pride,
Like wheels on iron pins,
We tire in eternal treadmills,
Always wandering, and always here![29]

Everything is such a rock—a rock that you kick to see if it exists, and to discover that you only experience yourself. The ennui (I understand Sartre's nausea) with entities that you can only reach by way of your own perception. Everything *is*, *is*, *is*. One cannot escape the essence—the fatigue, the boredom of it, the multiplication of it. And when I say *essence*, I mean *being* that underlies every manifestation of existence, thought, life, etc., etc.

And then there are Ontic Empires that constantly form. You know how they arise, you know how they endure: consistency, similarity, power, hierarchy, collections, duration, traditions, communities (linguistic, of interest, cultural, racial, etc.), shared experiences, wholes, convergences, etc., etc. *Lika barn leka bäst*[30] says the Swedish proverb, and it contains not only the wisdom of the nation but also the conviction of the existence of a generational community. This, *already*, is the Ontic Empire of a Generation, but proverbs themselves constitute an Ontic Empire of their own—the English equivalent of the proverb above is: *birds of a feather flock together*. There is an infinite number of self-organizing Ontic Empires in various dimensions: diachronic and synchronic, individual and collective, private and official, unknown and well-exposed, ephemeral and lasting, weak and strong, accepted and negated. One is always under the rule of some Empire, but one is also at the helm of another Empire. I think I have explained these relationships in my writing.

A friend advised me to expand individual fragments from *OE* into larger forms. It might be possible; many books would come out as a result. But I believe that anyone could do this on their own; I only want to show the paradigm, not to fill it with specific cases, not even to carry out the entire repartition. You may have many remarks, questions, additions—you may wonder why I am inconsistent, where this or that citation comes from, whose words I quote, etc. You know… I don't want to explain it to you, because if I do, I am afraid I will entangle myself in some argument that will certainly, more or less efficiently, justify my choices. But I think you do believe that I can reasonably explain my actions. I believe you can do it, too. Perhaps better than I could, perhaps differently, depending on your good (or ill) will. Trust me about one thing: I have renounced many entities (beings) with the view to conveying only the fundamental structure of my thinking—the state of my convictions. The thought of writing an "opus magnum" was abhorrent to me. Do not think I am bragging with what I have written, and please do not think I am arrogant when I tell you that masterpieces bore me to death. You know how they do… you reach for a masterpiece and you *always-already know*, for it contains everything about you, about life, about the world. You see "clearly, in wonder." But then it turns out that there is yet another masterpiece, and another, and all tell you the truth about you, about life, about the world. Indeed, you begin to wonder whether to expel poets from the republic, or who should remain in it. But of course, that thought is also repugnant to you. You think about your place in the republic and about your green tree of life. Yet there are other green trees, too. There is a forest, in which you found yourself on your life's journey.

So, allow me to explain my writing as an inclination towards asceticism, neoasceticism, as I would call it (again, words). If any fragment, any thought, any being from this writing resonates with you, you can consider it yours, completely yours, ours. In the humanities, the cult of the individual is still practiced. And yet, if you travel by train, car, or airplane, you do not think about who invented the wheel, the engine, the wing. Everyone invented it—it is the effect of *creatio humana*. So it is with art, with the humanities. A single guru, a "sender," is not enough; the sender alone does not suffice to account for arts' and humanities' existence. It also requires a code and a receiver. It needs a language that creates me more than I create it, and *you* are needed. And your consent to my loquacity. You know that to speak the truth, a few words will suffice, but behind the one who speaks these words must stand an Ontic Empire. The stronger it is and the more authority it possesses, the less one needs to speak; then speech can be contained in "yes" and "no." Małgorzata Szpakowska bravely said that the humanities are verbose—no one limits themselves to expressing their knowledge in one sentence. And since we must use this "high" language, the meta-language of myths, literature, quotations, names, references, allusions, thought shortcuts, metaphors, parables, symbols, etc., etc., let us remember that it is our shared culture that creates our expression. Perhaps it is better if I speak for myself, again—not in "my own words": through me flows a stream of beauty, but I am not beauty itself. It is very difficult, in fact, to free oneself from the common complex of Nero, that pitiful versifier and graphomaniac, who could not accept that the world did not appreciate him. "What an artist dies in me," he gasped at the moment of death, regretting not so much the loss of his Empire of Power as of the Empire of Art, in which, as he rightly assessed, no one acknowledged his greatness. Everyone probably has such an artist complex, everyone has a moment when they say their *Qualis artifex pereo*, for through everyone flows a stream of beauty. Yet, it is not necessarily something to be preserved, retained...

Andrzej Chojecki[31]

I received Andrzej's brief "2×2B" ("Twice to Be") note in the middle of my doctoral sabbatical, in my beloved wintry Stockholm, the city at the heart of the Garden of the Isles.[32] Although I remember it making my day, I did not respond immediately. In fact, I took a long time to meditate on it—and it was only when the chill of the Swedish February made me organically comprehend the absolute indispensability of the *institution* of the sauna that I found the key, which allowed me to properly collect my thoughts:

Dear Friend,

Please, forgive my autotranslatological inaccuracies—which, alas, all result from the pettiness of your generation of humanists. You, the seekers of the Formula, are so dreadfully infatuated with your Language of Definitions... And for some time now, impressed by the epiphanies of the 60s, you all have been pretending that you need it just about as much as a hole in a bridge.

You wrote your *OE* over ten years ago, likely because this very "discovery" caused you pain. And perhaps it is only the frustration with the impotence of conceptual language that can explain the extent to which you had to give up your freedom not to define, how cruelly (masochistically?) you had to beat yourself to compel yourself into self-enslavement, among the overwhelming cascades of vowels and consonants, in hope to achieve a synthetic perspective: a perspective that leads back to the helplessness of speech anyway, and, ultimately, produces yet another version of an already cancerous figment of the last few millennia: the OE of many faces. Furthermore, as I engaged in reading, I began to lose myself in this multiplicity, starting to almost trust it. The self-preservation instinct to follow the intuition of the existence of a non-verbal community uniting Us-Always-Others-in-the-cold prompted me (somewhat against my will) to enter your generation's discourse and plunge into seeking, uncovering, and always finding the implications of the OE in your cohort's terminology (which I inherited), in myths (both your own, and those rooted in the culture), as well as in literature. Do I know more about myself than before? Perhaps. I can describe—or approximate—myself in new/other (?) ontological categories emphasizing the cultural continuity of thought and of its development, in which light *I have become different*. Am I calmer? Certainly not. And that is good, because it is amidst the uncertainty that I found myself anew.

If he could know the direction, in which the reflection of the supposed ancestors of the Overman would evolve (and which your examples, drawn from various cultures and philosophies, both European and beyond, insistingly indicate) Nietzsche (whom—he had not decomposed long ago—your generation would have de-composed anyway) might never have chosen to join your "club," for, as he quietly whispers to me, he does not like it when suppositions extend beyond creative will—and *not* the will aimed to call deities (artificial beings) into existence. Instead, he murmurs, he does value the body—and, especially as Zarathustra, loves to dance. Perhaps, my friend, in the panic of the "ineffable," informing chapters 111 and 112 of the *OE*, one loses sight of the existence of the physical body that articulates this panic, forgets that institutions, such as *bastu*, were created to soothe it, and misses out on the fact that the durability of the structures that you have so excellently laid bare in your book depends solely on the temperature of the sauna and on the resilience of the body? I know, today, it is not easy to believe that "the non-verbal" and "the scientific" do not necessarily have to behave like oil and water. *Not easy*, but perhaps not impossible, especially when one chooses to ask a few critical questions: Since when has such a division been in effect? What is it that legitimizes it? Can the shaky conceptual edifice based upon such a mutually exclusive binary be reduced—without resorting to other concepts—to something so fundamental that it would precede conceptualizations? A non-verbal structure, notion-free, yet attainable?

Let's heat up the *bastu* and find out. Let's invite all the characters populating your book into the sauna—all of them, beginning with the theosophers of Jehovah and Shiva, through Confucius, Plato, Aquinas, to Husserl, Sartre, and finishing with the Errata and your own Readers (including myself). Let's knock ourselves out, splashing plenty of water over the red-hot stones—and then let's wait. When our guests can dream of nothing else—let them jump into an icehole and immerse in the wintry Baltic; protected with a layer of heat, and gently swayed by the water, the will relish the sensation of every muscle relaxing in delight. But after a brief while, the chill will have penetrated them to the bone: soon, they will dream of *nothing* but getting out of the icehole. Rushing back into the inferno of the *bastu*, they will enter the space of *jouissance*: before the layer of cold, still emanating from their bodies, becomes thin and eventually disappears, millions of subtle needles will turn the formerly unbearable heat into a caress. And then again, and again, and again. Each time, let's extend the time of their alternating immersion in the heat and in the cold *just a little*—after all, we are experts in tickling our egos with the specter of immortality! Tricking their pride (or their need of pleasure) we can have them endure *more heat* and *more cold*, we can have them balance on the verge of fainting. And then again, and again, and again.

What will happen? Very soon, our guests will no longer see their own nakedness, as if there had never been apples in the world. They will abandon their search for Formulas, or logical foundations of Being; the all-organizing sub-sub librarians or museum curators, hiding behind shelves, will lose their status of *éminence grise*. God, god, Gods, or gods will no longer be invoked, and soon will cease to matter altogether. Philosophers, theoreticians, and theologians alike will strive cover their mouths so that the heat, their source of delight, does not burn their own lungs—lungs independent of any OE. On the verge of fainting, they will feel a sensation reminiscent of a rising orgasm (the importance of which you elsewhere emphasized as D. Freya, pointing out to the frequent connection between Eros and Thanatos), but fulfillment will be hateful to them for they will lose egoistic, authoritative control over their own bodies—and what *they are* will take over what they *made themselves to be*. Anyone familiar with the *bastu* knows well that by the hot stove, every exhale is cool, but when one breathes too quickly—even the breath loses its cool.

It is mid-winter, Stockholm, I am buried in snow up to my ears. Speaking now is like breathing warmth into frozen hands, it warms me up. Therefore, I will not stop talking. However, when it is very cold outside, or when I am in a very hot sauna—I am silent. There, the world does not need any First Principle, no capital letters. The measurable alpha and omega here are marked by the Celsius scale. If you manage to push the entire panopticon of the OE onto the upper bench of the *badstuga*—then rest assured that Being will not trouble itself with mustering hope to find of an organizing formula for the OE of all beings, because all its elements will vibrate and oscillate faster than anything could describe it, let alone the fact that speaking, or describing anything, will be the last things anyone could think of. Here, in the sauna, one begs one's own body for the grace of a few steps—without words, concepts, or doubts.

Your "beautiful letter" is truly beautiful. Your *OE* is an act of masterful erudition, a victory over the phenomenon of the Tower of Babel, and yet a splendid defeat in the battle with language, and if I understand the term "ontic" correctly (as "possessing the character of real rather than phenomenal existence"), for me, personally, it also energizes a whole series of immediately arising, vivisectional, "juxtapositions," operating on the living body of the text:

— "OE" versus "sauna"
— "Human and divine *creatio*" versus "tiled pool/icehole and cold water";
— "*Homo/deus absconditus*" versus "*The Scream* by Edvard Munch—somewhere on the bridge between man and Overman, or perhaps when a person accidentally slips and leans against the stove in the *bastu*";
— "Jag söker döden utan att dö"[33] versus "three seconds too long on the upper bench";
— "OE must know the state of its possession" versus "the individual possession of the valve immediately activating the icy-cold shower." The rest, in this dimension, ceases to matter: wealth, power, memories, [...] gender, context. The inventory of the sauna, when I want to leave it, is the penultimate thing that will cross my mind. The last is its description;
— "Beings connecting in ontic chains try to link everything known with the unknown, they attempt to bind the OE of Everything" versus "long protein chains break down at high temperatures, decomposing into elemental parts. In the sauna, a compound sentence will be uttered in a series of simple sentences, separated by greedy gasps of air, or will be reduced to monosyllables—or to zero." The known heat does not require connection with the unknown, it only requires the neutralization with icy-cold streams of water, wonderfully flowing all over one's body. The rest can wait.
— "Be careful and cautious, for the Universe—God's Puzzle—might fall apart" versus "Be careful and cautious, for in the sauna, when it is properly heated, the Universe—God's Puzzle—might wonderfully weld together." And what then?

- "The IO seeks the perfect method of self-knowledge" versus. "Reduction of the OE in the sauna to the state of absence of concepts is a reduction to zero." In the stove, Ockham would burn his hand on his razor's blade, not to mention that he would not even notice anything that would need trimming. In the *bastu*, the method is an excess;
- "Being hopes that an organizing formula for the OE-of-all-beings will reveal itself to it" versus "In the sauna, being hopes that in three seconds it will still be able to hope for anything, hence it is interested only in maintaining a relatively stable organization of its own atoms." And even that hope is non-verbal and non-conceptual; the rest, entirely non-solipsistically, just in general—does not concern it at all. After all, in the *bastu* one begs for mercy without words.
- "Languages we know do not only describe and explore our OE. They are also an echo of, and a timid attempt to make contact with, the Unknown, with the Unknowable. [...] Each of our languages contains an attempt—and possibility—to communicate with God" versus "In the inarticulate wheezing of a person who can no longer endure the heat of the sauna and fights for life, there is an attempt, possibility, and necessity of communicating with Another, who stands in his way, blocking the door separating the sauna from the pool." The boundaries of the unknown in the sauna become quite sharp when they coincide with the limits of endurance, and these—with the boundary of existence. The "unknown" reaction of the Other to my request to make way quickly takes on concrete shapes. The same with good and evil: the dialectic loses its operativity here. There is only one exit—and it has nothing to do with the synthesis—or deconstructions—of the life/death opposition.
- "Each newly introduced ontic reduction order only increases the number of beings" versus "Introducing constructs is adding constructs." Regardless of intention. And in the claustrophobic sauna, the scattered atoms already take up too much space, until finally there is no room left for me. And I have to leave.

And so, driven by anxiety, in my attempts to resist entanglement in the vicious circle of cancerously multiplied beings, I could contemplate the entire Judeo-Christian philosophy of the OE in the context of the seemingly trivial sauna, reducing all OEs to the OE of such a situation, where there is no more room for any OE. After all, "even a fool can create entities in contact with the OE." Understand me, friend: it is only from such a vantage point that can I start looking at the universe without prejudice—and it is not a petty thought, "illiteracy complex," or the need to show off that drives me: I am well aware of my shortcomings.

Your deconstructions, however, suspend the meanings of fundamental approaches and concepts evolved in culture somewhere amidst their own multiplicity, and the fact that you undertake them probably indicates your own anxiety, caused, perhaps, by the entrapment of contemporary humanities in the quicksands of autotelism. All I can do now is learn a lesson from your doubts, and to question the accuracy of the worldview (not: the conclusions!) of both the philosophers you engage, and that of your own. You are aware, after all, of the legacy your generation has bestowed upon mine: the acceptance of the Master's worldview (OE, 13) is tantamount to the consent to the fact that, "[...] Everyone is right and No-one is right, and the Master's being right comprises the consent to all possible forms of being right." This must ultimately align with the essence of chapter 113 of the OE:

"Everything is language."
Everything is interpretation.
Everything is method.
Everything is a symbol.
Everything is self-deception.
Everything is I.
Everything is the Other.
Everything is God.
Everything is everything.
Is.
IS NOT.

If "Everything" is "right," then there are no longer any grounds for ethics, aesthetics, or even for the method, or for the philosophy of humanities in general, or such a philosophy will transmogrify into a monster constantly devouring itself and interested only in itself. This sounds enticing, but remains somehow untranslatable to the non-verbal (and, perhaps, "academic") reality, which nonetheless constitutes a large part of your world (worlds) and mine. I suspect that somewhere along the way, an error might have occurred: somewhere in the rush towards Everything, a detour sign might have been missed. Nevertheless, it seems there is no reason to worry. In the end, when someone leads us into a dead end, there remains one way out: to return to the crossroads. And when long contemplation there chills us to the bone—on we go to the sauna. It is from there that one should one step out naked into the snow, looking at the world without prejudice and without aim, and it is only from there that one should proceed to the observatory. And then, perhaps, one will see something.

Gratefully yours,
Paweł Jędrzejko

Surprisingly quickly, Andrzej responded:

> [...] There's something of Barthes in your perception, this corporeality confronted with thought, *jouissance*, where the body is not sure if it is satisfied but feels happy, pure, innocent—feels "itself"—it can enter the world, emerge (the sauna as a womb). [...]. The idea of the *OE* in the sauna is an excellent one—and it is not merely an idea, it is more: a desire to juxtapose raw existence with the tradition of thought. [...] You deploy an interesting juxtaposition-based reading—through the body and vulnerability, you want to engage with that ontic mass of which you ask if it would endure on the top bench of the *badstuga*. It's curious that ten years earlier, I used to go to the sauna almost every day in the same place as you do—most likely the same one. I really liked it; it was a rebirthing experience of sorts. [...] You do know that the sauna—the authentic Finnish sauna—once organized the lives of Finns? In the sauna, births were delivered, ablutions were performed—not just hygienic, but ritualistic ones, too. The sauna serves as a powerful metaphor—and I have not yet encountered it in essayistic literature, and, undoubtedly, it is something wonderfully original in the Polish academic literature, and it is not only a metaphorical concept. [...] The *OE* did not fare well with reviews; I know from other sources that people were afraid to review the book. R [...] told me that there is fascist thinking in *OE*, which meant that the book excludes possibilities for discussion and dialogue, disarms potential reviewers, and sets traps. There were many compliments, but that was not my concern. What irritated me the most were comments that praised my erudition: erudition is something akin to a circus—today, in the age of the Internet, this is becoming increasingly apparent. [...] My copy of the *OE* is so heavily annotated that a second volume of the book could successfully be derived on the basis of the marginalia alone. But in these annotations (you prompted me to review the book) I recently found a passage that would fit your juxtapositions [...]: "Perhaps what is most important lies between the heat of desert sand and the coolness of a lizard?"

Perhaps. Yet, in search for that which is "most important," Andrzej's misconceived "oneness" with "matter" questions the very idea of well-being by rendering the seeking subject void: the formula of a reflection purified in the heat of the *bastu* proved insufficient. Hell-bent to break out of the hegemony of the Ontic Empire altogether, Andrzej Chojecki wrote his last letter, followed only by the information from friends and family that the brilliant thinker, writer, and friend went ahead with his original exploratory plan, and effectively put himself "under erasure." Permanently. And I still cannot help wondering whether he would have chosen the path of self-annulment if he had had a chance of joining the crew of the *Oskard*.

Living in the Isles

I am well aware that I have a tendency to idealize Scandinavia, and I do know that the *bastu* with its philosophy is one of the reasons why I do; I do admit that I have a particularly soft spot for the islands of my favorite Nordic triad: Sweden, Finland, and Norway. Indeed, for nearly 30 years since I first arrived in the Stockholm Skärgården as an enthusiastic deckhand aboard a gorgeous mahogany Opal-type sailing yawl, I have always been coming back, deeper and deeper in love. And although the endless horizon and long swell of the open ocean always clear my mind, although long transoceanic cruising is precisely what my body craves after each long academic year, it is always to the "happy isles" of the colder seas that I eventually plot my course. Crossing oceans to get from one garden of the isles to another is, to paraphrase Ishmael, my substitute for Prozac and therapy. But the people of the isles, out of necessity organically attuned to the rhythms of the elements, seem not to need to *resort* to sailing. Being at home with the sea is an indispensable, and inextricable, part of their routines, which in themselves are aligned with the pace of their limitless skärgårdens.

Admittedly, wherever my journey takes me—be it Society Islands or the archipelagos of the Caribbean Sea, be it the Azores or Madeira, the Canary Islands or the Balearics, the islands of the Northern Adriatic or Southern Dalmatia, the Ionian Islands, Cyclades, Sporades, or the Dodecanese—I relish the atmosphere of the isles, always markedly different from that characterizing land-locked existence. Decades taught me to open up to the insular pulse. And even though I do realize that, at least initially, I am clearly "out of place," and that it always takes time for the locals to decide that I am trustworthy enough to "let me in," I always begin to feel "in place" once my own pace adapts to the rhythms of the place that I am in.

Each of these gardens is different. The heartbeat of some is dictated by the regularity of the oceanic ebbs and flows, sea currents, or seismic activity—but even in the islands unaffected by the tides, volcanic eruptions, or permanent winds, life, primarily, depends on the elements. The geomorphology of the land and seabed, seasonality of winds, currents, and temperatures, the duration of daytime, the amount and type of precipitation—all affect the vegetational cycles of their flora, reproductive cycles of their fauna, and the routines of its people. Life in the islands, especially secluded ones, where survival, despite the availability of 21st-century technologies, is organically inscribed into the fabric of the collective mentality of local communities. Their attunement to the rhythms of nature is not the question of *choice*: I feel that (both historically and today) it is a *sine-qua-non* condition of survival—a condition on which the islanders do not reflect philosophically, but which they intuitively embrace.

To a tourist, my assessment might sound largely exaggerated. After all, the folders at the travel agencies, postcards, and snapshots taken for family and friends, cropped to please the eye, are manufactured to impress, not to deter—I have yet to see a folder promoting a Caribbean vacation with photographs of Antigua, Barbados, or Guadeloupe in a hurricane. But hurricanes do happen with amazing regularity, and when they hit—boats sink, buildings get razed to the ground, and people die. I have seen thousands of folders promoting recreation in French Polynesia or in Hawai'i featuring the unbelievable colors of the reefs and the mind-boggling richness of their life, but I have never come across a folder featuring a sailboat shattered against the barrier reef, or one including the statistics of hospital interventions related to encounters with stonefish, cone snails, or fire corals, or number of skull fractures caused by falling coconuts, or people drowning in the ocean, because they neglected the current. But chameleon fish, inconspicuous snails, beautiful corals, tasty coconuts, and currents *do* regularly kill those unmindful, those arrogant, those careless, those unattuned. Even routine spearfishing may prove deadly: many an eager diver were attacked by sharks attracted to the blood of the would-be trophy, and many suffered from a severe food poisoning consuming fish caught within the lagoons. Charter skippers will routinely be warned of these dangers in the course of their orientation and so their awareness would expand, but *tourists*, as a rule, would be "taken by the hand": local guides, who acquire knowledge of the wildlife of the reefs and mountains alike along with language, and whose ability to swim and freedive develops along their ability to walk, who do not have to *learn* to be attentive and mindful to their environment, will accompany tourists wherever they go, warranting their safety, and thus relieving them of the need to watch out for themselves. After all, travel agencies would want no casualties to mar their reputation. But we, sailors, have to adapt quickly: the more "trained" in the practice of attunement we are, the more languages we know—both natural languages and the languages of the elements[34]—the less prejudice we harbor, and the greater our chances to avoid danger, which, even in paradise, may prove mortal. Epistemic humility and epistemic mindfulness—are key.

Our beloved Greece is no different, and Dariusz made the point very clear: a tourist, as a rule, thinks of Greece as a country of splendid beaches, infallible sun, must-see antiquities, and excellent food and wine. Much less often do travelers to Greece realize that between July and September the Greek archipelagos are affected by *meltemi*, whose force may reach gale levels; that catabatic winds may turn shelter into a trap; that water, in many islands, is a rare commodity that must be transported from the mainland; that every year people die in fires; that many arrogant lovers of mountain trails regularly suffer life-threatening sunstrokes and effects of severe dehydration; that sunburns, especially close to the sea, land tourists in hospitals. Lovely folders of the isle of Santorini almost always feature fabulous white-and-azure architecture, donkeys in decorative trappings, and elderly Greek ladies, who—clad in black despite the sweltering heat—smile at the camera while carrying heavy grocery bags up the very steep lanes of Thira are rarely interpreted as they should be. Many of the tourists bragging their physical fitness and regularly frequenting gyms would not be able to live up to the challenge that these inconspicuous Greek ladies do not even *consider* a challenge, because it has always been a *norm*. Stamina is a part of their attunement to the physical conditions of the island, and it would probably never cross their minds to think of it in any other terms but those of "just another Tuesday at home."

I could, of course, multiply examples. But in essence, whether in Greece or in the Azores, whether in the Canaries or in the Åland Archipelago, the rhythms by which the islanders live are attuned to the nature on which they depend, unlike many of the "mainland" urban professionals, often living in bustling centers, dependent (more and more inevitably) on virtual communications. Allegedly sheltered from the elements, they are easily fooled into pacing themselves in accordance with the rhythms of artificial—corporate, academic, or other man-made—clocks, further and further away from the rhythms of dusks and dawns, summers and winters, ebbs and flows.

Attunement, as I see it, and as Dariusz's observations confirm, affects the timeflow. On the islands, time it is often unrelated to the indications of chronometers, it resists the rigidity of flight schedules, or ferryboat timetables. A heavy gale, thick fog, or excessive swell might halt commercial traffic for hours, if not for days: fishermen, knowing that being furious at the gale is pointless, use that time to repair their nets, or socialize, or both. Tourists get anxious, frustrated, or angry. I have often observed Westerners, proud of their punctuality, mock the "mañana," or "siga-siga" philosophy, often practiced by the inhabitants of the islands. But far away from the mainland, not only timeflow is different; "I must" tends toward "I want," "I can't" to "I'll do it when conditions allow it," and "I should" to "I will be happy to if I can." Anger at someone's unpunctuality might meet with the "guilty" person's genuine surprise: in all sincerity, they might fail to understand the reason why someone should get upset at a few minutes, hours, or even days of delay. They will deliver. When the time is right.

Sailors understand it only too well. We are well aware that unlike train, car, or Flixbus operators, we entirely depend on the weather, and that our ETAs—Estimated Times of Arrival—are at best *informed approximations* that cannot be taken for more than their name would suggests: *estimations*, not *certainties*. In the course of our "Greek Amazement" experiment, the ETAs, calculated well in advance of the journey, proved to be Marcin Fabjański's major source of frustration, as—due to nautical conditions or the crew's fatigue—some of the meditation sessions, in the preparations of which he had invested countless hours at home, had to be postponed, or even cancelled. Similarly, in the course of my 2023 transatlantic cruise, 400 nautical miles away from the Azores some of my crew, heedless of my conservative advice, would use our satellite communicators to have their families book their airfares home for them. Although their choice to book early would be perfectly reasonable on land (after all, early bookings save money), doing it in the middle of the ocean is, to put it mildly, risky. Of course, the Atlantic rendered their attempts futile: all of them missed their (thriftily purchased) flights, because the damage of the mainstay, followed by its time-consuming repair at sea, delayed us by four days. Interestingly, in the face of the survival-dictated necessity to replace the broken stay hundreds of miles away from the nearest land, out of reach of any SAR unit, and in the aftermath of the complicated maneuver of a nighttime approach to the stern of a cargo ship, whose Greek captain graciously supplied us with much-needed reserve diesel fuel, the shared understanding of what *force majeure* really is proved profound enough for each of the "eager beavers" to accept their monetary losses without a single word of complaint. The incident taught them the lesson of mindfulness: the here-and-now is always in the state of flux. And, as all of them universally came to realize, much more than money is at stake when you are at sea.

Needless to say—when commercial clients come on board, the confrontation of their expectations and the endless possibilities of failure to follow the itinerary is always difficult: "I pay, I demand." Naturally, as commercial captains, we tend to do our best to make sure that the customers (not: crew) are beyond satisfied with their experience. Yet, in their air of entitlement, such clients often play their "paying customer card." In obviously adverse conditions, such attempts are, of course, ineffectual: we will find ways to mellifluously evade dangers. But when the customers insistence on actions on the part of the skipper (whom they usually treat with little more respect than they would a shuttle bus driver) coincides with safe, though possibly tiring, nautical conditions, and—against the better judgment of the captain, who understands weather forecasts—the ship goes to sea, clients, arrogantly pressing on a "timely transfer" to another port "on their itinerary," often end up seasick, or excessively fatigued, and every second of their self-inflicted torture is a lifetime. Time, as it seems, indeed is of the *essence*, in the literal sense of the phrase: becoming forcibly attuned to the elements, the "tourists" usually revise their schizophrenic split between the "immateriality" of time and the "tangibility" of its economically impactful flow. Yet, struggling with one's own frail body, one appreciates the gravity of every minute on board: time tends to expand or contract, depending on the intensity—and desirability—of experience. Although seasickness and possible bruising renders the "attractions-to-days" ratio inessential, when back on land, as a rule, the "tourist mind" represses it all, restoring ego to a central position.

The Philosophy of *Lagom*: Toward Balance

But even the "paying customers," should they choose to forgo their stereotypical idea of a vacation "under the palm trees," are mesmerized by the clock, by which Scandinavia ticks. Away from the "far away and long ago," fed by the primitivist fantasy of the "exotic" islands, and inspiringly close to familiar, Western "civilization," the skärgårdens of Sweden, Finland, and Norway infect visitors with the germ of the peace of mind. Myself, whenever I think of the Scandinavian "gardens of the isles," I experience a sensation of a curious mixture of longing and tranquility. Recalling countless moments of awe, I compile my notes, reprimanding myself for losing concentration.

As I retrace my Scandinavian cruises of the last 30 years, I wander off in my fantasies, imagining the spectacular beauty of Grinda, with its marina's wooden walkways, each plank meticulously trimmed to fit the shape of the rock, to which it is attached; I am reliving sailing the *Rawa* among the dangerous rocks at the entrance to breathtakingly gorgeous Käringsund; in my mind, I am docking the *Blue* at the pier sheltered by the red granites of amazing Havsvidden and gaze at the stars in the open air jacuzzi installed inconspicuously among the rocks; alive comes the dawn, painting the white hull of the *Mystic* pink in the marina of Rödhamn; I feel an upsurge of joy when I think of the shuffling of the snow off the deck and knocking ice of the rigging of the *Hi Ocean One*, in full view of the majestic mountains on our approach to the Norwegian port of Ålesund; I recall the splendor of the midnight sun reflected in the glaciers of the peaks towering over Balfsfjord, just a few miles off Tromsø. Words fail me, but I do realize how much I miss Scandinavia, and how wonderfully sailing in those dangerous, unspeakably beautiful, and limitless gardens of the isles transforms me.

I do not really need to talk to anyone to sense why this is the case, and that is *good*: people in the marinas are courteous, friendly, and helpful, but reluctant to the prospect of a recorded interview anyway. I let go after the first few attempts, my long experience of the Scandinavian respect for personal space tells me never to insist. Swedes, Finns, Norwegians—whether city dwellers spending a weekend on the water or the inhabitants of the islands—all appreciate the unique "camaraderie of the skärgårdens" based on the shared need to partake of the natural beauty of the garden of the isles: they smile, they are ready to chitchat, they will share tips and hints, but—as a rule—when they go sailing, fishing, or when they use the sauna, they do it alone (to collect their own thoughts), with closest friends or family (to enjoy quality time together). Far away from the logorrhea of the city life, they appreciate the tranquility of their own space and they respect the very same need in others. This is one of the faces of the grassroots practice of *lagom*, the fundamental concept of the Scandinavian philosophy of balance:

> The United Nations has released the *World Happiness Report* again this year [2019]. It involves the ranking of 156 member countries sorted by the degree of happiness of its citizens: the Scandinavian nations dominate the podium and seem to really have something to teach all the others. In particular, we have often heard about the *lagom* lifestyle, defined as the Swedish response to the Danish trend called *hygge*, which focuses on all those moments full of familiarity, warmth, harmony and conviviality. [...] *Lagom* in the Swedish language literally means *not too much, not too little, the right amount*, and is a philosophy that aims to balance, in every area and juncture of everyday life. [...] Lola Akerström, photographer, journalist and author of the book *Lagom, the Swedish recipe for living with less and being happy*, defines this incredible philosophy of life that is also a genuine way of dealing with

> routine. [...] The concept of *lagom* is based on a proud opposition to excess and on a rejection of originality at all costs, hence the first, fundamental advice that the Swedish lifestyle provides us with: balance as a true goal, the "enough" as the ultimate goal, to dispel the myth of being satisfied as a negative trend. Knowing how to grasp something positive in every situation, being able to learn from every experience, aspiring to live fairly and in harmony with the world. [...] In short, being able to find happiness even in hidden corners and live sobriety as a conquest rather than a condemnation, all this makes the *lagom* lifestyle not only relevant, but also essential in the frenzy in which we are forced to live every day.[35]

The term *lagom*, dating back to the Viking Era (ca. 8th–11th centuries), is believed to be etymologically derived from the phrase *laget om* ("around the team"), referring to the Viking custom of passing a mug of mead around and ensuring everyone enjoys enough of it irrespective of how little or how much of it was available. The custom, epitomized by the phrase, reflects a broader phenomenon: Scandinavia—both the peninsula and the gardens of the isles—with its long winters and short summers, is exposed to rough weather conditions for the most of the year, and land cultivation, especially in the inhospitable north, is particularly demanding. Historically, the survival of the individual very heavily depended on the reciprocality of the sense of communal responsibility: sharing and fairness being at the core of the concept of justice in the Scandinavian societies. Today, *lagom* translates into egalitarianism and community welfare. Rooted in the idea that "everyone should have enough, but not more than they need," the social welfare systems of Scandinavia—particularly Sweden—aim to provide an equitable distribution of resources but, in a larger scale, the balanced resource management affects these countries' commitment to sustainability and environmental protection, which largely relies upon individual awareness of the necessity to waste reduction by reducing *excess* consumption. Born out of the demands of life in the North, *lagom*, therefore, encourages living in harmony with nature, and thus can be recognized as a historical fundament of contemporary environmental ethic shared by the Scandinavians, but also as the root of the norms of social interactions in the Nordic countries. Grounded in the principle of *lagom*, the Scandinavian preference for humility and modesty (which, especially after the Reformation, aligned with the teachings of the church) goes hand in hand with the discouragement of excess, which, to a large extent, explains the character of the lifestyle to which the Norwegians, Swedes, and Finns adhere in almost every aspect of their lives. The quoted article provides further examples to illustrate the point:

> This principle of simplicity also applies to the kitchen, which is based on a concept of sustainability and the use of genuine ingredients. Swedes are also careful to limit the consumption of meat and not to buy food

> that results from practices that are unsustainable or even harmful to the environment. [...] Although it is an inner state, the *lagom* philosophy is composed of many concrete applications. The second piece of advice it gives us is to find time for both exercise and rest, both activities that purify the mind and help to deal with everyday life in a positive spirit. Finding time to relax, indulge in little joys or carve out a corner of peace to recharge, are all different solutions to get away from the anxieties or worries of work or family. The *lagom* lifestyle focuses heavily on the concept of "ritual," a simple activity that we dedicate to ourselves to try to intercept our deepest needs and live more consciously. [...] In short, *lagom* means harmony, moderation, utility and versatility, and all these indications also apply to the house, to which the fourth advice of the Swedish philosophy applies: choose a functional and logical furniture. The well-known furniture brand Ikea has dedicated to *lagom* an entire line of furniture, characterised by simple and clean lines, just respecting the priority given to practicality, convenience and resistance. [...] There are a few clear rules to organise your home in accordance with the *lagom* philosophy: first of all, place the reading corners or desks where you work near the windows, so as to take advantage of natural light and save electricity; then choose wood both in the furniture and in the coverings, both because it is a warm material and because it is processed with a low release of pollutants; as far as cushions and blankets are concerned, prefer handmade materials such as wool and cotton and always opt for neutral colours such as beige and dove grey, able to create a warm and serene environment. [...] In a house based on the *lagom* lifestyle, there is no shortage of plants, the real protagonists of Swedish philosophy: not only do they free the air from polluting substances, but they also create small corners of peace. It would be ideal to also have aromatic plants, to be able to flavour food with additions at zero kilometre, not to mention the garden and the possibility of eating something that has been grown in your garden or on your balcony. [...] A *lagom* house is therefore comfortable and welcoming, but also ecological and sustainable: getting rid of the superfluous and keeping only what makes us really happy, objects that have a history and are important because they convey emotions: this is the mantra of the *lagom* lifestyle. [...][36]

The households, gardens, vehicles, and boats owned by Scandinavians are neither flashy nor Spartan, which simultaneously warrants sufficient levels of comfort and facilitates communal interactions. Socially, the principle of *lagom* encourages individuals to fit in with the group and not stand out too much, which fosters a sense of belonging even in communities consisting of families of different levels of affluence. But this philosophy also affects preschool, primary, and secondary education, in which civic values promoting togetherness, fairness, and responsible attunement to the rhythms of nature are of paramount importance. Not too much, not too little, but enough for everyone; no excess and no waste—but also no austerity; diligence at work and commitment to profound relaxation; compassion for the others, but also respect for one's own personal space; partaking of the goods provided by the land and the sea, but also tending to both, protecting the integrity of the whole: everything one does, one does *lagom*.

> The last piece of advice that Swedish philosophy gives us is to abandon the (unattainable) goal of perfection: there is nothing more frustrating than aspiring to something unattainable. Wonder lies in the middle ground, in the moments of imprecision and imperfection that characterise our lives, in moments of forgetfulness and freedom. Once we have recognized the subjective proportions that allow us to reach a balance, it is only up to us to defend that space by focusing on small steps. True happiness is knowing how to immerse oneself in one's life with joy and preserve the strength to face every day with serenity, living and working with harmony, simplicity and awareness.[37]

It is not surprising that the unnamed authors of the article resort to particular vocabulary: *lagom*, manifest in every area of everyday life, is "an inner state," "a ritual of immersion in life," a "mantra," it involves "the purification of mind," the recognition of the fact that objects with which we live convey "emotions," it entails compassion in one's relation to family, friends, and community members, and—last but not least—is a principle of "proud opposition to excess and on a rejection of originality at all costs," a paradoxically transgressive principle, the adherence to which allows one to resist the schizophrenia of late capitalism. After so many years of exploring Scandinavia, I understand *lagom* only too well. Among others, it tells me not to insist on recorded interviews when I see that my interlocutors, albeit curious about my project, become uncomfortable when I produce the microphone. I understand, when they go sailing, they enter a peculiar comfort zone: the comfort, derived from full concentration on what could kill you—the rocks, other ships, or sudden gusts of wind that might make maneuvering dangerous—and then, in the harbor, transforming into a contemplative immersion in the rhythm of the isles, would be disrupted with long conversations with strangers. And so, instead of "putting them to work" on my project, I invite fellow sailors, who soak in the lush beauty of summer Skärgården, to a drink aboard my ship. And then, free from pressure, we talk. Or not.

Coda: In Search of Language

I sit in the cockpit of the *Rawa*, a 55-foot Bavaria Cruiser with her owner, my friend, Magnus Broden, gazing at the splendor of the sun setting over the island of Lidingö, in silence. My crew has already disembarked; we make the most of the moment before the next group arrives. Magnus's inner peace tempers my loquacity, and so tranquility enshrouds us both, along with the yacht, the marina, and the bay. People on neighboring boats sip their drinks; chatting softly, they do not disrupt the quietude: everyone shares the sunset, in sync with the atmosphere of the Skärgården. Soon, the season will be over; the autumnal colors will yield to the monochrome of the winter. Like Tove Jansson's Moominvalley, attuned to the rhythms of the seasons, the garden will soon hibernate—yet life will go on, concentrated around fireplaces in the *lagom* households, but also outdoors. Dressed properly, people will continue to fish, hike, or hunt; indoors and outdoors, the Swedes will continue to work, socialize, or enjoy the bliss of the *bastu*. And by that time, I will be back home, at my computer, collecting my thoughts, re-listening to the few recordings we were able to collect, and struggling with the matter of the language to translate my organic experience of attunment to the elements in the open ocean and in equally inclusive gardens of the isles into figures of speech: into a language compassionate enough to stir emotions both in those intuitively drawn to the great outdoors, and those preferring the bustle of city life, both travelers and tourists. Perhaps a language that would allow everyone to appreciate the world available to us—although nature does love to hide—in broad daylight—the world as an everyday gift, which may have gone unnoticed.

Gift

A day so happy.
Fog lifted early, I worked in the garden.
Hummingbirds were stopping over honeysuckle flowers.
There was no thing on earth I wanted to possess.
I knew no one worth my envying him.
Whatever evil I had suffered, I forgot.
To think that once I was the same man did not embarrass me.
In my body I felt no pain.
When straightening up, I saw the blue sea and sails.[38]

Notes

1 The source of the description: the section "About the place" of the official website of the Apennine School of Living Philosophy in Italy—https://selfoff.com (accessed: December 12, 2023).
2 The source of the description: the section "About the School" of the official website of the Apennine School of Living Philosophy in Italy—https://selfoff.com (accessed: December 12, 2023).
3 The source of the description: the section "What is Self Off" of the official website of the Apennine School of Living Philosophy in Italy—https://selfoff.com (accessed: December 12, 2023).
4 "Greckie Zadziwienie: warsztat filozoficzny na jachcie," Facebook event, hosted by Akademia Siedmiu Mórz, January 1, 2024, https://www.facebook.com/events/256229042604893?ref=newsfeed (accessed: November 15, 2023).

5 Marcin Fabjański, interview by Przemysław Poznański, "Szukajcie dostrojenia do wszechświata," *Zupełnie inna opowieść*, March 17, 2020, https://zupelnieinnaopowiesc.com/2020/03/17/marcin-fabjanski-szukajcie-dostrojenia-do-wszechswiata-rozmawia-przemyslaw-poznanski/ (accessed: April 25, 2024).

6 Marcin Fabjański, interview by Przemysław Poznański, "Szukajcie dostrojenia do wszechświata," *Zupełnie inna opowieść*, March 17, 2020, https://zupelnieinnaopowiesc.com/2020/03/17/marcin-fabjanski-szukajcie-dostrojenia-do-wszechswiata-rozmawia-przemyslaw-poznanski/ (accessed: April 25, 2024).

7 Michel de Montaigne, *The Essays of Montaigne, Complete*, trans. Charles Cotton, ed. William Carew Hazlitt, Project Gutenberg, May 28, 2001, https://www.gutenberg.org/ebooks/3600 (accessed: June 03, 2024).

8 Izba Morska przy Sądzie Okręgowym w Szczecinie, Orzeczenie z dnia 9 maja 2003, Sygn. WMS 46/2002 w sprawie zatonięcia jachtu Oskard w dniu 16 września 2002, in Janusz Zbierajewski, "Wypadki jachtów," https://wypadkijachtow.pl/oskard-2002-zatoniecie-zatoka-biskajska/ (accessed May 5, 2024). Translated by Paweł Jędrzejko.

9 —especially commodore Bodo Gorus and captain Andrzej Latusek, who personally took care of the logistics of our transportation back home.

10 This section is based on the more or less scattered reflections that I presented in the Polish language in my monographs: Paweł Jędrzejko, *Płynność i egzystencja: Doświadczenie lądu i morza a myśl Hermana Melville'a* (Katowice–Sosnowiec–Zabrze: BananaArt.pl/ExMachina/MStudio, 2008) and Paweł Jędrzejko, *Melville w kontekstach: Prolegomena do studiów melvillistycznych* (Katowice–Sosnowiec–Zabrze: BananaArt.pl/ExMachina/MStudio, 2007).

11 Herman Melville, *Moby-Dick, or The Whale*, ed. Harrison Hayford, G. Thomas Tanselle, and Hershel Parker (Evanston and Chicago: Northwestern University Press and the Newberry Library, 1988), 4.

12 Herman Melville, *Moby-Dick*..., 273.

13 Herman Melville, *Moby-Dick*...,158–159.

14 Herman Melville, *Redburn. White-Jacket. Moby-Dick* (New York: Library of America, 1983), 107.

15 Herman Melville, *Moby-Dick*...,159.

16 Herman Melville, *Moby-Dick*...,159.

17 Howard P. Vincent, *The Trying Out of Moby-Dick* (Kent, OH: Kent State University Press, 1980), 157. See also: Paul Schreiber, "Queequeg's Coffin and the Vortex. The Spatiality of Ishmael's Narrative in *Moby-Dick*," in *Proceedings from the Sixth Nordic Conference for English Studies*, Tromsø, May 25–28, 1995, ed. Gerd Bjørboude and Gøril Rague (Tromsø: University of Tromsø, 1995).

18 Herman Melville, *Moby-Dick*..., 572.

19 Herman Melville, *Moby-Dick*..., 4.

20 Nathaniel Hawthorne, *The English Notebooks*, based upon the original manuscripts in the Pierpont Morgan Library and edited by Randall Stewart (New York: Modern Language Association of America; London: Oxford University Press, 1941), 423–427.

21 Herman Melville, *Moby-Dick*..., 5.

22 In Swedish: "Jag söker döden utan att dö."

23 This section contains excerpts from my correspondence with Andrzej Chojecki in the Polish language, included in my text: "'Belle Lettre' o (I)mperium (O)ntycznym Sauny—Andrzejowi Chojeckiemu," *Er(r)go*, no. 4 (1/2002): 113–123.

24 Johann Wolfgang von Goethe, *Faust*, translated by A. S. Kline. *Poetry in Translation*. https://www.poetryintranslation.com/PITBR/German/FaustIProl.php (accessed: October 10, 2023). For an alternative translation consult the "classical" edition: Johann Wolfgang von Goethe, *Faust*, trans. and intro. Walter Kaufmann (New York: Anchor Books, 1962).

25 Lucius Annaeus Seneca, "Moral Letters to Lucilius, Letter 48: On Quibbling as Unworthy of the Philosopher," paragraph 12, https://en.wikisource.org/wiki/Moral_letters_to_Lucilius/Letter_48. (accessed: July 5, 2023).
26 Johann Christoph Friedrich Schiller, *Gedichte*. Aus: *Das Lied von der Glocke* (1799), https://www.friedrich-schiller-archiv.de/inhaltsangaben/das-lied-von-der-glocke-zusammenfassung-friedrich-schiller (accessed: July 5, 2023).
27 The Qur'an, 2:95 (al-Baqarah).
28 The Qur'an, 2:79 (al-Baqarah).
29 English translation—PJ.
30 Literally: "Similar children play best.".
31 Andrzej Chojecki, *IO* (Kraków: Wydawnictwo Literackie, 1995), 7–10; see also: Andrzej Chojecki, *W stronę niewiedzy. Antyontologia* (Warszawa: Graf, 1991). I would like to express my heartfelt thanks to Ms. Lucyna Chojecka, who consented to the reprint of the introductory letter to the *OE* in the article for the *Er(r) go*, translated here, and for her permission to include it in this book.
32 That major Scandinavian cities are worth philosophically exploring how cities *feel* is well testified to by the recent book by Mikkel Bille and Siri Schwabe, *The Atmospheric City*, which explores how people make sense of the feelings they get in and of urban spaces. The monograph, also based on ethnographic fieldwork, explores everyday life in Copenhagen, Oslo, and Stockholm, focusing on the atmospheric power of people, places, and phenomena. See Mikkel Bille and Siri Schwabe, *The Atmospheric City* (London: Routledge, 2023).
33 "I am looking for death, yet without dying.".
34 See, for instance, Tristan Gooley, *How to Read Water: Clues, Signs and Patterns from Puddles to the Sea* (London: Sceptre, an imprint of Hodder and Stoughton, an Hachette UK Company, 2016).
35 "5 Tips for Getting the Most Out of Lagom Lifestyle," Technogym Newsroom, https://www.technogym.com/us/newsroom/5-tips-lagom-lifestyle/#:~:text=What%20is%20Lagom%3F,and%20juncture%20of%20everyday%20life (accessed: June 5, 2024).
36 "5 Tips for Getting the Most Out of Lagom Lifestyle," Technogym Newsroom, https://www.technogym.com/us/newsroom/5-tips-lagom-lifestyle/#:~:text=What%20is%20Lagom%3F,and%20juncture%20of%20everyday%20life (accessed: June 5, 2024).
37 "5 Tips for Getting the Most Out of Lagom Lifestyle," Technogym Newsroom, https://www.technogym.com/us/newsroom/5-tips-lagom-lifestyle/#:~:text=What%20is%20Lagom%3F,and%20juncture%20of%20everyday%20life (accessed: June 5, 2024).
38 Czesław Miłosz, "Gift," in: *New & Collected Poems 1931–2001* (Allen Lane, The Penguin Press, 2001). https://www.scottishpoetrylibrary.org.uk/poem/gift-0/ (accessed: May 5, 2024). Interestingly, the metatags describing the poem in the Scottish Poetry Library website are: "gardening," "gardens," "happiness," "Polish," "Tools of the Trade," and "zen.".

6 *Bhāvanā*

The Thai Forest Tradition Revisited

I am engaging in an endeavor like many of those undertaken in the West today. I am embarking on a journey to Malaysia, my destination being the Sāsanārakkha Buddhist Sanctuary, a monastery near Taiping. My plan is to spend three months there, immersing deeply in meditation and embracing the unique environment of this traditional garden of awakening—a space designed to facilitate profound transformation, steering one away from path of life laden with suffering. This Buddhist monastery appears to be a living embodiment of Epicurus's ancient concept, realized in the 4th century BCE in Athens—a garden where not only flora but also individuals thrive, as they meditate to eliminate the fear of death, the root of suffering. Simultaneously, it seems to resonate with the Stoic ethos, with its pedagogy directed against human cognitive fallacies, illusions, and subjugation to unhealthy emotions. Yet, unique to Buddhist philosophy is how it survives: it is not merely through ancient texts that it keeps alive; it is alive, because it has been handed down the continuous lineage of practitioners across generations. Thus, Buddhism, which has garnered popularity in the West, remains authentic, and, unlike increasingly popular Stoicism, it does not require the "neo-" prefix.

Buddhism has always evolved in close connection with nature. According to legends, Buddha's birthplace was a forest; his enlightenment occurred under a particular fig tree; he delivered his inaugural teachings in a forest clearing, and even met his demise amongst trees. These events, dating back 2,500 years, continue to inspire to this day. Although Buddhist teachings on awakening have taken on various expressions and found homes both in large urban centers and immense monastery-universities (like the 5th-century Nalanda in India), the philosophy's forest origins persistently resurface. Charles S. Fisher describes several Buddhist traditions, including early Buddhism, Chinese recluses, Japanese Zen hermits, and the forest monks of Southeast Asia, emphasizing the importance of those responsible for their durability: a "narrow thread of practitioners who earnestly adhered to Buddha's counsel to renounce societal life."[1]

The Malaysian monastery to which I am headed, although it is a global training hub for monks, remains deeply rooted in the Thai Forest Tradition. This tradition, emerging in late 19th-century Thailand and Laos as a reaction

DOI: 10.4324/9781003328469-9

to the urban corruption of Buddhism, reinvigorated Buddha Gotama's call for a life in harmony with nature. Its initial form was a "mobile garden," a nomadic community of monks adhering to a strict rule of non-interference with natural processes and subsisting solely on alms. During its formative years, until the mid-20th century, these monks made groundbreaking observations on the human-nature symbiosis, perceiving health as transcending individual organisms and viewing the immune system as an open entity. As deforestation advanced, the monks eventually settled, establishing monasteries within garden-like environments near Thai National Parks. These monasteries, resembling vast gardens, are dedicated to awakening.

Opting for a Malaysian monastery, which, interestingly enough, is not located in a predominantly Buddhist country, I made a conscious choice. I wanted to experience the "garden of awakening" in its most unadulterated form—a philosophy school where both monks and laypersons seek profound understanding of their physical and mental nature. This monastery, unlike its Thai counterparts, is primarily supported by the local Chinese community and focuses on the practice of Theravada Buddhism, which, though traditionally linked to Southeast Asian nations like Thailand and Myanmar, is gaining traction among Chinese populations across Asia and in China itself, despite Mahayana Buddhism's historical prevalence there.

The Sāsanārakkha monastery, nestled at the outskirts of Taiping, occupies a site chosen with deliberate intent. It sits discreetly beyond a Chinese cemetery and adjacent to a dense jungle, as if shielded by two elements traditionally feared in Asian culture—wild jungle beasts and ghosts. This strategic location seems to protect the monastery from casual, frequent visitations. Despite its contemplative nature, the Sāsanārakkha Buddhist Sanctuary maintains a connection with the external world. It stands as an inviting open garden for both monks and meditators (yogis), who seek it as a sanctuary for their meditative pursuits.

My intention was to conduct interviews with the monks of this monastery to gather autoethnographic insights into their practices. However, this plan was hindered by a precept of the Theravada Buddhist monks, which disallows them to discuss their meditation experiences with the outsiders. Therefore, reconceptualizing my narrative, I have decided to base it on my own, personal meditation accounts, adapting the autophenomenographic method proposed by Jacquelyn Allen-Collinson[2] to my own purposes. During my three-month stay at the monastery, I experienced several unique states uncommon outside of meditation. The account below, however, focuses solely on those experiences that I found most pertinent to the investigation of health as an atmospheric phenomenon.

The Goal Is *Bhāvanā*

At the monastery's reception, I receive a cushion, a pillowcase, and a blanket. I will spend the next 90 days on a mattress in an empty room with a view of the mountains. The days will be strikingly similar. Wake-up call at 5:00 AM, followed by a communal meditation session for yogis in the designated hall. The second group session takes place in the evening, from 7:30 to 9:00 PM. Between these two events, which mark the beginning and the end of each day, my task is to remain mindful and meditate while sitting, walking, and lying down—continuously, including during the two meals: breakfast and lunch.

My aim is *bhāvanā*: cultivating the mind to a state where it ceases to generate self-induced suffering through an unhealthy engagement with projections and mental constructs. Periodically, usually every ten days, I meet with my teacher to discuss my experiences. My task is to deliver a comprehensive phenomenological account of what transpires within the sphere of my perception (encompassing both sensory and mental experiences) during formal meditation sessions and outside of them. Over the three months, the Austrian vice-abbot of the monastery, serving as my teacher, will come to learn, among other things, that I frequently experience hightened states of physical pleasure, known in Pali as *pīti*; that I can maintain focus on the object of my meditation for timespans ranging from five minutes to an hour and a half, with varying intensity, due to factors I am yet to ascertain, or that I consistently go through phases of what I can only describe as "mental discomfort," interspersed with occasional episodes of fear and anxiety.

During each interview, I receive minor corrections from the teacher regarding my meditation technique. This technique is based on one of the fundamental methods taught by the Buddha (and still practiced in contemporary Buddhism), known as *ānāpānasati*—the mindfulness of breathing. However, the vice-abbot agrees to also let me meditate using the four primary elements. I chose this form of meditation because it seems to connect a person most strongly with the natural environment.

The practice requires of the yogi to attentively observe the four elements—earth, water, air, and fire—as they are manifest in the various characteristics of matter (such as the pushing of the wind, the hardness of the earth, the heat of fire, or the flowing nature of water). This exercise fosters an exceptional acuity in the meditator's attention, leading to the realization that the world is not merely a collection of stable objects in space, but rather a vibrant interplay of dynamic processes. To put it in simpler terms, my aim was to immerse myself as deeply as possible in what I perceived as the experience of a plant, i.e., in existing as a convergence of various impersonal forces continually acting through me, sustaining my individual life. Most important in this context (and in the context of the philosophy of the garden in general), is the fact that meditation on the elements, as designed by the Buddha, aims to dissolve the barrier between the meditator and the environment. This objective was eloquently described by contemporary Buddhist nun and meditation master Ayya Khema:

> Gaining insight into ourselves as consisting of these elements helps us to realize that we are no different from our environment. No matter where we look, we find the elements of earth, fire, water, and air. As we fix our

> attention on this reality, our feeling of separation will diminish, giving us a greater sense of being part of the whole manifestation in this universe. We can feel embedded in this totality and no longer threatened by other people, or by natural or man-made catastrophes. We are part of the whole, the whole is part of us; there is no separation, no alienation.[3]

From this experiential and meditative realization, which transcends intellectual construction, health emerges not merely as a concern for a single organism but as something much broader and all-encompassing, environmental and atmospheric. While the meditation on the four elements was not originally devised by monks of the Thai Forest Tradition, it was adopted by them along with other meditative practices. These practices aim to cultivate, as Khema describes it, a "sense of being part of the entire manifestation in this universe."[4] Autobiographical accounts from monks within this tradition suggest that their most fundamental practice consisted in immersing themselves directly in the natural surroundings, with minimal artificial protection against environmental factors, except for absolute necessities (like mosquito nets). The goals of training in the Thai Forest Tradition, which have evolved to include advanced methods of meditatively embodying the natural environment, reflect the most essential pedagogical principles of the school.

Several fundamental features characterize the pedagogy of the Thai Forest Tradition: the willingness to sacrifice and surrender one's needs, desires, and old habits to the natural environment; the ability to bear loneliness; and fortitude. When imagining a stationary training of these virtues in a classroom, we think of lectures, reading, and meditative exercises. In the jungle, much like in the open ocean,[5] such training occurs spontaneously; it does not even require a teacher other than the environment itself. According to the Thai Forest Tradition pedagogy, meditation in nature is much more efficient than in a monastery. In a meditation hall, you must force yourself to be mindful, at least until mindfulness becomes habitual. In the jungle (like in the open ocean), mindfulness is necessary for survival. However, the monks' task was not merely to survive: it was much more sublime and demanding—to develop the mind (*citta*) to an extremely subtle level, at which awareness becomes tender and mental obstacles (*kilesa*) are eliminated.

The pedagogical approach once practiced by the Thai Forest Tradition monks has been significantly altered. The vast wild jungles of Thailand and Laos, which made this form of teaching viable before 1957 (when intensive deforestation started), no longer exist.[6] This change has led to a compromise with respect to the tradition's foundational principles: monks in Thailand have returned to monasteries, often established near Thai National Parks. Yet, although my stay at the Sāsanārakkha Buddhist Sanctuary is not a sojourn in a jungle, but rather in what is best described as a vast garden, from day one I am immediately immersed in evident natural rhythms. The most prominent of those is the rhythm of day and night. Although my *kuti* (meditation cabin) is equipped with electricity, I make a conscious effort to avoid using it. Consequently, my most vigorous activities, such as meditative walking and light physical exercises, are confined to daylight hours—much like the activities of the cicadas that chirp the loudest just before dusk. Another noticeable natural pattern is the variation in temperature, which is pleasant in the mornings and evenings, but uncomfortably hot around midday. I also quickly observe that certain flowers emit their strongest scents in the evenings and that rain is most frequent around the middle of the day. However, meditation itself does not seem to follow any discernible pattern; moments of mindfulness, or of getting lost in thought, can occur at any time of day, seemingly at its own volition.

The Pedagogy of the Buddhist Garden

The Buddhist literature of the Thai Forest School is rich with garden-inspired comparisons. These similes aim to facilitate the mental development of students. In the context of the Thai Forest tradition, machine-related, or other technologically-oriented similes would be entirely inappropriate: meditation is not a simple sequence of linear causes and effects but a complex, dynamic, and delicate phenomenon involving various, rapidly changing, and interdependent factors. A quotation from a book by the legendary master Ajahn Chah (1918–1992) illustrates it well:

> Right now we are sitting in a peaceful forest. Here, if there is no wind, a leaf remains still. When a wind blows, it flaps and flutters. The mind is similar to that leaf. When it contacts a mental impression, it, too, flaps and flutters according to the nature of that mental impression. And the less we know of Dhamma, the more the mind will continually pursue mental impressions. Feeling happy, is succumbs to happiness. Feeling suffering, it succumbs to suffering. It's in a constant flap.[7]

In the passage above, the term "Dhamma" signifies the teachings of the Buddha. When spelled with a lowercase "d," it refers to "nature." Buddhism describes the essence of Buddha's teachings as the law of nature. However, Buddha's teachings diverge from the conventional understanding of biology as a science that describes nature objectively, as if it were external to the cognizing mind. According to Buddha, humans are deeply integrated with nature, and his concept of biology encompasses the human mind. Therefore, as inferred from the passage, Buddhism explores the nexus between mental and physical phenomena. Perceiving this interplay with clarity requires meditatively honed attention. The analogy of the mind as a leaf stirred by the wind, symbolizing fluctuating mental formations, represents the mind non-linearly—as a convergence of simultaneous micro-events yielding a specific outcome. This pedagogical technique aims to induce a relaxed attentional state in meditators, enabling their activation of "peripheral attention"—akin to peripheral vision—to gain a broader, atmospheric, understanding. Repeated practice cultivates a habitual expansive mindset, leading to a more spacious perception of our position in the world, where the center of our perspective becomes less pronounced. However, such meditative training demands consistent effort, a notion aptly encapsulated by another garden-inspired simile.

> In meditation, you must continuously be attentive, just like when planting a seedling. If you plant a seedling in one place, then after three days you pull it up and plant it in another place, and after three more days, pull it up again and plant it somewhere else, it will just die and not grow up and bear any fruit. Meditation is the same. If you do a seven-day meditation retreat and after leaving it, for seven months you go around "soiling" the mind, and then come back and do another seven-day retreat where you don't speak and you keep to yourself, it's like the tree. Your meditation practice won't be able to grow and it will die without producing any real results.[8]

The parallels between nurturing plants in a garden to optimize their growth and the practice of meditation to enhance its efficiency are striking. The central premise is that neither process is entirely within our control. Our role is to cultivate favorable conditions for these processes to thrive. In gardening, we have limited influence over the weather, just as in meditation, factors such as the past *kamma* (Sanskrit: *karma*), or the potential accumulated in previous lives, may be beyond our control. It is noteworthy that in Pali texts, the pivotal moment in Buddhist meditation—the moment when previous efforts culminate in a degree of liberation from suffering—is referred to as "fruit" (*phala*). Elsewhere in his book, Ajahn Chah likens the human mind to a fruit swayed by the wind, using this metaphor to depict the human existential condition:

> When a fruit tree is in bloom, a strong gust of wind will blow some of its blossoms to the ground. Those that don't fall will eventually grow into small green fruit. But then another gust comes and some of them will fall, too. As for the rest, they will grow to become fruit nearly ripe, or even fully ripe, before they fall. And so it is with people. Like flowers and fruit in the wind, they, too, fall in different stages of life. Some people die while still in the womb, others within only a few days after birth. Some people live for a few years, then die, never having reached maturity. Some die in their youth. Still others reach a ripe old age before they die. When reflecting upon people, consider the nature of fruit in the wind—both are uncertain. Our minds are also uncertain. A mental impression arises, draws and blows at the mind, and then the mind falls—just like fruit. The Buddha understood this uncertain nature of things. He observed the phenomena of fruit in the wind and reflected upon the monks and novices who were his disciples. He found that they, too, were essentially of the same nature—uncertain! How could it be otherwise? This is just the way of all things.[9]

In another section of the same book, Chah includes a passage that offers guidance on how to adeptly navigate the existential reality as he described it above:

> Does anyone order the trees to grow the way they do? They can't talk nor can they move around, and yet they grow away from obstacles. Wherever it's cramped and growing will be difficult, they bent outwards. Trees by nature don't know anything. They act on natural laws, yet they do know enough to grow away from danger, to incline towards a suitable place. People are like this. We want to transcend suffering, and if that which we like and that which we don't like are suffering, we should then not go so close to them, not be cramped by them. When we incline toward the Buddha, suffering will lessen and eventually come to a complete end.[10]

Ajahn Chaa identifies a shared trait between human and plant intelligence: the innate capacity to evade the detrimental forces of the environment. In doing so, he highlights a fundamental characteristic inherent to all life on Earth and possibly beyond. A recurring theme in these similes, and many others employed by the Forest Tradition masters, is that of a profound trust in the natural order, juxtaposed with skepticism toward human constructs developed through enculturation and socialization. These garden similes aim to remind Ajahn Chaa's students of the ideal way of living, with which they may have lost touch amidst their entanglement in their daily activities. His advice to "be like a tree"—or to see oneself as a tree or fruit—subtly implies that we are, in essence, creations of our environment.

The simplest form of learning from the environment of the garden is to observe the processes occurring there and relate them on a one-to-one scale to ourselves. Just like the garden, we, too, are part of the grand process of life, and there is no reason why the rules dictating how, for example, fruit trees live, should differ from the rules that determine how we ourselves open and fold in the rhythm of nature.

> Here in this one pitiful apple tree we glimpse the gap in nature's show. Before its flowers are entirely gone and before the dead trunk is cut down, and a healthy sapling planted in its place, we may contemplate the contradictions in living things and wonder what conclusions to make about our own life and the whole fantastic mass of perceptions we call the world. What do we feel about the slow death of the tree?[11]

Monk Nyanasobhano, often referred to as "Thoreau of American Buddhism," further discusses the realization of natural laws taught by trees as a foundation for our own flourishing, encompassing both wisdom and health. We, too, are conditioned and transient, only slightly more durable than the states of weather. The longing for immortality and the fear of death often lead us to overlook this truth, as Nyanasobhano articulates it in a manner reminiscent of Epicurus. Certain natural phenomena, whether we wish to acknowledge them or not, have a way of revealing the truth through their suggestive presence. Immersing ourselves in the natural surroundings of a tree garden, sooner or later, we will inevitably begin to interpret and understand its signs.

> We can understand well enough that any form that occurs in this universe of cause and effect must come to an end when its causes disappear; but seldom do we see the rising and the following so closely conjoined as here, where death and life pivot in a single branch.[12]

The sight of a dying tree branch, for Nyanasobhano, serves as a wake-up call. However, it is only the beginning of the process: we are still far away from the solution to our existential dilemmas. The wake-up call must be followed up by the practice of Buddha's teachings. At the same time, other Buddhist teachers representing the Thai Forest Tradition perceive our immersion in the natural environment as more than just an alert; they see it as an intrinsically valuable lesson, with or without Buddha's teachings as an interpretive lens. One such teacher is Bhante Sucitto, the author of numerous books, videos, and audio talks available online. The way the environment shapes my experience during meditation is reflected in the meditative exercises he proposes. His foundational belief is that the human organism continuously communicates with the environment, receiving messages from all around. This environment determines the processes of our flourishing and healing, independently of our will. There is an intelligence at work outside our minds, which we often overlook or ignore, and flourishing involves nurturing relationships, including those with the environment.[13]

In his talks, Bhante Sucitto advocates for meditation techniques such as "graduated expansion," which aim to extend our field of sensitivity beyond our bodily limits and open our mind-body system to enlightening influences from the exterior. Ajahn Maha Boowa, a renowned Thai Forest Tradition master, shared a similar insight,[14] describing wisdom as an impersonal, intelligent force that operates within us, yet independently of our will. An excerpt from his autoethnographic account of the experience of pain illustrates it:

> Wisdom began to tackle this problem in earnest. Before I found myself cornered like that with no way out, I never imagined that wisdom could be so sharp and incisive. It went to work, relentlessly whirling around as it probed into the source of the pain with the determination of a warrior who never retreats or accepts defeat. This experience convinced me that in moments of real crisis wisdom arises to meet the challenge.[15]

The notion of a non-personal intelligence accessible to individual organisms is not a modern (Ajahn Boowa) or contemporary (Bhikkhu Sucitto) Buddhist invention. It appears in ancient Buddhist metaphysics, which, for instance, informs teachings on the universe as a sphere where five universal laws (*pañca-niyāma* in Pali) operate, creating a natural order—and humans are subject to all five. Lily De Silva, in her book on Buddhism and ecology,[16] outlines them as follows:

1 Season laws: Governing the activities of the four elements—earth, water, fire, and air.
2 Seed laws: Regulating the functions of life forms, including plants.
3 Mind laws: Pertaining to the mental activities and instincts of humans and animals.
4 Action laws: Addressing the moral sphere of action, emphasizing the transition from instinctive impulses to purposeful, wholesome actions, vital for liberation from the cycle of rebirth.
5 Universal laws of cause and effect, a core Buddhist concept also known as "co-dependent arising."[17]

De Silva notes that "human beings exist within these natural laws on Earth's troposphere, which upholds the atmosphere, biosphere, psychosphere, and moral realm."[18] She adds that environmental pollution is an external manifestation of humanity's internal moral pollution, which has reached crisis levels.[19] According to Buddhism, the scholar further argues, moral consciousness, the human mind, the body, the external world of fauna and flora, and society are interwoven in a comprehensive cause-and-effect network, forming an eco-system that is psychologically sensitive and responsive.[20]

In the Sāsanārakkha Buddhist Sanctuary, the monks referred to the third and fourth laws, citing *kamma* (known in Sanskrit as "karma"), as a crucial factor affecting health. This suggests that their concept of health aligns with its "atmospheric" notion, as introduced in the third chapter of this book. While monastic rules ordinarily prevent the monks from discussing their meditative experiences, my personal experiences at the sanctuary offer some insight. Midway through my stay, I encountered a phenomenon that illustrates these principles in full clarity.

The Felt Space (An Autophenomenographic Account)

Countless spiritual books narrate the quest for the awakening, often portraying a path of commitment to meditation, challenging, yet inevitably leading to enlightenment—a permanent escape from suffering. However, there is a less romanticized version of that story: one you confront when you cease reading and embark on the journey yourself. This version unveils a reality of uncertainty, fluctuating progress and regression, disorientation, doubts about one's sanity, and wavering trust in both teachers and in the chosen meditation techniques. I encountered this raw, unvarnished truth years before my time at the Sāsanārakkha Buddhist Sanctuary, during a much earlier trip to Asia. The following account, reconstructed from my notes, captures this experience:

Shwe Oo Min Buddhist Monastery in Myanmar was my destination in 2002 and 2003. I spent two months there, following a three-month intensive retreat in another Burmese monastery near Yangon. About a month into my stay, my perception began to shift dramatically—it was not a subtle process: it was radical. Initially, the concrete wall at the end of my walking meditation route was flat and gray. About a month later, one day it started shimmering with colors and gaining depth.

But the most profound change occurred within me. I became a symphony of vibrations, an ebb and flow of energy. Then, one night, everything halted—the energy dance, and the "I" observing it. It was as if my existence paused, yet life continued.

This was new and powerful—a groundbreaking experience for me; the first. The last time when I experienced something I had never experienced before may have been when I was three. Back then, it might have been a new taste, a scent, a human action, or perhaps my first thunderstorm. The details escape me, but whatever it was, most probably I ran to my Mom, with such an expression on my face that she would have no other option but to hug me.

Yet there, in the Burmese monastery, there was no Mom. Only teacher monks, unflustered by what I was experiencing, and limited in their English. The monastery offered me no framework for integrating this revolutionary experience, nor did life post-monastery. I craved certainty about what I had experienced, to name it, to articulate it in any language to myself or others, but I never got this luxury. It is either "to be free" or "to be certain": these two states are mutually exclusive. In a particular sense, being free is to be uncertain. To be free is to be uncertain without fear.

This pivotal experience, although earth-shattering upon recollection, cast a shadow over subsequent retreats in other meditation monasteries. Its exact classification within Buddhist pedagogy eluded me. Subconsciously, I yearned to relive it. But it never happened eversince. This longing bred expectations, universally recognized in Buddhist teaching as impediments to the development of meditative insights. The experience I recount below—and which I analyze using autophenomenographic methods—occurred at the Sāsanārakkha Buddhist Sanctuary. While it did not evoke the same level of wonder as my experience in Myanmar, it was, beyond doubt, an outcome of my meditation practice in this garden monastery.

The autophenomenographic approach I employ, modeled after Jacquelyn Allen-Collinson's[21] proposal, is rooted in phenomenological tradition; it is a form of investigation focusing on the researcher's personal experiences. It involves five steps: (i) collecting first-hand descriptions, (ii) setting aside biases, (iii) an initial impressionistic analysis, (iv) a detailed re-examination, and (v) imaginative free variation, intended to discern the essence of the studied phenomenon. Yet, this final step poses challenges, both academically and from the Buddhist standpoint. Relying on imagination to determine a phenomenon's essence seems questionable, even if, as Collinson suggests, its role is to *refine* insights derived from the preceding steps. Furthermore, the very notion of *an essence* is a construct of Western metaphysics, not entirely aligned with Eastern phenomenology. Therefore, in light of these discrepancies, instead of the free variation I propose maximum immersion in natural surroundings, as untouched as possible, especially in the course of the "setting aside biases" step, or, if that is not feasible, I would argue that turning to the landscape of our bodies is the next best option. After all, central to Buddhist phenomenology, which advises cautious reliance on imagination, is the concept of *papañca*, or mental proliferation:

> In doctrinal usage, it signifies the expansion, differentiation, "diffuseness" or "manifoldness" of the world; and it may also refer to the "phenomenal world" in general, and to the mental attitude of "worldliness." […] The opposite term *nippapañca* is a name for *Nibbāna*.[22]

Proliferation, which skews our perception of reality, stands in stark opposition to the clarity of *nibbāna*, the ultimate cognitive state in Buddhism. Buddhist phenomenology, paralleled in Arthur Schopenhauer's philosophy, suggests that distorted perception of reality arises from craving, which itself is rooted in the illusory experience of self as an autonomous entity. The procedure I propose seeks to diminish the dominance of the self by subtly shifting focus, allowing the researcher to become acutely aware of the impersonal forces within his or her perceptual field.

My procedure comprises the following stages: (i) gathering subjective accounts, (ii) attuning myself to life's rhythms, permitting impersonal sensations to reveal themselves within the field of my awareness, (iii) re-reading these accounts as if they were authored by another, and (iv) re-writing these accounts, eschewing a self-centric viewpoint. This approach is underpinned by a mindset advocated by contemporary Buddhist master, venerable Pamojjo, trained, among others, in the Thai Forest Tradition, which emphasizes "avoiding to know in advance."[23]

Gathering Subjective Accounts

This particular experience unfolded over approximately a week. My notes have been slightly refined for clarity:

> *I shift the focus of my attention away from my body and place it in the space in front of me. This is where I exile my hyperattention.*[24] *Then, it seems as if I perceive reality from an elevated vantage point—an experience drained of identity.*
>
> *Allow the experience to strike the perceiving awareness.*
>
> *A day marked by struggle, trying to maintain attention within the body when it yearns to scatter in every direction. Relief comes when I let it go.*
>
> *Attention like a shiver.*

Within the light field of attention, events that aren't mine emerge. But as I write this, my attention contracts significantly. Let it do as it pleases. What will happen to my attention when I cease to give it tasks?

Attention calmly observes sensations, sometimes itself, and when a thought emerges, it ends in a sudden storm within the cloud of sensations urging the body to action.

Being has contracted to a state of painful encapsulation. Yet, there is something that can observe such a state—identify it, name it, study it. I practice mindfulness of the state of encapsulation.

Presence is the atmospheric backdrop, not necessarily containing self-derived influences.[25] *This backdrop of presence is soft and velvety, fairly uniform in intensity throughout the whole space—and it extends nine, maybe twelve feet around my body. Sometimes, it seems to emanate from my heart. Will it connect me to the forest around? Where does the zone of self-derived tensions end? How do relationships with people affect this zone? How do thoughts affect it? This sensation clearly contradicts materialistic metaphysics.*

Having navigated through the steps of (ii) attuning myself to life's rhythms, permitting impersonal sensations to reveal themselves within the field of my awareness, (iii) re-reading these accounts as if they were authored by another, I have reformulated the experience as follows:

As a result of intensive meditation and attention experiments, I encountered a state where the space outside my body felt more palpable. As the experience became more and more profound, I started feeling the space outside as if it were a part of me, as if the territory of my self expanded. This process was triggered with my decision to shift my focus outward from my body to the space in front of me. Gradually, I began to perceive the surrounding space with greater intensity, as if enveloped in superlight, imperceptible, yet substantial clothes. In the process, I experienced several instances of painfully constricted attention, from which I learned to maintain a certain distance. Thoughts disrupted the uniform intensity and smoothness of my attention field. Following their emergence, I felt a compulsion to act.

One afternoon, I became aware that my attention had expanded from the center of my head and chest in all directions. My tangible existence in the world was no longer confined to the space within my skin, which no longer felt like a barrier. I started sensing a space of about a dozen feet around my body in every direction as soft and pleasant. Most intensely present in the center of my chest, the sensation seemed to radiate from there, hand in hand with the lightness in my heart and body, sparcity of thoughts, and an absence of worry.

This sensation faded after three days. It did not bring a heightened sense of connection with the environment. Although the range of my

> *experiential field clearly extended beyond my skin, with its boundaries becoming indistinct, I did not feel a deeper connection to my surroundings. Interestingly, the experience was more pronounced inside my meditation cabin than outside. It did not yield any particular insight either.*

As a consequence of the phenomenographic procedure I employed, the second description is devoid of abstract conjecture. Based on this narrative, I can conclude that this was an evident instance of atmospheric well-being, aligning with Schmitz's definition already introduced in Chapter 3. Despite the expanded and somewhat indistinct sense of self, I maintained clarity in my experience. I felt a direct engagement with the objects revealing themselves to me, accompanied by minimal cognitive clutter or mental fog.

However, this experience did not alleviate the persistent fever I endured for many weeks at the monastery, unlike the experiences of Thai Forest monks during their meditations in the jungle. As the esteemed master Dhammādharo articulates it:

> Living in the forest, you also learn from the vapors that each plant exudes. Some plants are good for your health, some are bad. Sometimes, for example, when I've been feverish, I've gone to sit under certain kinds of trees and my fever has disappeared. Sometimes when I've been feeling well I've gone to sit under certain kinds of trees and the elements in my body have become disturbed. Sometimes I've been hungry and thirsty, but as soon as I go sit under certain kinds of trees, my hunger and thirst disappear.[26]

This account not only narrates the traditional understanding of health as a state devoid of illness but also delves into well-being. Its author details the healing of a specific malady—fever—by means of the vapor exuded by trees, providing a quasi-scientific rationale for the process. It suggests an elemental balance restored between a human organism and the forest environment, perceivable to a mind honed in meditation. This instance illustrates yet another dialogue between the mental and physical realms, often segregated in our post-Descartian world. Contrary to contemporary Western medicine, which frequently seeks physical causes for illnesses, Buddhist wellness practices often identify mental states as the root of physical ailments. Emotions like intense anger, hatred, or love agitate the mind, "splash into the body, where they mix with the various properties of the body—in the blood, for instance, which then flows to the various parts of the body, causing weakness and fatigue."[27] When such emotional turbulence stagnates in a specific body part, it manifests as illness.

One characteristic of the meditative experience described above was the clear revelation of the connection between thoughts and the space perceived as a part of myself. The emergence of a thought disrupted this space, compromising its uniformity, while simultaneously diminishing the sensitivity of attention and the overall sense of vitality. This observation appears to validate the feasibility of the phenomenologist's task, as outlined by Hermann Schmitz. This is how I interpret the philosopher's position:

> Schmitz offers his philosophy as a means to regain this lost sensitivity. He evokes the concept of "felt body," or *Leib* in German, the notion of corporeal communication, and the theory of emotions as spatial atmospheres. *Leib* is a dynamic and rhythmical process, manifesting itself, for example, as vigor and languidness. The world outside *Leib* is a collection of dynamically changing atmospheric fields, fluctuating in various forms. The aim of a philosopher, or a new phenomenologist, to keep to Schmitz's vocabulary is to feel these dynamics.[28]

The re-description of my experience, according to the procedure I adapted from Collinson, demonstrated that some of the results I expected of my long-term meditative seclusion in a place I could call a "garden" never materialized. One of these was the anticipated sense of connection with the natural environment. Although my field of perception clearly extended beyond my body and its boundaries became blurred, it did not lead to a weakening of the sense of separateness or to an increased sense of empathy with the surrounding trees, cicadas, or monkeys. Another effect I expected from my stay at the monastery was a heightened sensitivity to the rhythm of life, especially since *ānāpānasati* meditation, which I sometimes practiced, involves attuning one's attention to the rhythm of one's breathing.

> According to Schmitz, the often nonspecific, diffusely localized corporeal feelings operate most of the time in the form of a pulsating rhythm in the felt body constantly oscillating between corporeal expansion and contraction, regularly at work already in breathing. This oscillation he christens "vital drive." Corporeal expansion is a marked widening of the felt space in the region of one's body, most notably occurring in states of relaxation. Characteristic examples of corporeal expansion are the experience of beholding of a wide, beautiful landscape, the first breaths outside in fresh air after having been locked inside a cramped and stuffy room, or the pleasant relaxing of the felt body when gently gliding into a hot bathtub. The opposite pole of corporeal contraction is a marked narrowing of he felt body, often in states of sudden, unexpected change to one's bodily orientation—such contraction occurs in states of shock, in panic or moments of great focus and concentration. Usually, expansion and contraction are dynamically related—Schmitz speaks of a "dialogical character" of the vital drive.[29]

However, what Schmitz calls "nonspecific, diffusely localized corporeal feelings," which should arrange themselves into a certain rhythm of expansion and contraction, I perceived as a constant, seemingly radiating field. On the other hand, from descriptions like the one below, it appears that this rhythm is related to certain interactions such as the "experience of beholding a wide, beautiful landscape," whose effect can last for some time. Thus, the meditative expansion of the field of perception beyond the body could be considered in terms of Schmitzean phase of expansion, while moments when I felt "painful encapsulation" could parallel what he describes as the phase of contraction.

From the perspective of Buddhist meditation pedagogy, if Schmitz is right and expansion-contraction engage in a dialogue with each other, and one responds to the presence of the other, then perceiving events in this way enhances an important factor of awakening called *dhamma vicaya* (the investigation of the nature of reality) in Pali. The concentration of attention on an unchanging object gives rise to another factor of awakening: concentration (*samādhi*). In this context, in the next section, I present anecdotal stories about the effects of Buddhist garden pedagogy, recounted by four monks from the Sāsanārakkha Buddhist Sanctuary.

Living in the Garden

One notable characteristic shared by all early Buddhist meditation methods is their concentration on transcending egocentric perspectives. This is especially evident in the Thai Forest Tradition, as the forest monks' continual focus is on transforming personal experiences into non-personal ones. For instance, hunger would be interpreted as the movement of the wind element in the intestines, while tiredness would be perceived as a sensation of gravity's pull. The jungle serves as a teacher of non-personal attention, and Master Boowa even goes as far as to describe the forest as a university. For monks of the Thai Forest Tradition, nature was, and still remains, a vast mirror, in which they are reflected in an impersonal manner. Fisher suggests that this practice is ancient, designed by the Buddha himself 2,500 years ago to help his students understand their "inner workings only by dispassionately observing themselves as part of nature."[30] In contrast, we, contemporary people, tend to do the opposite: we interpret every event in a personal way, to which the popularity of selfies testifies.

The conversations with monks at the Sāsanārakkha Buddhist Sanctuary distinctly suggest that transcending the egocentric perspective was a key objective they sought to achieve through their practice. Each of them stressed that the reason to embrace monastic life was primarily fatigue with the self-centered lifestyle, prevalent in modern society. Prior to monkhood, all four of my interlocutors had successful careers; their lives were lives of accomplishment, filled with pleasures. All of them held good jobs or had prospect for lucrative positions, and all were appreciated as gifted in their particular fields—like the Mexican monk, who, before he chose the monastic life, gained fame and critical recognition as a composer. However, despite their respective achievements, all of them found their former lifestyles unsatisfying and unhealthy, which discovery eventually led them to the monastery. The four monks who shared their stories with me came from three different continents: Asia, Europe, and South America. Below are excerpts from some of my interviews:

A 40-year-old monk from France, interviewed on October 14, 2022:

> *Nature teaches you openness to whatever is happening. This reinforces the type of meditation known as open awareness. As an effect of this meditation, you reduce craving and, in the second step, also reduce self-centeredness. I cannot tell you how nature influences my own meditation, as we are not supposed to talk about our meditative experiences. I can tell that all desire to consume things that was in me before I became a monk totally evaporated, and exposure to nature was an important factor that produced this effect. And this is significant. Craving is ultimately a source of environmental catastrophe that we experience.*

A 48-year-old monk from Mexico, interviewed on October 17, 2022:

> *It is obvious to me that the garden around produces a positive effect on my health. But health is also a personal matter determined by kamma. A natural environment of the type present in this monastery and other forest monasteries where I stayed influenced my emotions and taught me humility, effort, awareness, and concentration. One particular insight one can learn from nature while observing it is that whatever arises—ceases. This is the essence of nature, revealed by it if we want to listen.*

A 37-year-old monk from Sri Lanka, interviewed on October 18, 2022:

> *When I meditated in a cave in Yala National Park, I had eye-to-eye contact with a four-meter-long tiger. I felt strong fear, but I realized from how the animal looked at me that it did not care about me at all. That was a lesson in the impersonality of life. Buddha promised that when in the forest, your meditation does not regress. I confirm it. You do not need to meditate at all. You just learn forest skills, and this makes you wiser. In the forest mind itself conducts the investigation. The health of the body and the mind is one of the investigation results.*

A 47-year-old monk from Austria, interviewed on October 20, 2022:

> *Nature [...] simplifies life.*[31] *I only had one duty—sweeping the area around my mosquito net, so I could see snakes approaching. I realized how much rubbish I had in my apartment in Austria. Although one can learn a lot from nature, I would not call nature a teacher; that would be anthropomorphizing it. You learn from it how things change: leaves fall, trees collapse, and bodies of dead animals rot. I was a forest creature, no different from other forest creatures. Meditation on elements broke the borders between myself and the world around me. My well-being was a part of inter-being.*

From the point of view of atmospheric health, it is important to note that all the monks acknowledged the significance of the influence of the environment on health, whether understood as *flourishing*, *well-being*, or even *physical well-being*. For the Mexican monk, wisdom and health blend into one: insight into nature is a part of flourishing in general, strengthening certain emotions, but also reinforcing concentration. It leads to the realization of the three characteristics of nature: *impermanence*, *unsatisfactoriness*, and *lack of self*. Hence, *kamma* (Sans. *karma*) is a central factor influencing health, which observation coincides with the claims of such Buddhist scholars as Peter D. Hershock, who asserts that "ignorance and habitual karmic dispositions render us deeply susceptible to compromised or eroded capacities for health and healing."[32]

***

I spent three months in the Sāsanārakkha Buddhist Sanctuary without any special insights. However, one moment struck me powerfully: a flash of realization that in life, What Hurts Is That Which Holds. Further, What Hurts and That Which Holds are the same. The arresting, solidifying, and encapsulating of experiences—is pain. Letting go leads to the absence of pain. It is not that clinging to something *causes* pain: clinging to something *is* pain—a symptom of illness.

The ancient Buddhist school of philosophy, *Abhidhamma*, studies experience in a different way than does science arising from materialist metaphysics. Unlike in "classical" physics, in the perspective of which life unfolds outside of us and demands an objective description, in *Abhidhamma*, in every millisecond it is co-created by virtue of observing, entangled, attention. *Abhidhamma* sees the swelling of each moment, its culmination, and, eventually, its fading. Each such moment leaves the landscape ready to receive the next one, like a wave that has carved the beach, welcoming the next.

Abhidhamma teaches us that there is nothing like a single moment of decision, but rather decisional ebbs and flows. It is their rhythm that determines why we act as we do, or why we do not act at all. Such teachings explain why we are usually inconsistent: our decisions are governed by an underlying pulsation that disrupts the rational strategy that compels us to hold a single, established position until the next substantive reassessment—propelled by new factors—changes our mind. The Buddha, however, goes further: as if satisfying Heidegger's need for *angst*, he says that not only there is no position to hold on to, but also that there is no one to take such a position. There is only a selfless rhythm of ebbs and flows.

> *Pattaya, Jomtien Beach, high tide. The South China Sea tries to grab a new piece of dry sand several times before finally spitting out a wave strong enough to do it. As if it needed to gather strength. The process of our decision-making has a similar rhythm but is more complex:*

> *whether our decisions are intelligent or not has already been determined before they arise, as if in the course of their swelling. Ancient Western and Eastern philosophies, like Abhidhamma, with their obsessive love of nature and distrust of human ego, would teach that the lesser my ambition in the crucial moments preceding the maturation of decisions, the better. When decisions are born within me, I am not needed at all.*

Coda: A Cross-cultural Concept of Health?

Peter Hershock distinguishes between health concepts based on the notion of disease, which allow for relatively precise "quantitative assessments of an organism's health," and those based on the concept of well-being.

> Conceptions biased toward the former assumption take health to be the natural equilibrium that is manifest when an organism's overall mode of functioning is not disturbed by disease or infirmity. Health, in other words, is simply the absence of disease or abnormality. It is the default mode of organic functioning. [...] Conceptions of health biased toward the centrality of well-being yield more qualitative approaches to assessing health that are rooted in recognition of the normative character of healthy functioning. Here, both individual subjective perspectives and culture are taken to have crucial roles in defining the meaning of health. According to such a view, health is best understood as the optimal state or functioning of an organism rather than a simple baseline status.[33]

On the World Health Organization website, we find a definition of health that assumes mental health is an integral part of general health ("there is no health without mental health"), and the definition of health itself reads: "Health is a state of complete physical, mental, and social well-being and not merely the absence of disease or infirmity."[34] The official definition of the most important international organization focusing on health thus agrees with what Hershock defines as "the optimal state or functioning of an organism." However, "optimal" means something completely different in the context of the functioning of contemporary societies in various fields, and in the philosophical context of flourishing. The latter context assumes that the optimal state is a state of enlightenment (Buddhism) or *ataraxia* (e.g., Epicureanism). Such a state must first be striven for by the philosopher, and this is precisely the purpose which the practices created by each school of philosophy serve (in the case of Buddhism, for example, these are meditative practices). Such a state is not a commonplace state in social life: its achievement requires that certain social norms should be questioned.

Further, Hershock writes that "health is not a state to be achieved and maintained—the model of homeostasis; it is a distinctively creative quality of interrelatedness."[35] Such a concept of health is atmospheric, reflecting the influence of various forces acting on both sides of an organism's body wall. Yet, health also possesses a significant social dimension, which the metaphor of the "garden" aptly describes:

> What are we not doing with this time and attention? The deep penetration of the media into our day-to-day lives marks a catastrophic export of attention from our families and local communities. This is true regardless of program content. Each hour spent consuming mass media commodities is an hour not spent attending our own immediate situation or actively and jointly improvising changes in its meaning. Like gardens, families and communities left untended will eventually go to ruin. As attention is systematically exported from unique and local patterns of relationship, largely through the consumption of both subsistence and media commodities, there occurs a depletion of precisely the resources needed locally if a population is to meaningfully resolve its own problems.[36]

In this context, what the Descartian—Western—mindset distinguishes as two separate realms is, in fact, interwoven. Acts of will can directly interact with elements—such as water. Ignorance is detrimental to health as it disrupts the delicate balance of life's processes:

> Early Buddhist teachings on the "four nutriments" (*ahara*)—material nutriments (food, water, air), sensory contacts (through sight, hearing, taste, touch, and smell, but also thinking), volitional complexes, and consciousness—suggest that the long-term effects of habit formations can be conceived as analogous to those of malnutrition. Ignorance and habit formations effectively restrict the degree to which we are nourished by our situation. Just as there are material diets that can be physically debilitating, there are patterns of experience—sensory contact, volition, and consciousness—that bring about an analogous compromise of emotional, cognitive, and social functioning, resilience, and coping ability. Health depends upon realizing a proper range, quality, and balance among all four nutriments.[37]

Meditation in the garden is based on opening up to impersonal forces that act within the meditator's field of attention. It is a continuous learning of impersonal, natural processes. This form of meditation frees one from the illusion of materialist metaphysics, the belief that an objective world is being displayed before us. The meditator learns that the way reality manifests itself depends on the quality of their attention. Buddhist meditation is a continuous training of attention, to make it sensitive and habituated to focusing on natural processes occurring in the body and mind, rather than processing images and thoughts.

> Ajahn Maha Boowa repeats many times in his writings that wisdom operates in us as a non-personal force of attention. Attention has its strategies, sometimes attacking its objects intensively, sometimes

> retreating, as if it wanted to rest and regain the strength, but it is always self-independent. At a certain level, meditation develops by itself, in the form of attention independent of us, runs around in the body with a great speed, and investigates each nook of our flesh, bones, and bodily liquids. Its acting is caused by forces independent of our will, not necessarily local, perhaps forces of the whole universe. In environmentally oriented Buddhism, conversion takes place thanks to nature and against concepts[38]

Likewise, in a Buddhist monastery human flourishing occurs as a result of the process of maturation. Interestingly, while describing the idea of "good life," William James once resorted to the concept of a "meaningful stranger." A similar, albeit not identical, concept comes to mind when we consider the Buddhist approach to good life, which can be summarized in terms of a "meaningful surrender." I flourish in the Buddhist monastery-garden by allowing nature to influence me as much as possible. It is nature that makes crucial decisions regarding the course of healing. Although, for obvious reasons, earlier in the Thai Forest Tradition, it was the forest that used to serve this purpose, it is the garden that facilitates the evocation of such an experience best. It is clear that describing the process of flourishing is unattainable in mechanistic language or the language of cognitive science, just as it is impossible to efficiently use these languages to render the essence of a flower blooming or a vegetable ripening. This is why we need to consider a new description of humanity, a new anthropology: the anthropology of *kepos*.[39]

Such an anthropology (and its language) often departs from the world of "medium dimensions" in which we live our lives. It refers us to micro-events present in our body and mind, which can only be noticed by way of intensive meditative training. Engaging in it, we abandon the dramaturgy generated by the world of medium-sized things and the dramaturgy produced by our thinking process in favor of what I call a *submorphic dramaturgy*. We no longer observe objects placed in space, nor become entangled in thinking about them, but, instead, shift our focus to micro-events in our body and mind. These micro-events contribute to a constant process; they have no boundaries, and they allow no stagnation. An example of such a "microscale" focus is the meditation on the four basic elements, where we observe the fluidity, hardness, or the spreading of heat or cold within our organism. In such a context, it becomes clear that, the postulated anthropology of *kepos* might certainly embrace the effects of applying meditative techniques present in the Thai Forest Tradition, which are very well defined—as is the process of developing gradual insight in the Buddhist tradition *sensu largo*:

> For wisdom or insight to arise, the meditator must learn to suspend the normal constructive, synthesizing activity of the mind responsible for weaving the realms of immediate sensory data into coherent narrative patterns revolving around persons, entities, and their attributes. Instead,

the meditator must adopt a radically phenomenological stance, attending mindfully to each successive occasion of experience exactly as it presents itself in its sheer immediacy. When this technique of "bare attention" is assiduously applied, the familiar world of everyday perception dissolves into a dynamic stream of impersonal phenomena, flashes of actuality arising and perishing with incredible rapidity.[40]

Notes

1 Charles S, Fisher, *Meditation in the Wild: Buddhism's Origin in the Heart of Nature* (Winchester, UK; Washington, USA: Changemakers Books, 2013), 27.
2 Jacquelyn Allen-Collinson, "Intention and Epoché-in Tension: Autophenomenography, Bracketing and a Novel Approach to Researching Sporting Embodiment," *Qualitative Research in Sport, Exercise and Health* 3, no. 1 (2011): 48–62, https://doi.org/10.1080/19398441.2010.541484.
3 Ayya Khema, *When the Iron Eagle Flies: Buddhism for the West* (Somerville, MA: Wisdom Publications, 2000), 108.
4 —as mentioned in her talks available on various websites, such as www.dharmaseed.org (accessed: April 10, 2024).
5 See Paweł Jędrzejko's autoethnographic account in Chapter 5.
6 Kamala Tiyavanich, *In the Cool Shade of Compassion: The Enchanted World of the Buddha in the Jungle* (Boulder, CO: Shambhala, 2018).
7 Ajahn Chah, *A Tree in a Forest: A Collection of Ajahn Chah Similes* (Taoyuan: Yuan Kuang Publishing House, 1994), 49.
8 Ajahn Chah, *A Tree in a Forest: A Collection of Ajahn Chah Similes...*, 69.
9 Ajahn Chah, *A Tree in a Forest: A Collection of Ajahn Chah Similes...*, 123.
10 Ajahn Chah, *A Tree in a Forest: A Collection of Ajahn Chah Similes...*, 190.
11 Bhikkhu Nyanasobhano, *Longing for Certainty: Reflections on the Buddhist Life* (Somerville, MA: Wisdom Publications, 2002), 36.
12 Bhikkhu Nyanasobhano, *Longing for Certainty: Reflections on the Buddhist Life...*, 37.
13 See, e.g., the *Dharma Seed*, https://dharmaseed.org/talks/68169/ (accessed: April 10, 2024).
14 It appears that upon achieving the status of a "legend," a Buddhist monk may disclose their meditation experiences to the public.
15 Ñanadhammo Boowa, *The Path to Arahantship: A Compilation of Venerable Ācariya Mahā Boowa's Dhamma Talks About His Path of Practice* (Wat Pa Baan Taad: Ampher Meuang, 2005, 19.
16 Lily De Silva, *Essays on Buddhism Culture and Ecology* (Dehiwala: Buddhist Cultural Center, 2001).
17 Lily De Silva, *Essays on Buddhism Culture and Ecology...*, 38.
18 Lily De Silva, *Essays on Buddhism Culture and Ecology...*, 38.
19 Lily De Silva, *Essays on Buddhism Culture and Ecology...*, 93.
20 Lily De Silva, *Essays on Buddhism Culture and Ecology...*, 94.
21 Jacquelyn Allen-Collinson, "Breathing in Life: Phenomenological Perspectives on Sport and Exercise," in *Routledge Handbook of Qualitative Research in Sport and Exercise*, ed. Brett Smith and Andrew Charles Sparkes (London: Routledge, 2016), 315–330; Jacquelyn Allen-Collinson, "Intention and Epoché-in Tension: Autophenomenography, Bracketing and a Novel Approach to Researching Sporting Embodiment," *Qualitative Research in Sport, Exercise and Health* 3, no. 1 (2011): 48–62, https://doi.org/10.1080/19398441.2010.541484.

22 Nyanatiloka Thera, *Buddhist Dictionary: Manual of Buddhist Terms and Doctrines* (Kandy: Buddhist Publication Society, 1997), 145.
23 Pramote Pamojjo, *To See the Truth: The Teachings of Venerable Pramote Pamojjo* (Chonburi: Suan Santidham, 2013), 6.
24 The idea of "hyperattention," stemming from my introspection training, has evolved—in my perception—into a descriptive concept rather than one that serves as an analytical tool. I use it to describe the sensation of heaviness in the area where I concentrate my attention, often filled with distinct emotions, typically desire or aversion.
25 Several years ago, I learned to distinguish between self-derived forces and free forces in my field of attention. The former I experience as an energy flow triggered by some desire, as if emanating from within me. The latter are clearly perceptible forces of which I am not the agent—such as the force of gravity acting on my body or within it. This distinction has become intuitive for me; it often becomes substantiated along with experience, before any reflection.
26 Lee Dhammādharo, *The Words of Ajaan Lee*, Kindle edition (publisher not provided, 2011), loc. 6961.
27 Lee Dhammādharo, *The Words of Ajaan Lee*, Kindle edition (publisher not provided, 2011), loc. 9011.
28 Marcin Fabjański, *Embodied Nature and Health: How to Attune to the Open-source Intelligence...*, 28.
29 Hermann Schmitz, Rudolf Owen Müllan, and Jan Slaby, "Emotions Outside the Box—The New Phenomenology of Feeling and Corporeality," *Phenomenology and the Cognitive Sciences* 10 (June 2011): 245, https://doi.org/10.1007/s11097-011-9195-1.
30 Charles S. Fisher, *Meditation in the Wild: Buddhism's Origin in the Heart of Nature* (London: Changemakers Books, 2013), 740.
31 This particular monk had even given up a meditation cabin to live for several years in the forest under a mosquito net.
32 Peter D. Hershock, *Buddhism in the Public Sphere: Reorienting Global Interdependence* (London: Routledge, 2017), 47.
33 Peter D. Hershock, *Buddhism in the Public Sphere: Reorienting Global Interdependence...*, 40–41.
34 World Health Organization, "Health and Well-Being," Global Health Observatory, https://www.who.int/data/gho/data/major-themes/health-and-well-being (accessed: June 29, 2023).
35 Peter D. Hershock, *Buddhism in the Public Sphere: Reorienting Global Interdependence...*, 42.
36 Peter D. Hershock, *Buddhism in the Public Sphere: Reorienting Global Interdependence...*, 98.
37 Peter D. Hershock, *Buddhism in the Public Sphere: Reorienting Global Interdependence...*, 47.
38 Marcin Fabjański, *Embodied Nature and Health: How to Attune to the Open-source Intelligence...*, 144.
39 *Kepos*—(Κῆπος) in ancient Greek: "garden," or -in the literary usage—"bosom.".
40 Nyanaponika Thera, *Abhidhamma Studies: Buddhist Exploration of Consciousness and Time* (Boston, MA: Wisdom Publications, 1997), XXVI–XXVI.

Part Three

Toward a New Metaphysics

7 *Friluftsliv*

The Cultural Practice of Attunement

The practice of meditation within the Thai Forest Tradition, leading to the lived experience of one's profound connection with nature, the mindful, yet organically perceived, never-ending journey in the mountains and isles of Greece, the process of learning to attune to the rhythms of the elements in the open ocean or in secluded islands world-wide—all share a common denominator: they are all *processes* involving the cultivation of a heightened state of awareness that transcends quotidian perceptions. Each of these contexts, as we have demonstrated, is conducive to the suspension of the mind's habitual narrative-building, and each allows one to experience the immediate flow of sensations, unfiltered through preconceived notions. "Bare attention," whether attained by way of practice or induced by the here-and-now circumstance, reveals the transient nature of phenomena, facilitating the emergence of insight—and wisdom. This immersive state of mind enables one's direct interaction with the environment that can transform one's "inherited" perceptions and evoke a sense of spiritual wholeness. Such language permeates Hans Gelter's beautifully written "Friluftsliv: The Scandinavian Philosophy of Outdoor Life," which appealingly bridges the practice of *lagom* with the organic experience of attunement to the rhythms of nature:

> We are in a splendid remote wilderness—the Wind River in northern Yukon. Crystal clear water sparkles around us with the marbled river bottom several meters below, giving the sensation of our canoe gliding in open air. The strong current and our synchronized paddle strokes carry the canoe down this Arctic river with a force that creates a deep shiver of pleasure. The breathtaking big sky above us, the river valley bordered by magnificent mountains, and the sensation of undisturbed wildlife surrounding us causes a deep emotional storm of happiness within, filling my eyes with tears—a spiritual, almost religious feeling I often experience in nature. This landscape absorbs me so completely, entering through all of my senses and directly touching my limbic system. This gives me a sensation of a total integration with this land; a strong feeling of being at home in a place I have never visited before.

DOI: 10.4324/9781003328469-11

Sensing myself as part of the landscape I experience the processes and evolution of this place unfolding itself inside my consciousness. I get a strong feeling of knowing the ways of things around me.

The power and intensity of this feeling has always surprised me, as well as the fact that not everybody experiences the same feelings. Trying to discuss this feeling with people who have never experienced it is difficult, as they cannot relate such strong feelings to nature. The feeling of being a part of the river or the mountains seems too spiritual to most people. Many of today's urban people have lost this ability to experience nature in a subjective way, seeing the landscape in an objective and disconnected way. The ability to be absorbed by a place is a state of mind, a skill that needs training. Many modem people have lost this ability to give the landscape free access, through open senses, to the limbic system. This limbic system makes up the functional centre of human emotions and memory. Sensory information enters the brain here and the higher centre of integration in the cerebrum consults the limbic system for memory retrieving and memory processing. By electrical stimulation of the limbic system hallucinations, religious experiences, out-of-body experiences, and near-dead experiences can artificially be created, indicating this system's importance for the experience of reality. Thus it seems as if there is an ability to let the landscape interact directly through open senses with the limbic system when interpreting the world, which can create a strong emotional reaction—a reaction we may experience as spiritual or religious. This has nothing to do with the endorphin and adrenaline rush one can experience in adventure activities.

This spiritual feeling of connectedness to the landscape is probably the deep experience in Arne Naess's philosophy of Deep Ecology [...]. Naess, himself a mountaineer and outdoor person, proposes that a deep experience of nature creates deep feelings leading to deep questions and a deep commitment for nature [...]. This may result in a paradigm shift in one's way of viewing the world. In Scandinavia we would say that this deep experience of the landscape is the essence and reward of a lifestyle we call *friluftsliv*. The word translates to "free air life" meaning a philosophical lifestyle based on experiences of the freedom in nature and the spiritual connectedness with the landscape. The reward of this connectedness with the landscape is this strong sensation of a new level of consciousness and a spiritual wholeness.[1]

The Essence of *Friluftsliv*

The essence of *friluftsliv*, literally translated into "free-air-living," is thus not merely about being-in-nature, but about experiencing it deeply and personally. It involves a mindful engagement with the natural environment, fostering a sense of belonging and gardener-like stewardship. *Friluftsliv* is thus a philosophy epitomizing a holistic approach to outdoor living, which, as will be shown, contributes significantly to the well-being understood in a plethora of dimensions. And although countries such as Canada, New Zealand, and Japan—promoting outdoor activities and nature-based lifestyles—tend toward the embracement of the therapeutic benefits of one's profound presence in nature, the global dominance of the neoclassical paradigms of thought continues to stand in the way of individual and communal physical health, mental clarity, and social cohesion. Interestingly, the paradigm shift, for which Gelter calls, is precisely what our reflections seek to energize. Yet, although—like the author of the article—we do realize that grassroots philosophies of attunement are, have long been, and probably will be spontaneously practiced in numerous locations world-wide, the paradigm shift that would bring down the hegemony of the Descartian-Newtonian model, that has effectively cut the capitalist, production-oriented, greed-driven world off its original roots would have to be a veritable philosophical coup. While in many locations—especially ones particularly affected by the cycles of nature—this organic connection between individuals, communities, and the elements remains strong, the mentality of an average Westerner remains disconnected from the fundamental principles of life. It must be realized that

> The cultural roots of *friluftsliv* in Scandinavia come from the self-image of Scandinavians as a nature loving people. This image is partly based on these countries' unpopulated landscape, where even urban people have free nature very close by for recreation. This self-image is also reflected by the unwritten law of "Allemansrätten" ("everyones-right") in Sweden and Norway that allows everyone access to the land, even private property. The image has its historical origin in the long history of living in this cold Ultima Thule where skis were already used for hunting 5200 years ago. The romantic "back-to-nature" movement in the 18th century, as a reaction against urbanization and industrialization, strongly influences Scandinavian culture. Through music, poetry, and art, this natureloving image was introduced to the upper-class society. Successful Scandinavian explorers like Fritjof Nansen, Sven Hedin, Roald Admunsen, Adolf Nordenskiöld, and others strengthened this image. But the upper class had no natural connection to nature; they weren't hunters, fishers, or farmers. Therefore *friluftsliv* became a way to realize the ideas of romanticism, to reconnect with nature and the old Scandinavian outdoor tradition. To guide the people back to nature *friluftsliv* was organized and developed by the worlds first tourist organizations (1868 in Norway and 1885 in Sweden), and later (in 1892) the Swedish outdoor organization "Friluftsfrämjandet." Their goals were to foster people's good health through skiing and other nature experiences to better cope with the urban and industrial development. During wartime, *friluftsliv* was used to develop and foster strong people for the defence—similar to the origin of the

> Anglo-American Outward Bound [...]. This self-image of a nature-loving people was also reinforced in the 1930s during the building of the Swedish socialistic "folhemmet" (folk-home). The increased disposable time for the working class had to be used for healthy recreation in nature, and *friluftsliv* was the way.
>
> Today a strong commercialization creates a never-ending flow of new consumption-lifestyles for outdoor recreation. Activities and equipment now overshadow the original goal of *friluftsliv* to be close to nature. For many people nature is becoming an arena [...] to test oneself and the equipment. This commercialization excludes many from *friluftsliv* today because of the high price of gear, the long journeys to "the right places," and the expertise needed for many activities is too high. This new trend in *friluftsliv* is in strong contrast with the essence of the word first used by Henrik Ibsen in an 1859 poem ["On the Heights"]. He was sitting in a cottage, looking into the stove, and said "[...] this is *Friluftsliv* for my thoughts."[2]

Although we cannot be certain whether or not Henrik Ibsen indeed coined the phrase,[3] we do realize that it is his particular application of the notion that gave it a philosophical dimension, thus connecting the idea of *friluftsliv* to the ideas underlying the "transgressive" trends in Europe and in America, that strove to overthrow the hegemony of the paradigms imposed by the Age of Reason. And so Hans Gelter is right in observing that:

> [t]he word *friluftsliv* was thus first used to describe a thought, an idea about life. Before Ibsen, the word "Frilufts-painting" had been used by Theodore Rousseau and others in poesophy (poetry and philosophy) of the European deep Romantic Movement. In a meeting with the Norwegian Tourist Association in 1921 Nansen [...] talked about *friluftsliv* as a philosophy and as an alternative for youth to avoid "tourism," a superficial acquaintance with nature. He spoke about the ability to co-operate with nature's powers and the joy of being in nature. He believed that free nature was our true home and that *friluftsliv* was our way back home.[4]

As can be seen, the sense of the distinction between "tourism," implying as "superficial acquaintance with nature" and *friluftsliv* being the philosophy of attunement, which clearly underlies Dariusz Kubok's reflection, is deeply ingrained in the culture of Scandinavia, with Norway and Sweden pioneering in the institutionalization of the latter. Interestingly, personal reflections and stories of individuals who practice *friluftsliv* offer insights into how vivid this philosophy remains in the day-to-day practice and how powerfully it has affected the legislatures of the Scandinavian cultures. Many Norwegians—like Robert DeVries and his partner Paola, our hosts, whose attic and basement have been transformed into hiking gear storages, and whose yard features a rather impressive assortment of kayaks and SUPs—describe *friluftsliv* as a source of joy: the simple act of spending time in nature, whether hiking in the mountains, fishing in a fjord, or enjoying a quiet moment by a forest lake, provides a profound sense of fulfillment and connection. Like Robert, everyone we encounter recounts experiencs of hiking in the Norwegian mountains, almost always using language that emphasizes the sense of freedom and exhilaration they feel upon reaching subsequent summits. Breathtaking panoramic views, the crisp mountain air, and the sense of accomplishment combined, attract people of all ages to the mountains. But the love of the country and the fascination with the sublime magnificence of its nature also motivates them to contribute to the maintenance of the mountain trails, shelters, and huts, established by Den Norske Turistforening (The Norwegian Trekking Association)—one of the oldest outdoor organizations in the world—irrespective of whether, or how often, they use them.

Den Norske Turistforening (DNT) plays a pivotal role in promoting *friluftsliv*. As Hans Gelter emphasized, established in 1868, the DNT originally aimed to attract the garden-bound upper classes, but the implementation of the Romantic program of "the return to nature" effectively yielded a much more all-encompassing result. The DNT made nature more accessible to the general public, encouraging exploration and appreciation of Norway's natural landscapes. Initially, the association built a few cottages in the mountains, providing safe havens for hikers and nature enthusiasts. These cottages, strategically located in scenic, and often remote, areas, allowed romantically driven *aficionados* to experience nature firsthand without the need for extensive planning or equipment. Over time, however, the DNT expanded its network, establishing thousands of cottages across Norway. The expansion of the DNT has effectively democratized access to the outdoors, ensuring that today people from all walks of life can enjoy Norway's natural beauty. These cottages are often equipped with basic amenities such as beds, kitchens, and firewood, making them suitable for families, solo adventurers, and groups alike. By reducing the logistical barriers to outdoor recreation, DNT has played a crucial role in embedding *friluftsliv* into the fabric of Norwegian society.

"Everyman's Right" and Trust

Importantly, however, as all of my interlocutors stress, these cottages—like much of Scandinavia in general—operate on trust: visitors are expected to pay for the food they consume and charcoal or wood they burn leaving money in a money box at the cottage, or scanning a QR code, and they are trusted to keep the facilities clean and operational for the next user. This system reflects the high level of mutual confidence that characterizes Scandinavian societies: in many other countries, such facilities might be susceptible to vandalism or theft, but in Norway, Sweden, or Finland the communal respect for these shared resources—and for one another—is profoundly ingrained in local mentality. The sense of confidence in others is based on the long-cultivated communal responsibility, anchored, as our discussion on the origins of the philosophy of *lagom* highlighted, in historically (or perhaps genetically?) shared experience of survival in the North.

Our Norwegian friends explain that although the overall situation is markedly changing, in rural Norway, a high level of trust continues to permeate daily life. People often leave their homes and cars unlocked, and communal resources are shared freely. Confirming our intuitions, Robert and Paola attribute this level of confidence to the historical conditioning of the development of small, close-knit communities, where mutual reliance was essential. However, they do stress that this tradition faces challenges posed by urbanization and increasing migration. These phenomena, although enhancing beneficial aspects of diversity, and conducive to the overall growth of affluence, are also responsible for the present day concurrence of alternative cultural norms (which do not necessarily coincide with the traditional Scandinavian values) and, especially in urban centers, for the increase in crime rate levels.

The above notwithstanding, rural and wild areas still constitute the majority of the territory of Scandinavia, and therefore the sense of trust and community remains alive among its population, affecting the steady growth of the variety of forms in which people interact with nature. *Friluftsliv* is no longer tantamount to individual enjoyment of the outdoors: it has also become a label for organized group activities such as hikes, skiing trips, or fishing excursions, further fostering social bonds and reinforcing the collective identity of the Scandinavians, centered around a shared appreciation for the natural world. Furthermore, the high level of trust typical of the rural communities extends to interactions with visitors—both "travelers" and "tourists." After all, the "The Right to Roam" is "Everyman's Right," and the Scandinavians to not restrict it to themselves alone.

The Norwegian *Allemannsretten*, Swedish *Allemansrätten*,[5] and Finnish *Jokaisenoikeudet*, are the cornerstones of *friluftsliv*: "The Right to Roam" is a historic custom law, allowing the public to access and enjoy natural areas, even on private lands, provided they respect both the environment and property owners, adhering to the principle "Leave No Trace." Enshrined in Norwegian law and universally respected in Sweden and Finland, it reflects a deep-seated Scandinavian belief in communal access to nature. "Everyman's Right" permits activities such as hiking, camping, and foraging for wild berries and mushrooms, fostering a culture where nature is seen as a shared resource for all to enjoy and profit from. As our host, Robert DeVries expressed it in an informal conversation, "the whole area around us is ours, even if it is somebody else who owns it." In stark contrast to countries where land access is restricted and natural areas are often privatized, "The Right to Roam," based on mutual trust and respect, promotes inclusivity and sustainability. Intuitively, it encourages people to explore and engage with nature and with one another, further fostering a collective responsibility for its preservation. Such "open access" to space, as one can imagine, has significant implications for public health: encouraging outdoor activities, it promotes lifestyle that values both physical and mental well-being, to which numerous reports and academic studies available in the web services of the Norwegian umbrella organization Norsk Friluftslif or the Swedish Environmental Agency (Naturvårdsverket) testify.[6]

Importantly, "Everyman's Right" is not only a cultural norm that shapes the way Scandinavians interact with their environment; it embodies a profoundly anchored philosophy of mutual respect and coexistence determining the cultural practice in which individuals are free to roam and enjoy nature, provided they do so responsibly. It is also a *legal* framework, which encourages a deeper connection to the land and instills a sense of stewardship and care for the natural world. The practice of "The Right to Roam" has proven to generate measurable economic and social benefits: it supports local economies by promoting tourism and outdoor recreation, attracting visitors who contribute to the sustainability of rural communities. Simultaneously, it enhances social equity by ensuring that access to natural spaces is not limited by economic status or property ownership, allowing everyone to benefit from the therapeutic and recreational values of nature. And despite demographic transformations, the atmosphere of respect for the natural environment and the property of others makes the newcomers observe the principles shared by the locals. This compassionate relation is evident in the state of the natural environment: well-maintained trails, pristine campsites, the absence of garbage, and the ecologically oriented facilities in the marinas. Visitors (whether hikers, campers, or sailors) are welcome, provided they observe the rules of *lagom* and the ramifications of "The Right to Roam." And if they do—they are readily integrated into local traditions, such as communal bonfires, fishing festivals, and events organized by hiking clubs. Such "community of values" strengthens the bonds between residents and newcomers, but also allows the visitors to learn the Scandinavian ways "by osmosis," which, in time, might make it easier to transplant them to continental Europe, or the industrially advanced regions of the Americas.

Upbringing and Formal Education: "Open Air at School"

Unsurprisingly, *friluftsliv* has become an indispensable element of the Scandinavian family life and education. Paco Kaliciński and his wife Agnieszka—musicians, master chefs, sailors, and, as all of the masters of STA ships and small boat skippers alike will confirm, unofficial ambassadors of Poland in Flekkefjord, Norway—highlight the role of *friluftsliv* in their own homestead experience. Spending weekends at remote mountain huts or cabins, disconnected from the distractions of modern life, the whole family bonds while exploring nature together, teaching their children the values of respect for the environment and, simultaneously, self-reliance. On the one hand, Paco and Aga emphasize the importance of these experiences in fostering a sense of adventure and curiosity in Norwegian children—the values which they would then carry into adulthood. On the other—they tell me of how the general recognition of the significance of such exposure to nature translates into the integration of *friluftsliv* into the Norwegian education system. From a very young age, schoolchildren are encouraged to spend time outdoors, regardless of the weather; and thus, rather than with rulers, crayons, and compasses, parents are advised to equip their children with foul-weather gear and towels, as outdoor play is a fundamental part of the school day. This practice, known as *uteskole* or "outdoor school," fosters resilience, independence, and a deep connection to nature, which benefits have long been recognized by the Norwegian Ministry of Climate and Environment, whose initiative "Friluftsliv i skolen" is described thus:

> *Friluftsliv i skolen* (*Outdoor Life in Schools*) is part of the Ministry of Climate and Environment's initiative to ensure that nature is increasingly used as a space for activities and learning. Norsk Friluftsliv—the Norwegian Outdoor Life Organization—aims to facilitate the use of nature by schools as an extended classroom for health-promoting activities, deeper learning, and social interaction. Norwegian Outdoor Life has long advocated for outdoor activities and outdoor schools to become a more significant part of children's school lives. In the project Outdoor Life in Schools, we collaborate across the education and outdoor sectors to enhance competence and model teaching in nature. The Forum for Outdoor Life in Schools is led by the Norwegian Outdoor Life with a mandate from the Ministry of Climate and Environment. The forum brings together actors across the education and outdoor sectors for competence exchange and coordination of a joint effort to strengthen nature as a space for learning and activity in primary education. The Outdoor Life in Schools project was established to achieve one of the main goals in the government's action plan for outdoor life, namely that "Nature should increasingly be used as a learning and activity area for children and youth." We believe that by using nature as a learning space, Norwegian schools can actively engage with the overarching goals and subject plans in the Curriculum Knowledge Promotion 2020. By using nature as an extended classroom in teaching, it facilitates deeper learning, the development of social skills, and activities that strengthen physical and mental health. Therefore, Norsk Friluftsliv works to ensure that all students experience community, well-being, and mastery in nature's classroom.[7]

As the website of the "*Friluftsliv* i skolen" initiative demonstrates, outdoor education in Norway is now *institutionalized*. Far beyond just play, it is based on structured learning activities that take place in natural settings. Subjects such as biology, geography, and physical education are often taught outside, offering the students hands-on experiences that enrich the learning process and transform it into a fascinating adventure. This approach not only enhances academic understanding of the world, but also promotes physical health and builds the sense of togetherness, as outdoor education proves conducive to the development of social skills and teamwork. Group activities such as hiking, orienteering, and outdoor games require cooperation and communication, helping children develop important interpersonal skills. The shared experiences of overcoming challenges and discovering new aspects of nature foster a sense of camaraderie and community among students, and allows the children, regularly interacting with nature, to develop a lifelong appreciation for the environment and a commitment to preserving it. But the benefits of integrating *friluftsliv* into education do not end there: physical activity (which, of course, is essential for the healthy development of children) combined with regular exposure to natural environment has been shown to reduce stress, improve concentration, and enhance mood. Furthermore, learning in nature fosters a sense of curiosity and creativity, as students engage with their surroundings in dynamic and meaningful ways, which jointly contributes to the overall growth of the communal and individual sense of well-being. Hence, the institutional position of the Norwegian government is to undestand *friluftsliv* as a *child's indispensable right.*

Friluftsliv and Health

The "*Friluftsliv* at School" initiative may, in fact, be perceived as a preventive measure: attuned early, their culturally rooted proximity to the natural rhythms enhanced, the children grow into adults equipped with mechanisms allowing them to defend themselves against the stringencies of the digital era. The above notwithstanding, for those who might need it anyway, nature therapy (ecotherapy or green therapy) is institutionally promoted. The practice of attunement to the rhythms of the biotope with which one interacts leverages the therapeutic effects of natural settings to improve mental health, and thus activities such as forest bathing, mindfulness walks, and nature-based meditation are being integrated into mental health treatment programs with increasing frequency. These practices, as it goes without saying, align with the principles of *friluftsliv*, a grassroots philosophy emphasizing the healing power of nature.

The mental health benefits of *friluftsliv* are well documented: spending time in nature has been shown to reduce stress, anxiety, and depression while enhancing mood, focus, and overall well-being.[8] The calming effects of natural environments, combined with physical activity, contribute to a holistic approach to mental health. Personal accounts from individuals who have experienced the mental health benefits of *friluftsliv* provide compelling evidence of its impact. One of our interlocutors described how regular hikes in the forest helped them cope with work-related stress and anxiety—the tranquility of the natural surroundings and the rhythmic movement of walking would enter them into a meditative state, each time allowing them to clear their mind and gain perspective. Another person, sharing their journey of overcoming depression through *friluftsliv*, recounted how spending time in nature allowed them to discover a sense of solace and hope, which helped them to reconnect with their own, now extended, non-egotistic self, and thereby find meaning in life. Stories like these, and many others, underscore the profound transformative effects of nature on mental health, highlighting the importance of integrating *friluftsliv* into daily life both at the level of personal decisions, community practice, and state administration rulings.

Everyday Practice of "Living in the Open Air"

Deep in their mental and physical makeup, the Scandinavians' commitment to *friluftsliv* today is visible not only in their fascination with the beauty of their respective countries and care for the rural and urban areas alike (even in major cities, community gardens and green spaces provide opportunities for residents and visitors to connect with nature and each other—very much in accordance with the principles of *friluftsliv*), but also in their long-term life choices. In their driveways, one often sees recreational vehicles (RVs) or campers; in their backyards—more often than not—one gets to see boats. Propelled mechanically or naturally, of every imaginable type, these larger or smaller vessels allow even those who are not professional fishermen or island dwellers a massive degree of independence in roaming the gardens of the isles, fjords, or inland lakes. Whenever they can, they leave their daily routines behind and sail off to places immune to the artificial hassle of the major cities, or stringencies of work. Some, like Sævar Hansson and Valdis Brynjolfsdottir, originally from Iceland, take a step further: they tie their professional life to *friluftsliv*. Infatuated by the glorious nature of Norway, this warm, nature-loving couple build a welcoming, friendly, and unassumingly beautiful camping site on the shores of the Rokosjøen lake, featuring a few cottages, a small sauna, and a number of plots, where you can put up your tent or park your RV. One evening together made it clear that both Sævar and Valdis see the profundity of their connection with the universe around them in terms of both their lifetime experiences of interconnection and transcendence: attuned to Iceland, whose rhythms, out of necessity, require mindful observation of every possible symptom of danger, attuned to the seasonal pulse of the Rokosjøen's nature, both our hosts recount their experiences of "transparence," moments beyond time, yet moments of concentrated presence. On the one hand, these experiences have proven conducive to their search for the understanding of the essence of their immersive experiences in light of hermetic philosophy—beyond categorial limitations. On the other, their family business is a palpable manifestation of the translatability of their contemplative insight into living practice: they established the campsite to practice *friluftsliv* every minute of their lives. Yet, as opposed to institutions capitalizing on "pop-mindfulness," they indeed deliver to others what they promise to deliver—a space conducive to one's attunement of the pulsing-life-within to the pulsing-life-without—or, at the very least, a wonderful space for relaxation, recreation, and contemplation.

Stewardship as a Policy

As is already clear, the atmospheric philosophy of *friluftsliv* has imperceptibly transformed Scandinavia into a healing garden: simultaneously attuned to its own biological clock, it is also tended to by those who share it. Norwegians, Swedes, and Finns are deeply committed to environmental stewardship, a value instilled in every Scandinavian child from an early age. This commitment is visible not only in the already emphasized meticulous cleanliness of natural areas, but also in the widespread practice of recycling and waste segregation. The societal norm of maintaining and respecting natural spaces contributes to Scandinavian countries reputation as a global leader in environmental conservation, which, in itself, is sustainable owing to environmental education—a key component of the *friluftsliv* philosophy. Schools and community organizations regularly offer programs that teach children and adults about sustainable practices, wildlife conservation, and the impact of human activities on the environment. These long-term initiatives, supported by local and state governments alike, foster a sense of responsibility and encourage proactive measures to protect natural resources.

The commitment to environmental stewardship extends to policy and infrastructure as well. Norway, Sweden, and Finland have implemented numerous initiatives to reduce carbon emissions, promote renewable energy, and protect biodiversity. Biodiversity offsetting and ecological compensation,[9] practical implementation of ecosystems thinking,[10] sustainable development[11] and sustainable consumption,[12] environmental legislation,[13] and more specifically, policymaking concerning everyone's "right to free air" (*Friluftslivpolitiken*[14]), although historically—in one way or another—they have always been practiced by Scandinavian communities attuned to the demands of the land, sea, and weather, upon which their lives depended, are now institutional efforts.

These efforts are supported by robust legislation and significant public investment, reflecting the high priority placed on environmental sustainability. One notable example of Norway's dedication to sustainability is its transition to renewable energy sources. The country has invested heavily in hydropower, wind energy, and electric transportation, significantly reducing its reliance on fossil fuels. Norway's capital, Oslo, has been recognized as a European Green Capital for its efforts to promote sustainable urban living and reduce its carbon footprint. Norway's environmental policies are often developed through a collaborative process that involves government agencies, non-governmental organizations, and the public. This inclusive approach ensures that diverse perspectives are considered and that policies are widely supported: public participation in environmental decision-making, banking on the "traditional" sense of collective responsibility, further enhances the sense of communal agency and empowers individuals, who feel motivated to personally contribute to sustainability efforts.

Friluftsliv and "Civilization"

While *friluftsliv* is often associated with remote wilderness areas, it is equally relevant in urban settings. Urban *friluftsliv* involves finding and creating opportunities for outdoor activities and nature experiences within cities. This can include parks, green spaces, community gardens, and urban trails. City planners and architects in Norway and other Scandinavian countries increasingly often incorporate elements of *friluftsliv* into urban design. Green roofs, vertical gardens, and pedestrian-friendly streetscapes create pockets of nature within the urban environment. These green spaces provide residents with accessible opportunities for relaxation, recreation, and (re)connection to nature. Like in all other dimension of the stewardship to nature in Scandinavia, community initiatives—grassroots phenomena rooted in tradition and energized with education—play a crucial role in promoting urban *friluftsliv*. Local groups organize outdoor activities such as nature walks, cycling tours, and gardening projects, fostering the sense of community even further, and further encouraging residents to engage with their surroundings. Even a cursory look at the schools and childcare centers in Scandinavia will allow one to appreciate that projects like "Friluftsliv i skolen" work: as a rule, schools incorporate nature-based play areas and outdoor classrooms, ensuring that children have regular contact with nature even in densely populated areas. Needless to say, urban *friluftsliv* offers numerous benefits, including improved air quality, reduced heat island effects, and enhanced social cohesion. Owing to the fact that the municipal governments, city councils, and townhall adminstrations fully recognize the importance of the principles of *friluftsliv*, funds invested into integrating nature into urban environments, effectively can create healthier and more sustainable living spaces. Such spendings, in Scandinavia, are not considered a burden: they are deemed an indispensable investment in the future.

While traditional *friluftsliv* emphasizes simplicity and disconnection from modern technology, the digital age has introduced new dimensions to this practice. Technology can both complement and challenge the principles of *friluftsliv*. On one hand, digital solutions enhance safety, navigation, and accessibility for outdoor activities; on the other, excessive screen time and reliance on technology can detract one from the immersive experience of nature. The use of smartphones and GPS devices has made it easier for people to explore remote areas and navigate challenging terrains, and apps that provide weather forecasts, trail maps, and safety tips have become valuable resources for outdoor enthusiasts. Social media platforms allow individuals to share their *friluftsliv* experiences, inspiring others to venture into nature and, fostering a sense of community online, they promote the person-to-person interaction on trails or in the isles. However, as all of our Scandinavian interlocutors—parents of children—stress, a growing awareness of the need to balance technology use with the principles of *friluftsliv* can be observed; more and more practitioners of the outdoor exposure advocate for "digital detox" periods, where individuals disconnect from their devices and rely on preparation, observation, and physical maps or charts, to fully engage with the natural environment. *Not too little, not too much, just enough*: this mindful approach to technology, aligning with the philosophy of *lagom*, helps preserve the essence of *friluftsliv*, allowing individuals to experience nature in its purest form.

Friluftsliv, as a philosophy and as a way of life, offers a proposition for individuals, societies, and governments seeking to foster well-being among their own citizens—caught in the rat race, and more and more evidently affected by all of the "civilization diseases" recognized by contemporary medicine—by way of mindful insight into our ever-existing, yet long forgotten, place in the garden of nature. Its emphasis on simplicity, sustainability, and communal access to natural spaces provides a model for harmonious living in an increasingly complex and urbanized world. Scandinavian governments, who embraced *friluftsliv* as the fundament of their policies, highlight its profound impact on physical, mental, and social health:[15] as we face global challenges such as climate change, biodiversity loss, and social fragmentation, their example offers a path forward. From the mountains of Norway to urban parks, from the Stockholm Skärgården to the Åland archipelago, from Nordkapp to Vyborg, the effects of the actual practice of *friluftsliv* encourage us to slow down, mindfully immerse ourselves in nature, and cultivate a sense of wonder and gratitude. Such a lifestyle values nature, promotes sustainability, and fosters community bonds, and—although it does not warrant immunity to all existing pathogens[16]—it can become a fundament for a more resilient and harmonious world that the future generations will call theirs, identifying with it at the fundamental level of their own, garden-like, extended, atmospherically conditioned, yet too often enslaved bodies that, against all obstacles, are primed to flourish.

Notes

1 Hans Gelter, "*Friluftsliv*: The Scandinavian Philosophy of Outdoor Life," *Canadian Journal of Environmental Education* 5 (Summer 2000): 79–80.
2 Hans Gelter, "*Friluftsliv*: The Scandinavian Philosophy of Outdoor Life"..., 79–80.
3 In his paper "The Role of Friluftsliv in Henrik Ibsen's Works," Petter Erik Leirhaug explains that "*On the Heights* is a long epic poem. In total it consists of 387 lines. We follow a young nameless man who leaves his home and the neighbour farm girl whom he loves. He secures his little house, takes his rifle, and heads for the mountains. First, his plan is to return, with a splendid reindeer pelt for his mother, and some as well for his sweetheart. But then he meets a stranger who urges him to leave all human ties behind and stay in the mountains. At this stage, in the middle part of the poem, is the stanza where Ibsen's use of "friluftsliv" for the very first time appears: 'I den øde sæterstue al min rige fangst jeg sanker; der er krak og der er grue, friluftsliv for mine tanker. (HU XIV:395) / In the lonely mountain farm, My abundant catch I take. There is a hearth, and table, And friluftsliv for my thoughts.'" Petter Erik Leirhaug, "The Role of Friluftsliv in Henrik Ibsen's Works," (paper presented at the Henrik Ibsen: The Birth of "Friluftsliv," A 150 Year International Dialogue Conference Jubilee Celebration, North Troendelag University College, Levanger, Norway, September 14–19, 2009) https://norwegianjournaloffriluftsliv.com/doc/172010.pdf (access: June 5, 2024).

4 Hans Gelter, "*Friluftsliv*: The Scandinavian Philosophy of Outdoor Life"..., 79–80.
5 Bertil Bengtsson, *Allemansrätten – Vad säger lagen?* (Stockholm: Naturvårdsverket, 2023).
6 See the Norwegian websites of the Norsk Friluftsliv: https://norskfriluftsliv.no and the Swedish websites of the Naturvårdsverket: https://www.naturvardsverket.se (accessed: February 2, 2024).
7 Norsk Friluftsliv, "Friluftsliv i Skolen," https://friluftsliviskolen.no/om-prosjektet/ (accessed: August 5, 2023).
8 See, for instance, Anthoni Oisin, "Friluftsliv: The Norwegian Prescription for Health and Happiness," *News-Medical*, https://www.news-medical.net/health/Friluftsliv-The-Norwegian-Prescription-for-Health-and-Happiness.aspx (accessed: June 7, 2024).
9 See, for instance: Mikael Karlsson, Henrik Josefsson, Tim Schnoor, and Charlotta Zetterberg, *Utbytbar natur? Etiska och rättsliga aspekter på ekologisk kompensation: Slutrapport från ett forskningsprojekt*, Rapport 6997 (Stockholm: Naturvårdsverket, 2021); Ingemar Jönsson et al., *Ekologisk kompensation som styrmedel i kommunal planering*, Rapport 6992 (Stockholm: Naturvårdsverket, 2021); Lena Bergström et al., *Ekologisk kompensation i kustmiljön: Hur kan man uppväga förluster av biologisk mångfald och ekosystemtjänster i samband med mänsklig verksamhet i kustområdet?* Rapport 6994 (Stockholm: Naturvårdsverket, September 2021); Lena Bergström et al., *Ekologisk kompensation i kustmiljön: Hur kan man uppväga förluster av biologisk mångfald och ekosystemtjänster i samband med mänsklig verksamhet i kustområdet?* Rapport 6994 (Stockholm: Naturvårdsverket, 2021); Jonas Nordström et al., *Ekologisk kompensation: Upptag och integrering bland svenska aktörer och kvantifiering av de samhällsekonomiska effekterna*, Rapport 7008 (Stockholm: Naturvårdsverket, September 2021); *forskningsprojekten Systematisera ekologisk kunskap för att effektivisera ekologisk kompensation och När kan ekologisk kompensation bidra till att bevara biologisk mångfald och ekosystemtjänster?* Rapport 6995 (Stockholm: Naturvårdsverket, September 2021); for comparison, see also: Ryszard Mikosz, *Wprowadzenie do prawa środowiska* (Katowice: Wydawnictwo Uniwersytetu Śląskiego, 2021).
10 See, for instance: Sofia A. Wikström et al., *Ekosystemansatsen – praktiska erfarenheter från svensk havs- och vattenförvaltning*, Rapport 6934 (Stockholm: Naturvårdsverket, August 2020).
11 See, for instance: Lena Bergström et al., *Effekter av havsbaserad vindkraft på marint liv: En syntesrapport om kunskapsläget 2021*, Rapport 7049 (Stockholm: Naturvårdsverket, 2022); Tomas Pärt, *Våtmarker i odlingslandskapet: Restaurering och anläggning för ökad mångfald av våtmarksfåglar*, Rapport 6933 (Stockholm: Naturvårdsverket, November 2020); Antonia Nyström Sandman et al., *Grön infrastruktur i havet: Landskapsperspektiv i förvaltningen av Sveriges marina områden*, Rapport 6930 (Stockholm: Naturvårdsverket, June 2020), Rasmus Kløcker Larsen et al., *Omtvistade landskap: Navigering mellan konkurrerande markanvändning och kumulativa effekter*, Rapport 6908 (Stockholm: Naturvårdsverket, January 2020);
12 See, for instance: Louise Sörme et al., *Matavfall i butik – Analys av samhällsförändringar*, Rapport 6901 (Stockholm: Naturvårdsverket, November 2019); Elin Röös, Sarah Säll, and Emma Moberg, *Effekter av en klimatskatt på livsmedel*, Rapport 6965 (Stockholm: Naturvårdsverket, 2021), or Karin André et al., *Drivkrafter för hållbar konsumtion på lokal nivå: Svenska kommuners roll och möjligheter*, Rapport 6960 (Stockholm: Naturvårdsverket, February 2021).
13 See, for instance: Magnus Hennlock et al., *Styrmedel för livscykler: En integrerad modellansats mellan cirkulärekonomiska modeller och livscykelanalys – Slutrapport*, Rapport 6961 (Stockholm: Naturvårdsverket, 2021).
14 See, for instance: Naturvårdsverket, *Uppföljning av målen för friluftslivspolitiken 2023*, Rapport 7123 (Stockholm: Naturvårdsverket, December 2023).

15 Stefán Ólafsson, "Well-Being in the Nordic Countries: An International Comparison," *Icelandic Review of Politics & Administration* 9, no. 2 (December 2013): 345–372, https://doi.org/10.13177/irpa.a.2013.9.2.5.

16 Erica A. Yarmol-Matusiak, Lauren E. Cipriano, and Saverio Stranges, "A Comparison of COVID-19 Epidemiological Indicators in Sweden, Norway, Denmark, and Finland," *Scandinavian Journal of Public Health* 49, no. 1 (February 2021): 69–78, https://doi.org/10.1177/1403494820980264; see also: Swapnil Mishra, James A. Scott, Daniel J. Laydon, Seth Flaxman, Axel Gandy, Thomas A. Mellan, H. Juliette T. Unwin, Michaela Vollmer, Helen Coupland, Oliver Ratmann, Melodie Monod, Harrison H. Zhu, Anne Cori, Katy A. M. Gaythorpe, Lilith K. Whittles, Charles Whittaker, Christl A. Donnelly, Neil M. Ferguson, and Samir Bhatt, "Comparing the Responses of the UK, Sweden and Denmark to COVID-19 Using Counterfactual Modelling," *Scientific Reports* 11 (2021): 16342, https://doi.org/10.1038/s41598-021-95699-9.

8 Walking through Atmospheres

If ancient Greek philosophers—like Aristotle—were right in their claim that nature itself desires for us to flourish, and if modern medicine aims for the same, then why do we not bloom? Why is it that we witness a veritable epidemic of cancers and mental disorders in the modern world? What stands in our way toward our flourishing, our health? We believe that, among other places, the explanations of the present state of affairs should be sought in the area traditionally located in the realm of philosophy: in metaphysics. Many diagnoses, verbalized in a series of philosophical essays that, owing to their popularity, seem to resonate with what readers in the West intuitively feel, suggest that the cause of modern humanity's failure to flourish lies precisely there. One such diagnosis, proposed by Byung-Chul Han, a German philosopher of Korean origin, in the essay titled *The Palliative Society. Pain Today*,[1] seems to have address the essence of the problem with particular acuity.

The author focuses on the interrelation of pain and health, which as he argues, has become an "obsession" of contemporary humanity. According to Han, health—understood as a physical state, expressed in the upkeep of high bodily efficiency—has become the *telos* of modern Western man. It has supplanted not only the concept of flourishing known to the ancients as one's entering into harmony with the universe, but also the grand Christian narrative of the salvation of the soul. This *telos* has turned the human into a non-metaphysical being, devoid of a vision broader than the limits of his or her own body, and, to make things worse, treating that body as the center of the universe. The contemporary obsession with health is narcissistic, Han asserts, and it is one of the pillars of what he calls the *palliative society*. In his view, we can refer to modern society as palliative, because its members, contemporary people, are mortally afraid of pain. On the one hand, this widespread fear, which the scholar terms algophobia, leads us to cutting ourselves off from the natural environment, which in itself may be the source of pain. On the other, it is conducive to shifting our focus solely to our own bodies, effectively anesthetized by drugs. The result? Permanent anesthesia.

DOI: 10.4324/9781003328469-12

Suppressing Pain

The fear of pain has a broader context. In our attempts to eliminate all traces of negativity from our surroundings we have, at least allegedly, become a society of positivity. Even psychology—in its extremely popular, modern form—has transmogrified into positive psychology. Today, we wish to engage only with what is positive, which attitude appears to be subordinated to the imperative of productivity. In the modern world, productivity is both the goal of our actions and the context that gives these actions a positive connotation. Han adds that "for the neoliberal ideology of resilience, traumatic experiences should be seen as catalysts that increase performance."[2]

Furthermore, as the scholar suggests, the contemporary psyche can be described as the "sterile self." Several aspects of this concept are important for understanding health. First, such a self is detached from the environment, which, beyond its role in affirming their uniqueness, has no value for modern individuals. Yet, such an affirmation requires constant exposure: we need to present ourselves in a way that would garner the approval of others, whose feedback would allow us to develop the impression that we are liked. Consequently, what we cannot expose—erodes in value. Hence, we lose interest in the depth of our inner selves; the shadows in the chiaroscuro of our personality that once constituted our identity fade in relevance.

At the same time, the contemporary Western self is tied to the body as an individual organism, whose efficiency—attested to, for instance, by the recognition of its attractiveness in the eyes of others—determines our value. Such a self, utterly removed from the dramaturgy of the life process, is entirely non-atmospheric. It no longer adheres to the laws of nature, *phusis*, which the Greeks defined, among other things, as cycles of birth, maturation, and eventual decay. The natural environment for the sterile self becomes the rhizomatic world of digital networks, whose main principle is not *phusis*, but the algorithmic score of "likes" granted by individuals whose attention is almost entirely absorbed by social media—the greater the number of likes, the greater one's value. In such a neoliberal society of positivity, we are expected to be both happy and pain-resistant. But, as Han observes, "[w]hat has been forgotten is that pain purifies. It has a cathartic effect. The culture of the likeable and the agreeable lacks any opportunities for catharsis."[3]

It is easy to see how the sterility of the self emerges as a spiritual problem. Not because it belongs to the realm of religion, but because it concerns flourishing, and thus falls within the field of interest of philosophy a spiritual exercise in the sense in which Pierre Hadot conceived it. Without encountering pain, spiritual relief, defined in various philosophical therapies (e.g.: those evolved in the traditions of Buddhism or Stoicism) as the development of the habit of disattachment, is impossible. If Han is right, nothing could be more foreign to the modern human than philosophy understood as transcendence of the self (as it is in the Buddhist context), or as the attunement to Allnature (as it is in the Stoic tradition). And, by the same virtue, nothing could be closer than positive psychology and various forms of corporate mindfulness, aimed at making us the ever-smiling, productive employees.

Contemporary Western medicine, which—in greatly simplified terms—assumes that the doctor removes the body's defects, and the patient does not even need to know how it happens, corresponds to the mentality described by Han. The sterility of the patient's self also means that it identifies with the general outline of its body, which must meet the requirements of aesthetics and efficiency, and which—using Han's term—is fully "transparent." The Western doctor operates on material "brought to light," in contrast to the Chinese healer described earlier in this book, who, through conversation, delves into the shadows, seeking to find what is hidden beneath the surface, what is invisible to the naked eye (including the eye of a medical device). Before prescribing means to restore the body's harmony—herbs, massages, acupuncture, further conversation, change of life attitude—he seeks equilibrium expressed by the famous *tai chi* symbol: a black tear fused with a white one, where both contain elements of the opposite color. Light and shadow form a three-dimensional whole, perhaps less sharply defined than the printout of test parameters obtained with modern medical devices, perhaps partially intuitive, but much more holistic. It must be remembered, however, that such a less sharp, but much deeper image of health does not characterize Chinese medicine alone:

> Viktor von Weizsäcker describes the primordial scene of healing as follows: "When a sister, still very young herself, sees her little brother in pain, she senses what to do before knowing anything: her hand finds its way; she wants to caress him where it hurts—thus, the little Samaritan becomes the first doctor. A pre-knowledge about a primal effect is unconsciously at work in her. That knowledge guides her urge towards her hand and leads the hand to perform the soothing caress. For that is what the brother will experience; the hand will soothe him. Between himself and the pain slips the sensation of being touched by the sisterly hand, and the pain retreats before this new sensation." We are currently moving further and further away from this primordial scene of healing. The experience of healing care in the form of the sensation of being touched and addressed is becoming increasingly rare.[4]

This, however, does not change the fact that health remains a transpersonal and complex phenomenon: it cannot be reduced solely to the hard parameters of Western medicine. Among its components is the sensation of another person's kindness, but also the overall atmosphere, in which the healing process takes place. Health, as a process of flourishing, that can be tended to and cultivated, depends on the atmosphere, and therefore the garden—with its complexity, multi-aspectuality, simultaneity of openness and closeness—seems to be an adequate model to describe this phenomenon.

Micro-Geography of Health

Han observes that Friedrich Nietzsche described his times using a set of concepts, into which the idea of a palliative society would fit perfectly well. Palliativity, in Nietzschean perspective, manifests itself in the withering of the sense of life. In the philosopher's view, life has lost its vitality: it has become "comfortable survival." The opposite of vitality—which assumes tapping into liveliness—is permanent anesthesia, including mental numbness. Such a state makes genuine experience (*Erfahrung*) impossible. Furthermore, as Han writes, our world consists of a series of lived events (*Erlebnis*). The difference between experience and a series of lived events is simple: only the former way of experiencing the world leads to change and triggers a transformation that may culminate in catharsis. The latter may amuse, but do not transform.

If Han's diagnosis is correct, modern man has lost the ability to heal. Transformation is necessary in the healing process and in maintaining harmony, or health. The enemy of health is elimination—or limitation—of metabolic openness, which physically manifests as cancers that feed on stagnation. Many of the cancer cells thrive in the oxygen-free environment, and thus, one might pose, atmosphere, quite literally, is a key component of what we call health. Health arises from what Han, following Heidegger, calls a "pre-reflexively disclosed world," a world atmospheric in its nature.

> Heidegger points out that moods [*Stimmungen*] possess a world-disclosing power. The world as pre-reflexively disclosed by a mood precedes intentionality, the aiming at an object: "The mood has already disclosed, in every case, Being-in-the-world as a whole, and makes it possible first of all to direct oneself towards something."[5]

The will to heal does not have to assume the form of a verbal (or mental) utterance, consciously produced by a person, but it still defines one's direction. The intention to heal, or perhaps more precisely, the intention to flourish (or to wither), can be non-subjective and atmospheric; it may manifest itself as an *impersonal* intention of the natural environment of which other people are part, and which we perceive via moods. Yet, we can sense these moods only when we learn to wait. In contrast to "waiting for," Han conceptualizes this process as "waiting in," thus vesting it with relaxed attentiveness, or with what the Stoics called cheerful acceptance.

Such a form of waiting occurs in nature, as evidenced by the wildest forest of the European lowlands, the Białowieża Forest, located in northeastern Poland. Wandering through its most inaccessible and biodiverse corners, we do not hear the cacophony of bird calls, which we so often encounter in city parks. The forest is quiet. Peaceful. It is as if the actors in these places, which people seldom reach, did not need to put on a spectacle. Everything in the forest seems to conserve energy. The hazel grouse will take flight from the forest floor only when we inadvertently approach it within a couple of feet. Herds of deer will start fleeing only when we walk up really close, and when a bison suddenly crosses our path it feels like an avalanche coming down the mountain; the suddenness of this event is astounding, and the noise seems immense because the massive animal emerges from almost perfect silence. Such a forest is bio-diverse, offering thousands of subtle variations of stimuli—tactile and olfactory—that, in the environment of a contemporary Western city, are limited in scope and number. The forest is replete with energy, pulsing with readiness, and full of potential. It is alive—but whatever transpires within it takes proper time. The world of the forest stands in sharp contrast to the instantaneous, yet lifeless world of the Internet, where things happen quickly, but where no life energy, dissolved in the ever-shifting constellations of pixels and electronic sounds, is accumulated. The variety characterizing digital worlds may be described as the variability of lifelessness—variability, which, due to instantaneity, is devoid of rhythm.

Describing health as an atmospheric phenomenon rather than a property of a single organism changes the definition of the presumptive subject of that health—the human being. We can define such a subject as an open system and by that virtue we could describe it, for instance, by analogy to a reptile, whose internal states depend on the environment to a much greater extent than seems to be the case with the humans. In the spring, reptiles seek warm spots because the increase of their blood temperature depends on proper exposure to the sun, which, in turn, determines whether—and when—they will break out of winter lethargy and replenish their bodies with vitality. Today, when doctors increasingly often prescribe long forest walks instead of pills, we should perhaps adopt the reptilian strategy more often, seeking out places where our well-being flourishes.

This is not a new idea at all—and we are not even alluding to spas, which, owing to their healing microclimates, have attracted patients for centuries. An interesting example of such a "human-reptile" was Friedrich Nietzsche, who spent most of his adult life moving from place to place in search of balance. Usually, he would spent winters in Italy or France, and summers in Switzerland. But it was not only the benign temperatures that the philosopher sought: it was also about other environmental factors affecting his psychophysical state, some of which remain quite elusive—such as the unique light from above the Alps that he observed in Turin.

> He seems to have been a man without skin, someone who is supersensitive: always in danger of surrendering to external influences but, simultaneously, someone who is able to perceive forces not seen by other people, such as a "certain energy in the air"[6]. Nietzsche noted that down in the Italian city of Turin, where he was writing *Thus Spoke Zarathustra,* and divided his day into three activities: writing, eating, and walking in the mountains. "He told the Danish critic Georg Brandes that he had conceived the whole Zarathustra Book I in ten days out walking in the mountains"[7]

Nietzsche sought not only countries or cities where living would be comfortable, but also *special* spots—and he would often find them. One such spot was a small peninsula on the lake in Sils Maria, where he even considered building a house, though the plan never came to fruition. Other such special *loci* were, for example, a site by a solitary pine tree in Sorrento, or the area around a seaside rock near Genoa and its immediate vicinity—places to which the philosopher would frequently return while in Italy. It is in such locations that he would discover, or perhaps co-create, his own "gardens"—spaces that kept him alive and, even if physically he did not flourish to his utmost, these gardens allowed his philosophy to blossom fully.

> Speaking of Aurora, he declares that "almost every phrase in the book was meditated upon and slipped out in that tangle of rocks near Genoa, where I was alone and intimate with the sea." Thus, if Aurora, as the title itself wants us to understand, is a writing that is meant to be illuminating, this will also be due to the clear sky and sea air of which Nietzsche, an indefatigable walker, used to stock up on his daily strolls through the city's streets and along its seafront.[8]

Nietzsche's habits could illustrate Rosario Assunto's thesis, who claims that being in a real garden provides us with the opportunity for a metaphysical experience. For Nietzsche, every place he reached had to almost immediately become a life-protecting garden, one that would nurture his vitality just as the sun nurtures the vitality of a reptile. Whenever this failed to happen, he would quickly flee. Over the years, he learned to avoid places that would not foster his well-being, hence his continuous wandering between Switzerland, France, and Italy. For Nietzsche, vitality is a fundamental metaphysical category, which he often identifies with life itself. The German philosopher sees vitality as the sole and ultimate spiritual fulfillment, thus forerunning contemporary empirical research, such as the study carried out by John Trigwell, Alexander J. Francis, and Kathryn L. Bagot, who observe that

> [there is a] direct empirical support for a relationship between nature connectedness and spirituality, making a significant contribution to the environmental literature. Further, it has extended previous literature examining hedonic well-being by revealing spirituality as a mediator of the nature connectedness—eudaimonic well-being relationship across various well-being dimensions. Together, these findings provide empirical support for ecological-self theory as a useful framework for understanding the relationship between humans and natural environments, providing a foundation for a range of exciting directions in future research.[9]

One conclusion from such research is obvious—the sterile self does not *exist*. And if it did exist, it would be devoid of all intelligence.

Paesaggio Factor

In the 19th century, Friedrich Nietzsche and Ralph Waldo Emerson journeyed from their respective northern countries—Germany and Massachusetts, USA—to Mediterranean Italy, a region crucial in shaping the Western identity. One could expect that their travels through the historic areas, foundational to Western thought, would reinforce Platonic and Christian themes in their writings. Paradoxically, the effect was quite the opposite: they moved away from the Platonic concept of an independent soul toward an acknowledgment of human beings as defined by their vitality and integration within the natural, biological order. The Italian language, landscapes, and people they encountered unexpectedly influenced their metaphysical sensitivities, modifying their perceptions and provoking a transformation in their philosophical perspectives.[10] This shift led them to reconsider previous ideas, and abandon others, embrace new insights, and refine—or elevate—views, to which they would hold on since.

The respective experiences of Emerson and Nietzsche exemplify how the elements of nature can reveal themselves in thought through a form of osmosis—via the atmospheric absorption of the Italian landscape's richness, characterized by its ancient landmarks, the southern light (as Nietzsche described it), and the evocative scents of pine trees and Mediterranean soil. Centuries earlier, the same impressions had shaped the metaphysical sensitivities of Marcus Aurelius and St. Benedict. However, for such an osmosis to be truly effective, it requires active philosophical engagement. An awareness of humanity's co-existence with the rest of nature must be developed, which, in turn, demands adopting a critical attitude toward what Nietzsche referred to as the "herd prejudices." Such criticism requires the suspension, or even elimination, of the most fundamental prejudice—that the self is the center of the universe. While this was challenging in Nietzsche's era, it may be even more difficult today, especially if Han's diagnosis that modern human attention is compulsively focused on the body—is correct. This habit of self-centered focus hinders the development of atmospheric perception, which Schmitz defined as "unbound occupation of surfaceless space in the region of what is experienced as present."[11]

It seems that alienation from nature is deeply embedded in the structures of the Western psyche, more so than in the cultures of the Far East or non-Western Indigenous traditions. This process may have begun with Plato and his concept of the soul, which exists independently from matter and becomes corrupted through its association with the material world. This motif was adopted by Christianity and later cemented by Descartes, who divided the universe into mental and material substances, a duality further reinforced by Enlightenment philosophy, fostering such notions as, for example, Kant's transcendent self. Today, our neoliberal ego, an inheritance from this tradition, is often blamed for the exploitation of the planet and the environmental crises characterizing the Anthropocene.

Atmosphericity, including the atmosphericity of health, possesses not only spatial characteristics but also temporal ones. Although the insubstantiality of the atmosphere—including the insubstantiality of the very center of experience, which, in Western metaphysics, has been traditionally defined as "psyche" or "soul"—might seem threatening, it also presents an opportunity:

> The self is not only psychosomatic but also processual. It weighs, as the increased sensation of heaviness while walking up in Nietzsche's description shows. Thus, the self changes not only throughout the course of life but also in the act of moving our body and acting with it against the force of gravitation. It is not a weightless psychic bubble that walks through the landscape but a force (as in "physical force") generated from the act of walking. It changes and assumes a new form due to what can be named—as we are still near Turin—a *paesaggio* factor: the self-constituting force of the environment. It seems that the illusion of self-separateness is felt only by people with thick skin and abandoned by those more sensitive, especially by Nietzsche himself, the man without skin. The landscape around us does not modify our thoughts; it is present in what seems to exist "within us" as our thoughts. We are open systems, and the felt degree of our separateness is determined by our sensitivity of attention, not by frontiers existing objectively in the world. Thoughts do not originate from the soul of the walker but are established, modified, and sublimated as a result of a play of sourceless forces.[12]

It seems that the habit of wandering is what allows one to more easily grasp the self-constituting force of the environment, irrespective of the here-and-now cultural conditions, or historical circumstances. The discoveries of Ajahn Lee Dhammādharo, the legendary Thai meditation master who spent years wandering through the jungles of Southeast Asia, seem to confirm it:

> Some kind of trees make themselves quiet in ways we can see: We say that they "sleep." At night, they fold up their leaves. If you go lie under them, you'll have a clear view of the stars in the nighttime sky. But when day comes, they'll spread out their leaves and give a dense shade. This is a good lesson for the mind: When you sit in meditation, close only your eyes. Keep your mind bright and alert, like a tree that closes its leaves and thus doesn't obstruct our view of the stars.[13]

What if contemporary Western medicine—which is based on a model where patients visit doctors in sterile offices and receive treatments (often pharmacological) to remedy certain "defects" in the body—took this lesson? What if it were to be inspired by the plethora of other discoveries made by the masters of the Thai Forest Tradition, or drew constructive conclusions from Nietzsche's unique "therapy"? What if it factored in the investigation of the impact of the environment on health in the light of an atmospheric model? Finally, what if it included metaphysical interventions in its repertoire? Let us consider such a scenario.

Caring about Well-being: A Transgressive-Metaphysics-Based Procedure

An interesting example of a practice that impacts the practitioner's metaphysical stance is the New Presence practice, recently initiated in Poland. The New Presence Paths project aims to design and chart naturalistic and philosophical trails that could optimally cater to human health (and, by that virtue, his or her well-being), as well as to enhance ecological sensitivity based on recent research on the human-nature relationship. The trails are not physically marked, although they may follow existing paths. On these paths, there are designated spots where practitioners, following provided instructions, can perform exercises to enhance the sensitivity of their perception, as well as sites conducive to philosophical reflection on human co-existence with nature. The paths are designed to offer enriching, mindset-changing experiences to those who walk them. Additionally, by encouraging a deeper appreciation and understanding of the natural environment, they contribute to wilderness protection.

Being in nature enhances what contemporary scientific literature refers to as "eudaimonic well-being," which is described as a lasting sense of meaning, purpose, personal growth, and moral enrichment. This concept aligns with what was historically termed "human flourishing," a phrase, to which we purposefully return in this book. When designing the trails, insights from ancient sages complement latest scientific findings, such, for instance, as those arrived at by Andrew J. Howell, Holli-Anne Passmore, and Kimberly Buro, whose study on meaning in nature demonstrates that nature, spirituality, and eudaimonic well-being are strongly correlated.[14] Examining this relationship more closely, the three authors discovered that the sense of life mediates the relationship between being-in-nature and well-being. The reduction of everyday life stress, along with the fascination and awe that can arise in wild environments, can lead to "peak" or "transcendent" experiences, characterized by "a sense of unity with the universe, absorption in the present moment, and a sense of timelessness [...]."[15]

The more biodiverse the trail we choose is, the greater the chance of attaining the intended effect.[16] Time matters, too: the longer the contact with nature, the better the results; a walk on the trail should last at least 30 minutes. Furthermore, eudaimonic well-being is most intensely enhanced by the sublime—by "large natural objects, such as mountains or vistas, by natural events such as storms, or by objects with infinite repetition such as fractals, waves and patterns in nature."[17] Such elements direct our attention to what is much larger than ourselves, which is why the designers of the New Presence Paths always take them into account.

The New Presence Paths are easy to navigate, offering views that enhance transcendent experiences and opportunities for deep philosophical reflection connected to specific locations. They aim to activate all five senses, which—research shows—radically improves our health. The project assumes that after walking such a path, cortisol levels in our bodies decrease, and blood pressure stabilizes. At the same time, the sense of life's meaning and ecological sensitivity increase. We return home healthier and more empathetic toward other people and other sentient, or feeling, beings. The excerpts from the description of one such path (located near Wrocław), including instructions for those who might wish to traverse it "philosophically," illustrate the procedure:

> Behind you is the city. You may even hear its hum from here. The world of concrete, asphalt, noise, and haste. Probably also other, more pleasant things. That's where your story unfolds. Think that no matter where you came from, that story is now behind you. Joys and sorrows, problems and struggles, all are now behind you. And think that you leave them there. You don't take them with you on this walk. They won't disappear, even those you would sincerely wish to. They will wait for you. Take a determined step forward. Let your sole strike the path firmly. Together with this step, you leave your old self behind. Ahead of you is nature—wilder and wilder. At the crossroads of the paths, turn right, following the green trail.
>
> [...]
>
> After about two hundred steps, where the green trail turns left, on your right side you will see an extraordinary phenomenon—the uncountable tree, and you will be able to perform the third exercise related to it. The uncountable tree stands right by a metal fence where the park canal ends. It consists of three trunks, perhaps joined together underground, or perhaps not. With a relaxed body, a free mind, and a will filled with cheerful acceptance, without expecting an answer, ask yourself a question and let it resonate in your mind: How many trees do I have before me: one, two, or maybe three? Or maybe I just don't know. Maybe I don't need to know, because the world I'm delving into, the world of nature, is not entirely countable. It's not a warehouse with a precisely determined number of objects that can be counted and cataloged, but a wild transformation. And numbers are not a feature of the world but our projection, which the mind imposes on this world. The philosopher Simone de Beauvoir wrote about a certain experience that frees one from seeing the world as a warehouse of things, "an abundant disorientation in which true life begins." Perhaps right now, in you, this state experienced by philosophers—the state of abundant disorientation—appears.
>
> [...]

> Just across the stream, there is a fork in the paths—go left until you see an old willow with a hollow trunk on the right side by the pond. It is right by the water. Reach it and enter the hollow, if you wish. Here you will perform the fifth philosophical exercise: the contemplation of space. Ask yourself the question: what is *the interior* and what is *the exterior*? Do one and the other exist objectively, or are they just roughly outlined by some shape (in this case, a tree)—essentially, only an approximation? Is the interior/exterior distinction a feature of reality, or just a projection of our mind?
>
> [...]
>
> When you move forward again, let your gaze wander among the trees and bushes. Do not rush. On the left side, among other trees, you will see a fallen birch. Approach it closer to perform the sixth philosophical exercise: the contemplation of old age. Look closely at the fallen trunk. It seems dead, yet fungi grow out of it. The trunk is covered with moss, green and fresh, as if it were only just beginning its adventure with life. So is the phenomenon before you. Are the fallen birch and all that grows on it and out of it old or young? Alive or dead? Look closely. And allow for the possibility that maybe youth and old age are not opposites, and that death is not the negation of life. Before you is the dance of life, momentarily frozen, like in a snapshot. Was it standing over such a tree that the Emperor and Stoic Marcus Aurelius noticed that "Death is one of life's actions"? Do you yourself fall under the truth of this record?[18]

The training that takes place on the New Presence Paths involves, on the one hand, developing mindfulness through honing one's awareness of sensory stimuli, and, on the other hand, elements of philosophical reflection, including those derived from the Stoic school of thought. For instance, the path in Krokowa near Gdańsk, Poland, has been designed around the philosophical discoveries of Johann Gottlieb Fichte, who spent several years as a family tutor in the nearby castle at the end of the eighteenth century. Walking the path, the practitioners reflect on quotes selected especially to encourage the adoption of metaphysics alternative to that of Plato's:

> "Where does this tree end and begin? My mind can determine its edges, give it identity. But when I get closer—the fungi, fresh twigs growing from the tree, and the rot break this illusion. At what point does the tree end and where does it begin? When can it be said that it is already a pile of punk and has ceased to be a tree? At what second of the existence of this phenomenon? Can I observe my inner self as impartially as I now look at the decaying tree trunk?"
>
> ***Fichte:***
>
> "I am surrounded by objects which I feel compelled to recognize as wholes existing for themselves and mutually distinct from each other.

> I attribute to each element properties and features by which I distinguish it from others; this plant has such a form, that one another; this tree has such, and another differently shaped leaves."
>
> [...]
>
> "Is the moss that has spread over the tree trunk part of it or something external? You can sit on the moss, on the high escarpment (to your left) overlooking the bottom of the ravine.
>
> A fairly steep path leads to it. It is a good place to observe the chiaroscuro. Our eyes habitually focus on what is bright. What will happen if we consciously direct them for a few minutes to what is in the shadow?"
>
> [...]
>
> Or maybe you can't sit down because of the cold, dampness, or insects? This is also a lesson about the life process. In nature, there is not much room for stillness. Maybe you are already tired and long to sit on a bench. You won't find one in the Sowi Jar. This allows you to ask yourself the question: What is it that wants to sit down—the mind or the body? Maybe you think there should be at least one bench here. But why change the order already created by nature?
>
> [...]
>
> ***Fichte:***
>
> "Nature is bound by the whole in every moment of its duration; at every moment, each particular part of it must be as it is because all others are as they are; and you could not move a single grain of sand from place to place without thereby changing something in all parts of the immeasurable whole, though it may be imperceptible to your eye."
>
> [...]
>
> "Cold. Does it pour into me immediately or slowly? In jumps or evenly? Is it an external force that attacks me or is it what connects me to the environment?"
>
> ***Fichte:***
>
> "I myself, together with everything I call my own, am a link in this chain of raw natural necessity." Am I the object or the subject of this continuous change? Is it only by describing in my thoughts what is happening around me that I exclude myself from the process of nature? Ask yourself these questions, but do not rush to answer.[19]

The effectiveness of pathways designed to enhance eudaimonic well-being still needs to be thoroughly studied, but the findings that have already been arrived at confirm that such a procedure facilitates interventions based on the atmospheric health model. Personal development is achieved by means of continuously relating the ongoing thought process (and consequently, the allocation of attention) to the world pulsating on both sides of one's skin, as well as by shifting attention between the worlds that, according to the dominant Cartesian view, allegedly belong to two separate realms: the mental and the perceptual. The account quoted below illustrates such a process, simultaneously supporting the thesis that the environment influences our emotions, thoughts, and perceptions.

> Apennine path, along the Simbrivio river, my students and I come across the body of a wild boar torn to pieces by a pack of wolves. The wolves ate the flesh, but left intact the boar's blueish-purple stomach with dangling intestines. What a picture! Here was the process of life revealing itself to us, a perfect classroom setting for my eager philosophers. Now we could apply philosophical fascination, except that our first reaction was to close off our senses. We held our noses and turned our heads away. This revulsed "don't want to see" response explains why we love to pay attention to the more controlled pixel and narrative worlds. There are no putrid smells there, nor views of evisceration.
>
> On our way back home, after having meditated on the rocks over the stream, we looked again at the remains of the wild boar, this time with curiosity, although from a safe distance. Between the two encounters, I asked my students to attune their senses to the rhythms of our environment: water splashing in a small waterfall, waves spreading at the base, twigs twitching in the wind, various smells, the sun coming out from behind a cloud, sudden warmth on the back of the neck. We did not start any eco-friendly conversations: we listened to the Earth's conversation. The activation of our senses changed our attitude towards the carcass of the boar, and perhaps something inside us as well.[20]

Evidently, the human self is not a sterile bubble that we transport from one landscape to another. Certain premises indicate that the self is a *process*, pulsating rhythmically at a frequency of five to eight times per second.[21] It is possible that while meditating by the Simbrivio stream in accordance with the instructions for the Buddhist contemplation of the elements—fire, water, air, and earth—we experienced a fragmentation of the sense of self, leading to its partial deconstruction and subsequent reconstruction. The self deconstructs and reconstructs several times per second, but we rarely have influence over the effects of this phenomenon. As we have observed earlier in this book, elemental exchange between a single human organism and the wild environment is always at play, and a mind trained in meditation can sense it clearly. This pulsation fits perfectly within the atmospheric model of health, where fire, water, air, and earth interact not only with one another, but also with the two elements from Empedocles's set: love and strife. This also coincides with what Buddhist meditation masters, like Ajahn Lee Dhammādharo, seem to believe: hatred or love, let us recall, "splash into the body, where they mix with the various properties of the body—in the blood, for instance, which then flows to the various parts of the body, causing weakness and fatigue."[22] Similarly, Hans Georg Gadamer believes that we do not *produce* health, because health is a natural condition, from which follows that treatment is tantamount to the restoration of the undisturbed natural processes.

Toward an Atmospheric Health Model

Atmosphericity presupposes the interpenetration of substances, which, in Cartesian dualist division, exist in different states of matter and seem to have fixed boundaries. Life is a theater of interactions about which we know little, as our science excels at capturing a static image of the world but struggles with its dynamics. The super-metabolism known as life may never be fully describable in the language of words or numbers, yet, from the perspective of the cultivation of health, this is not at all necessary—just as it is not indispensable to attain the complete knowledge of all the processes occurring in a garden to be able to cultivate it.

Books by practitioners—such as *The Intuitive Gardener: Finding Creative Freedom in the Garden*[23] written by Marilyn Raff, a psychologist representing the Jungian tradition—are often manifestos proclaiming the impossibility of efficient gardening based on dry information. In the practice of modern Western medicine, which assumes a division of roles between the expert doctor and the passive patient, decisions regarding treatment methods are increasingly often made automatically: such decisions are, in fact, conclusions from data accumulated by devices. Even as we write this book, research is underway to develop sophisticated shirts equipped with sensors that would monitor various health parameters. Such shirts, designed to be worn daily, like regular clothes, are supposed to send information on patients' vitals to hospital data banks in real time. In *Science Daily*, we read that MIT scientists have just made a breakthrough in this field.

> The electronic sensors consist of long, flexible strips that are encased in epoxy and then woven into narrow channels in the fabric. These channels have small openings that allow the sensors to be exposed to the skin. For this study, the researchers designed a prototype shirt with 30 temperature sensors and an accelerometer that can measure the wearer's movement, heart rate, and breathing rate. The garment can then transmit this data wirelessly to a smartphone.[24]

At this point, a disclaimer needs to be made: we are not postmodern Luddites. The product described in this article represents a notable advancement in medical convenience. However, even if its application becomes widespread, its use may present certain drawbacks. When technology is uncritically entrusted with the ultimate authority in determining the state of the patient's health, it enhances the risk of diminishing the personal responsibility for the outcomes of the critical judgment, traditionally exercised by physicians. The elimination of profound, compassionate relation between doctor-healer and patient-multidimensional being, may lead to further decline in our sensitivity to our bodies and to further loss of the "gardener's intuition" in our cultivation of health. This, in turn, may render us even more passive in terms of our participation in the healing process, dulling our perception of bodily signals and weakening our sense of what is and what is not healthy for us. After all, every form of critical engagement, including such that sometimes allow patients to be partners in dialogue with their physicians, is fueled by access to information.

If we accept—as the ancient philosophers did, and as we do in this book—that flourishing is driven by nature itself, and that, individually, it is conditioned by our ability to attune to it, then the MIT smart shirts, paradoxically, may prove to be a hindrance to our flourishing. While this new invention can undoubtedly save lives and be an excellent solution in critical situations, if it is not implemented hand in hand with the adoption of a holistic attitude to healing, it may further disrupt the synchrony, which the atmospheric health model assumes as indispensable. In this model, human consciousness, which contracts and expands, encounters a similar pattern in the natural environment, resulting in healthy resonance. The physicians' sole reliance on the data provided by such smart shirts may lead to our further alienation from the "primitive present," which, let us recall, is the "present" that contains a rich life "full of situations, from the meaningfulness of which no individual meanings can be isolated."[25] The patient's description of his or her own feelings, already less and less important in medical diagnostics today, may end up being omitted altogether.

As we remember, according to Schmitz, the atmospheric "primitive present" is an incompletely conscious core of our identity. Atmospheres affect what the philosopher calls "self-ascription." The primitive present is thus a stabilizing factor in the fluid world, in which modern humans find themselves—a world deprived of metaphysical structures that once determined its order. The primitive present also bestows depth upon our identity, protecting it against the "sterility" of self. Seen from this perspective, health may be construed as tantamount to self-ascription by attunement to natural rhythms. Yet, as we have declared in the introduction to this book, none of us would forgo the life-saving potential of modern devices in hospitals and ambulances. Therefore, the health model we describe in the final chapter acknowledges the ambiguity of the existential situation in which the Western medical paradigm, along with the development of medical technologies, has placed us. Emerging from Platonic/Descartian metaphysics, this paradigm posits an insurmountable boundary between the spiritual and the corporeal. In search of an alternative, in the following sections we outline a different model: one that seeks to blur these boundaries. Fusing both realms, the garden collapses the Descartian binaries, thereby embodying a metaphysics that we subsequently define as "transgressive."

Notes

1 Byung-Chul Han, *The Palliative Society: Pain Today*, trans. Daniel Steuer (Medford: Polity Press, 2021).
2 Byung-Chul Han, *The Palliative Society: Pain Today*..., 1.
3 Byung-Chul Han, *The Palliative Society: Pain Today*..., 1.
4 Viktor von Weizsäcker, "Die Schmerzen," in *Der Arzt und der Kranke: Stücke einer medizinischen Anthropologie*, Gesammelte Schriften, vol. 5 (Frankfurt am Main: Suhrkamp, 1987), 27, quoted in Byung-Chul Han, *The Palliative Society: Pain Today*..., 26.

5 Byung-Chul Han, *The Palliative Society: Pain Today…*, 43. It is helpful to notice, that the etymology of the word "Stimmungen" is telling. The prefix "Stimm-" relates to "voice" or "sound" (as in "Stimme," meaning "voice"), emphasizing a sense of resonance or attunement. "Stimmung" is thus derived from the verb "stimmen," which means "to tune" or "to harmonize." In its popular use, the word reflects the idea of bringing something into a state or condition, where it "fits" everything else, much like tuning a musical instrument to achieve harmony within the orchestra. The "pre-reflexively disclosed world," as one can conclude, is thus disclosed by way of attunement, and the non-verbal, atmospheric, pre-reflexive knowledge, gives us a possibility to "direct ourselves towards something." Intuitively, as we think, the primary "something" to which we spontaneously direct ourselves, is health.

6 Lesley Chamberlain, *Nietzsche in Turin: The End of the Future* (London: Pushkin Press, 1997), 38.

7 Lesley Chamberlain, *Nietzsche in Turin: The End of the Future…*, 40; Marcin Fabjański, "Walking-Derived Metaphysics in Nietzsche's *Thus Spoke Zarathustra*," *Forum Philosophicum* 29, no. 1 (2024): 29–41, https://doi.org/10.35765/forphil.2024.2901.02.

8 Giovanni Pulina, *Guida filosofica dell'Italia* (Bologna: Diogene Multimedia, 2018).

9 John Trigwell, Alexander J. Francis, and Kathryn L. Bagot, "Nature Connectedness and Eudaimonic Well-Being: Spirituality as a Potential Mediator," *Ecopsychology* 6 (2014): 241, https://doi.org/10.1089/eco.2014.0025.

10 See, for instance, Massimo Venturi Ferriolo, *Oltre il giardino: Filosofia di Paesaggio* (Torino: Giulio Einaudi Editore, 2019).

11 Hermann Schmitz, *New Phenomenology: A Brief Introduction…*, 139.

12 Marcin Fabjański, "Walking-Derived Metaphysics in Nietzsche's *Thus Spoke Zarathustra*," *Forum Philosophicum* 29, no. 1 (2024): 29–41, https://doi.org/10.35765/forphil.2024.2901.02.

13 Lee Dhammādharo, *The Words of Ajaan Lee*, Kindle edition (publisher not provided, 2011), loc. 6950.

14 Andrew J. Howell, Holli-Anne Passmore, and Kimberly Buro, "Meaning in Nature: Meaning in Life as a Mediator of the Relationship Between Nature Connectedness and Wellbeing," *Journal of Happiness Studies* 14 (2013): 1681–1696, https://doi.org/10.1007/s10902-012-9403-x.

15 Gwyneth Hamman and Itai Ivtzan, "30 Minutes in Nature a Day Can Increase Mood, Well-Being, Meaning in Life and Mindfulness: Effects of a Pilot Programme," *Social Inquiry into Well-Being* 2 (2016): 34–46, https://doi.org/10.13165/SIIW-16-2-2-04.

16 Emma Wood, "Not All Green Space Is Created Equal: Biodiversity Predicts Psychological Restorative Benefits from Urban Green Space," in *One Health: The Well-being Impacts of Human–Nature Relationships*, ed. Elizabeth Brymer et al. (Lausanne: Frontiers Media, 2017), https://doi.org/10.3389/978-2-88945-384-3.

17 Annabel Pritchard, Miles Richardson, David Sheffield, and Kelly McEwan, "The Relationship Between Nature Connectedness and Eudaimonic Well-Being: A Meta-analysis," *Journal of Happiness Studies* 21 (2020): 1145–1167, https://doi.org/10.1007/s10902-019-00118-6.

18 Marcin Fabjański, "Ścieżka w Lesie Mokrzańskim," *Ścieżki Nowej Obecności*, Fundacja Nowa Obecność, http://sciezki.org/sciezka-w-lesie-mokrzanskim/ (accessed: May 27, 2024). Translated by Marcin Fabjański.

19 Marcin Fabjański, "Ścieżka w Sowim Jarze," *Ścieżki Nowej Obecności*, Fundacja Nowa Obecność, http://sciezki.org/sciezka-w-sowim-jarze (accessed: May 27, 2024). Translated by Marcin Fabjański.

20 Marcin Fabjański, *Embodied Nature and Health: How to Attune to the Open-source Intelligence…*, 5.

21 Rick Hanson and Richard Mendius, *Buddha's Brain: The Practical Neuroscience of Happiness, Love & Wisdom* (Oakland, CA: New Harbinger Publications, 2009).
22 Lee Dhammādharo, *The Words of Ajaan Lee…*, loc. 9011.
23 Marilyn Raff, *The Intuitive Gardener: Finding Creative Freedom in the Garden* (Golden, CO: Fulcrum Publishing, 2002).
24 Massachusetts Institute of Technology, "Sensors Woven into a Shirt Can Monitor Vital Signs: Comfortable, Form-Fitting Garments Could Be Used to Remotely Track Patients' Health," April 23, 2020, story by Anne Trafton, https://news.mit.edu/2020/sensors-shirt-monitor-vital-signs-0423 (accessed: May 22, 2024).
25 Hermann Schmitz, *New Phenomenology. A Brief Introduction…*, 75.

9 Toward the Metaphysics of the Garden

Our previous analyses, substantiated by our autoethnographic insights, have guided us toward a new model of health, founded on the garden-inspired metaphysics. In this section, we synthesize our findings to outline a general model of the garden as a revised conceptual system expressing a new understanding of health. Health thus understood is a concept that transcends the egological notion of the human being as radically separate from the world; instead, health is conceived of a communal good, "common-wealth," which cannot be confined to any single, individual organism. This idea is not rooted in haphazard, isolated sets of beliefs. On the contrary, it is grounded in a holistic view, expressed as the metaphysics of the garden, aiming to systematically articulate the theoretical foundations of health understood as an atmospheric phenomenon. The following description of these rudimentary structures is not merely *inspired* by the garden, but, fundamentally, constitutes an outline of the proposed metaphysics. Aware that the detailed presentation of the vast ramifications of its potential would require another book, or even book series, in the sections below we will focus on its key concepts and principles, hoping to inspire and encourage other scholars, scientists, and medical professionals to follow suit.

As previously indicated, profound understanding of health demands that we adopt a metaphysical stance where "every atom belonging to me as good belongs to you." Encapsulated in Walt Whitman's famous phrase, such garden-inspired metaphysics triggers a revision of ethics, which logically transmogrifies into "atmospheric ethics." Consequently, the communion of humanity and the universe inevitably fosters a profound sense of union with others, which is the *sine qua non* condition of common/wealth. The health of the common/wealth requires attention to the rhythms of animate and inanimate nature and mindfulness in actions concerning the community; neglect in these respects will inevitably impact one's own well-being.

What Is Metaphysics?

Yet, before we delve deeper into the essentials of our model, we must begin by clarifying our understanding of the term "metaphysics," which, when juxtaposed with terms such as "garden" and "health," may seem controversial

DOI: 10.4324/9781003328469-13

or incomprehensible—if not altogether provocative. That, of course, is not our intention. As is well known, "metaphysics" has a long history and has been understood in a variety of ways. Historically, the term is associated with one of Aristotle's principal works, whose title—*Metaphysics*—was (probably) ascribed to it by Andronicus of Rhodes. Unable to categorize this particular set of Aristotle's texts into any thematic group, Andronicus labeled them as "those that follow physics" (τα μετα τα φυσικά—"those after physics"). This designation is not merely classificatory: Alexander of Aphrodisias argues it is substantively understandable, because the order of things variable, undergoing motion—the object of the insights of physics—precedes the order of the causes and principles (ἀρχαί), which are the objects of metaphysical reflection. Thus, "metaphysics" follows "physics," according to an order relative to us (πρὸς ἡμᾶς). The distinction lies between things that are *prior by nature* (absolute sense) and those that *are prior to us* (cognitive perspective). Thus, one should start with what is *prior to us*, namely *physics* (sensory experience), and proceed to *metaphysics*. Importantly, neo-Platonic thinkers, followed by many others throughout the history of philosophy, interpreted the prefix "μετα" not as "after" but as "beyond" ("ὑπέρ"), at the same time absolutistically monopolizing that which is *prior by nature*. Consequently, metaphysics became a reflection on what is *beyond the physical* and *primary by nature*. Importantly, Aristotle indicates that the reflection later termed metaphysical encompasses the fundamental principles of holistically understood reality.

It is noteworthy that the debate on the subject of metaphysics, highlighting its dual interest in the totality of reality and the suprasensory dimension of it, is already present in the writings of the thinkers of the antiquity. Some, indeed, merge these two understandings into one, but others do not. Modern and contemporary definitions of metaphysics reveal a variety of its potential senses. For instance, in his essay "What Is Metaphysics?" Martin Heidegger explores the content of borderline experiences (anxiety, boredom). Henri Bergson contrasts analysis and intuition; the former presents the object from the outside, while intuition, proper to metaphysics, penetrates the structure of the object. Ludwig Wittgenstein's terminal view posits that metaphysics is inexpressible as silence.

Aristotle referred to metaphysical considerations as first philosophy (πρώτη φιλοσοφία). Commentators on his legacy attempted to define the status of this field of reflection, seeking, above all, to determine in relation to *what* metaphysics is "first," and on what grounds this primacy could be established. However, without delving deep into this centuries-old debate, and leaving the endless, and ongoing, disputes concerning the grounds of the historical distinction between metaphysics and ontology aside, it is possible to state that the "first philosophy" concerns what is *fundamental*, in some respect *primary*, and *prior in relation* to something else. Let us then attempt to formulate the most general, core meaning of the term "metaphysics" as we use it here, avoiding adherence to any specific, historically concretized, stances. Metaphysics as we propose to understand it addresses what is most general, thereby providing the most comprehensive view of the world. As "first philosophy," it seeks to grasp what is most fundamental and primary. Attempts to identify these fundamentals may lead to defining that which is ultimate, or most essential, in our thinking about the world. Metaphysical reflection should also be seen as foundational for other types of reflection on particular aspects of reality.

Transgressive Metaphysics

In Chapter 1, we formulated our premises and presuppositions, which, now, in the light of the autoethnographic material analyzed in subsequent sections and in the face of ensuing philosophical reflection, need to be revised and reorganized in order for us to efficiently outline the essential structure of our proposed metaphysics of the garden. Sketching out its key categories and essential principles, we see our efforts as groundwork, or perhaps propedeutics, upon which more profound, and more multifaceted insights concerning being-in-the-world as a source of *eudaimonia* could be developed further. Needless to say, such an outline is also necessary to ensure that the assumptions and premises underlying the stance toward the world that we propose are not justified solely by our own, fervent "wishful thinking," or our nature-loving readers' enthusiastic acceptance of what we present as fundamental. This is why, having explained our understanding of "metaphysics," we need to shift our focus to "transgression," the explication of which brings us back to the garden, whose "commonplace" definition we now revise on the grounds of our findings.

Let us recall the etymology of the "garden," to which we referred at the onset of our reflections: the noun "garden" derives from the Proto-Indo-European root *gher-, meaning "to grasp, enclose." Traditionally, the garden is thus conceived of as an enclosed, or fenced-off, space—but, as the overview of the painterly, architectural, and poetic representations of the garden indicate, the "boundary," the very etymological determinant of the concept. Has always been problematic. It is not only that the "boundary," which only seemingly "solidifies" in the early modern era, ultimately "dissolves" in the context of the "Garden of the Isles" (whose only limit, as we have demonstrated, is the limit of the open sea), or that its sense is further questioned by the legal recognition of the "Right to Roam" (*Allemansrätten*). The fact that its perceptions—and definitions—fluctuate is only one of the premises motivating our reflections. The key, however, is that the garden, in which external (spatial) and internal boundaries concur, evades the simple conceptualization as a "place set apart" and "demarcated in space": in its multifaceted dynamics, the garden, fundamentally, "operates" on boundaries. We already know that the garden is not merely "a part of a larger whole" but is itself a whole, within which distinctions fade. The essence of the garden lies neither in its extraterritoriality in relation to what is conceived of as "non-garden," nor in its separateness. Instead, it seems to lie in its liminal dynamics, which, primarily, dissolve seemingly clear thersholds, blur differences, and cancel oppositions. The "liminal nature" of the garden is chiefly revealed in its operations on its internal boundaries, the effects of which (only superficially) affect its external limits. The garden's workings are workings "in-between," epitomizing the reality of being-in-the-world, interpenetration, processual interconnectedness. Thus, the metaphysics of the garden so conceived cannot be any other but *transgressive*, as it transcends and abolishes the internal boundaries of being in defiance of the hegemony of historical narratives, constantly struggling to bring them to sharp focus—and optical illusion to trick the unmindful eye.

Traditionally established clear distinctions are evidently helpful in organizing our understanding of the world, but—if they prove inefficient, or inoperational—they can turn out to be cumbersome obstacles at best, or, at worst, sources of a plethora of our problems, and—upon closer reflection—risks. It goes without saying that the more fundamental these distinctions are, the more impactful their consequences. As follows, rudimentary dualisms and oppositions are most deeply encoded in culture, and therefore the effects of either accepting or rejecting them—are crucial. The metaphysics of the garden focuses on diagnosing "liminal" distinctions, upon the reconsideration of their validity, which, potentially, may lead to the revision of their status, including the possibility of their ultimate abolishment. Thus, the transgressiveness of the garden, as it is conceived in our project, gives rise to a broader model of a new, qualitatively transgressive metaphysics.

At first sight, such a statement might raise immediate objections—or even outrage. But upon reflection, one will easily recognize the fact that transgressive metaphysics does not aim to surpass all boundaries, respecting none: this, of course, is impossible. Our thinking about the world cannot do without *any* categorization, without *any* distinctions. Thinking banks upon our separative capacities as its indispensable and fundamental precondition. Thus, speaking of transgression, we do not speak of *blurring all boundaries*: we only point to our capacity of eliminating those that, unfounded in the natural rhythms, prevent humans from reaching the fullness of life and hinder flourishing. It is true that the metaphysics of the garden weakens the significance of certain distinctions and oppositions or abolishes them altogether, yet, as such, it remains discursive, and therefore must accept *some* categories, even if their status should not be any other than that of "working categories." Such metaphysics, transgressive though it is, is also affirming, hence positive, leading to the acceptance of certain resolutions. The above notwithstanding, these resolutions must be subject to continuous critical reflection lest we should become enslaved by the illusion of ultimate certainty, claims to which indicate authoritarian dogmatism. The critical dimension of such metaphysics does not exclude areas of indeterminacy, variability, and ambiguity. Transgressive metaphysics of the garden is not a rigorous science catering to unrealistic dreams of early 20th-century "positivist" thinkers.

Such metaphysics considers the garden as a model of resistance to certain established oppositions, primarily those separating self from non-self, soul from body, and nature from culture. These and other oppositions will be discussed below, but at this point it should be noted that even though it contests these distinctions, the adoption of such a metaphysical perspective is not tantamount to abandoning the human perspective. Metaphysics is a human enterprise—not just a theory, but also an activity. We cannot step outside the structures of human thinking about reality, which, of course, does not necessarily imply axiological anthropocentrism, or the appropriation of the whole for the sake of immediate human needs. Metaphysics, as a human creation and project concerning the totality of what is, includes that which is non-human, and, as such, it need not collapse into anthropocentrism. Therefore, in our thinking, we reject all three fundamental types of anthropocentrism:

i epistemological anthropocentrism, which only accepts the human perspective, while rejecting or marginalizing ways of perceiving the world beyond human perception as epistemologically flawed or unnecessary;
ii ontological anthropocentrism, which assumes that humans are the center of the universe; especially, we reject radicalized forms of this view, which assume that the world's structure favors humans or that the ultimate goal of the existence of universe is the human, embodying the culmination of its purpose;
iii axiological anthropocentrism, which holds that only humans possess any value.

The metaphysics of the garden is transgressive, holistic, and non-anthropocentric in the aforementioned sense, yet—indubitably—it is a product of human thought and, as such, remains subject to all resulting limitations.

Deep Ecology

Engaging in considerations of fundamental issues concerning the totality of what exists, we ask the most primary, "first," or, one could say, "deepest" questions. Often, discussions of the most profound matters lack philosophical tools and critical attitude, unnecessarily tending toward ideological confrontations, often violent in nature. Given the fundamental character of the problems we discuss, it would be an error not to return to Arne Naess's remarks on deep ecology, already alluded to in our presentation of Hans Gelter's organic experience of *friluftsliv*. Returning to Naess, we do not intend to analyze or critically evaluate the premises of deep ecology as a social movement, but to emphasize the transgressive dimension of his thinking about the world.

The thinker contrasts shallow ecology, focused on environmental pollution and resource depletion, with deep ecology, which seeks to diagnose the sources of the climate crisis. He believes that overcoming the crisis will be possible by questioning the fundamental theoretical assumptions causing it. Hence, a transformation at the very foundations of our thinking about the world is necessary. In a way, our proposal is driven by similar aspirations: the metaphysics of the garden is an attempt to revise the very rudiments of our thinking about the world. According to Naess,[1] the dominant paradigm comprises a set of beliefs, including the conviction that humans exploit nature because the purpose of humanity is to intensify consumption and production alike. Technocracy goes hand in hand with anthropocentrism, finding its expression in the idea that "man comes first." Nature is viewed merely as a storeroom of resources, so even if we care for it, it is because we fear resource depletion. Deep ecology, meanwhile, assumes the shift away from the *man-in-environment* image to the *relational, total-field image* of the connections in the universe.[2] Furthermore, Naess advocates for biospherical egalitarianism, the equal right to live and blossom, life-form diversity, anti-specialization, and self-realization.

In his 1986 article "The Deep Ecology Movement: Some Philosophical Aspects,"[3] Arne Naess answers the question "What Is Deep Ecology?" referring to eight principles.[4] Although analyzing or evaluating them, or systematic reflection on the movement remain beyond the scope of this book, it is worth noting that two of the eight principles strongly correlate with our findings:

1 The well-being and flourishing of human and non-human life on Earth have value in themselves (synonyms: intrinsic value, inherent value). These values are independent of the usefulness of the non-human world for human purposes.
2 Richness and diversity of life-forms contribute to the realization of these values and are also values in themselves.

Notably, both principles speak of values *in themselves*: the well-being and flourishing of human and non-human life, and the richness and diversity of life-forms *are* such values. Moreover, Naess emphasizes the relational nature of these values in their absolute (not relative) sense. In our perspective, too, well-being and flourishing become central values in our reflection on human life. Importantly, like Naess, we embrace these values in their absolute sense—we do not define them with regard to any utilitarian or pragmatic ends: well-being and flourishing do not *serve* any external purposes. Humans do not flourish in order to work more efficiently, but because flourishing is an autonomous and autotelic value. Limiting our insight into Naess's principles to issues relevant to human life as an atmospheric phenomenon, we could nonetheless ask two questions that, in a sense, shed some light upon the author's way of thinking and its underlying metaphysics:

1 do all forms of non-human life have equal value (e.g., cancer cells)?
2 why limit the validity of the principles to Earth alone? Do they not apply beyond it?

Values related to diversity—currently the subject of numerous inquiries—are of central importance today. Biodiversity, DEI (Diversity, Equity, Inclusion), diversity of thought, etc., are the "hot" subjects in contemporary academia and beyond. Fostering hetero*geneity* prompts further *genesis* of diversity, which, in itself, enriches differences, increasing the number of varied entities. Such heterogeneity may also be perceived in terms of the structural non-homogeneity of the universe. Incentives fostering diversity are now of common concern, propelling cross-sectional activities, whose objectives range from biodiversity to social diversity to educational diversity.[5] The richness of diversity reveals its multifaceted utility, but it is not only this dimension that matters. Naess clearly states that diversity is a value *independent* of the benefits it brings, or, in other words, the functional benefits of diversity are not a necessary condition to recognize diversity as a *value in itself*. Of course, diversity thus understood is not an isolated value: since it stimulates multiplicity, then connecting this multiplicity to the unity of experience or, depending on the adopted assumptions, with the unity of the universe—is essential. Regardless of how one views it, all diversity manifests its fertility in the face of the awareness of wholeness-as-unity. What is important is to beware of stances potentially sanctioning radical relativism, which could lead to the impoverishment of diversity: understanding values related to diversity as values-in-themselves is linked to *relational*, not *relativistic*, understanding of constituent elements, and such thinking excludes potential tendencies to reduce diversity.

Life Quality

Diversity-oriented thinking enriches the world, caters to the heterogeneity of entities and contributes to the abundance of their variety, thereby serving as a safeguard against the unpredictable effects of the hegemony of anthropocentrism, disregarding the fact that living organisms, including humans, co-exist and interact in dynamic ecosystems. Alas, currently, efforts to create conditions for biological diversity to develop often seem to boil down to *narratives* presenting biodiversity as essential for human well-being as its upkeep supports economic and social development. Such an approach seems shortsighted and dangerous, as it reduces the essence of the human being to utilitarian, or even consumerist, dimension, which reveals its proponents' axiological blindness. Humans are not merely elements of a sociobiological process: humans are beings who self-actualize, find fulfillment, and flourish within the space of possibility and freedom. And this aspect of human existence is captured by Naess's seventh principle:

> 7. The ideological change is mainly that of appreciating life quality (dwelling in situations of inherent value) rather than adhering to an increasingly higher standard of living. There will be a profound awareness of the difference between big and great.[6]

The distinction between "life quality" and "standard of living" is crucial, as the latter category belongs to the anthropocentric narrative, where the human—as center and ultimate purpose of the universe—exploits all entities to continually raise his or her "standard of living." In a race toward the highest possible living standards, *life quality* is either overlooked, or gets subordinated to the requirements of the race, which brings to mind Henry David Thoreau's distinction between "profit" and "benefit," which he often emphasizes throughout his writings. What matters in the mercantile-economic perspective is, of course, the former, yet the sense of human life cannot be reduced to profit alone. A simple forest walk yields no profits, but benefits our inner life and the *whole*, in which we live. Criticizing materialism—and praising authenticity—in his *Life Without Principle*,[7] Thoreau clearly suggests that in profit-driven activities, we are not irreplaceable; there are always "spares" in the process of the exchange of goods. Yet, at the deepest level, what truly benefits me (e.g., a walk, contemplation, a kiss, a song, or writing a book) cannot be transferred to another—life is the one thing, in which no one can replace me. With analogous observations among its premises, the metaphysics of the garden, understandably, focuses on life quality, which is a *benefit*, not a *profit*, and which is grounded in the experience of one's inextricable connectedness with the surrounding world. Thoreau, like may other philosophers whom we engaged thus far, teaches that this union is not "personal" but "natural"—it "belongs" to the whole. Life is participation in nature, and the Concord thinker expresses it beautifully in his journal entry, where he asserts his profound belonging to the world in the wholeness of its diversity of forms:

> Must be out-of-doors enough to get experience of wholesome reality, as a ballast to thought and sentiment. Health requires this relaxation, this aimless life. This life in the present. Let a man have thought what he will of Nature in the house, she will still be novel outdoors. I keep out of doors for the sake of the mineral, vegetable, and animal in me.[8]

Metaphysics of the Garden and Deep Questions

Naess's term "deep ecology" primarily signifies the depth of formulated questions and the profundity of philosophical analysis. It emphasizes the need to rethink, and, consequently, zetetically inquire into the assumptions—and reasons—of particular actions. Deep ecological reflection must reference philosophical dilemmas and questions that inherently probe what is the most fundamental. Understood in such a way, *deep ecology* is essentially *first ecology*, and, in terms of its content, also *first philosophy*, or metaphysics. Metaphysics, as deep zetetic inquiry into what is *first* and *fundamental*, is particularly needed today, as these essentials have been obscured by the towering structures of anthropocentrism, hegemonic dualism, technocratism, pragmatism, and, most recently, naïve faith in the power of the surrogate artificial intelligence, the uncritical use of which renders philosophical discourse more and more shallow.

Furthermore, the necessity, or even inevitability, of metaphysical reflection—or in other words, the necessity of the inquiry into the issues of first philosophy—becomes all the more urgent as we realize how much we need an antidote to the advancing process of scientific specialization. The institutionally sanctioned fragmentation of expertise results in the fact that the reflection on the universe-as-a-whole has come to resemble a jigsaw puzzle: a collection of disparate—albeit often valuable—fragmented discourses, missing a common fundament of holistic thought. Contemporary scientific reflection tends to emphasize—and isolate—quantitatively understood cause-effect relationships, approaching issues of life quality with suspicion, or, at best, heroically attempting to quantify it. While the parameters of the standard of living are quantitatively measurable, the key determinants of life quality—balance and flourishing—resist reduction to parameters. Health, well-being, *eudaimonia*—like a garden—do not adhere to the principle of permanent quantitative growth, the causes and mechanisms of which we would like to understand in full, in order to be able to control it. Understanding the phenomenon of life quality is understanding the state of equilibrium,[9] which requires insight into the dynamics of the entwinement of a plethora of factors, awareness of our own cognitive limitations, and acceptance of possible areas of uncertainty—and, not infrequently, unknowability. The metaphysics of the garden, as first philosophy, abandons claims to decisiveness, unambiguity, or ultimate answers. Transgressive by necessity, such metaphysics embraces ambiguity and wildness, to which we dedicate more space in the sections below.

Intimacy: The Case of the Fenner Typology

In their recent book *The Art and Philosophy of the Garden*, David and Ethan Fenner assert that

> [e]very garden is unique; to use the philosophic term, every garden is autographic. It exists as a thing that cannot be replicated. Just as true, gardens as a kind are special insofar as they represent best among all things, the intersection and interplay between humans and nature.[10]

The authors further reinforce this observation with the claim that "gardens are special because of the intimacy of the intersection."[11] Our remarks on the metaphysics of the garden thus far have shed quite a lot of light on the "intersection and interplay," but the issue of the "intimacy"—the word recurs in the writings of Thoreau, Nietzsche, and many other thinkers seeking peace in attunement to natural rhythms—is a key challenge. To adequately address intimacy, it does not suffice to propose a "new" approach, if it is still clad in old conceptual garb, pulled out of the musty closet of the classical oppositional metaphysics. To understand intimacy, one must *become* intimate, yet to embrace it, one must address the fundamentals, that which is truly primary, that which is essential to the metaphysics of the garden. Although, considering the scope of this book, we realize that a more profound insight into the intimacy in the human/nature relationship might not be feasible, even a brief engagement with David and Ethan Fenner's work allows us not only to substantiate many of our autoethnographic observations, but also to indicate yet another, zetetical, dimension of the transgressive metaphysics of the garden.

Translating the intimacy into language is not an easy task. It is not surprising that to demonstrate why "intimacy" renders gardens "special," one could be tempted—like David and Ethan Fenner—to begin by distinguishing "gardens" from "non-gardens," the procedure, which, in the context of the liminality of the discussed phenomenon, may prove futile. The initial passage reads as follows:

> Certainly, we can experience [such intersections] when we hike through "untouched" nature or interact with nature through agriculture, but gardens are special because of the intimacy of the intersection. They exist as artistic co-creations of nature and humans, a series of back-and-forth actions and reactions, where both parties have something to add. Humans, through deliberation and design, express themselves in the creation of a place that cannot be said to be only nature or only artifice; these places become something new as vision leads to expression, expression to cooperation, and cooperation to instantiation.[12]

In the context of the above quotation—but especially in the light of our debate on the gardens as operating on boundaries—it becomes clear that attempting to distinguish the "garden" from the "non-garden," one will inevitably face the problem of the exclusivity or inclusivity of the definition. Definitions, of course, may differ depending on the variables taken into account, and the Fenners's "horto-lexicographic" experiment clearly testifies to this claim. The authors first distinguish two basic meanings of the term "garden": (i) as any particular garden, and (ii) as a set of objects (type, abstract kind, or ontological category).[13] Driven by the intention "to create a comprehensive inventory of all garden types,"[14] the authors provide examples of various attempts to define gardens based on different criteria. In Chapter 2, "Defining 'The Garden,'" they ultimately list ten constitutive features of all *true* gardens, proposing that any object possessing each of these traits can be considered a garden. And so gardens are: (1) spatially three dimensional, (2) geographically bound, (3) human scaled and perspectivally bound, (4) dynamic and time-dependent, (5) nature interactive and ecologically bound, (6) unbounded in time and lifespan, (7) designed, (8) human interactive and requiring maintenance, (9) rule governed, and (10) aesthetically engaging.[15]

Of course, it is not our intention to discuss the Fenners's "typological" proposal or its premises. Yet, in the light of what we already know, the requirement of the simultaneity of the manifestation of all ten "constitutive" traits disqualifies most of the spaces (physical, metaphorical, or imaginary) that we have discussed—as "true gardens." Although at first sight that should not be a major problem—after all, in a scholarly debate we may or may not agree on a definition—the assumption underlying *The Art and Philosophy of the Garden* renders it impossible not only to address the phenomenon of intimacy, but also to explain the uniqueness of the phenomenon of the *Skärgården/Skjærgården/Eyjagarðurinn*—spaces, whose inhabitants themselves dubbed them "the Garden of the Isles" for a good reason, most evidently adopting a very different, non-arbitrary, point of departure, and a very different concept of aesthetics. A brief mental experiment demonstrates the shortcomings of the "ten constitutive traits":

The gardens of the isles—not only those in Scandinavia, but also those in Greece, Polynesia, Thailand, or in the North Atlantic—are not *three*, but at least *four*-dimensional. They are "geographically bound" inasmuch as they are located within the political borders of a given country, but even that is not "binding" in the case of the islanders' attitude to geographical boundaries whenever they must seek shelter in a gale. Such gardens are neither "human scaled" nor "perspectivally bound"; they are "dynamic," yet "time-dependent" only in relation to the rhythm of biological cycles; they are inevitably "nature interactive," and equally inevitably—as ecosystems—"ecologically bound," they are "unbounded in time and lifespan," only on condition that one chooses to think in terms of a human lifespan, and not in terms of geological eras; they are not "designed," although they are tended to; they harbor man-made structures, and—in some instances—are home to man-demarcated natural reserves; they are obviously "human interactive," and, although they do not "require maintenance," maintenance is welcome; they are "rule governed" if we understand the "rules" in terms of the rhythms of nature and legislature that prevents their devastation, and finally—they are like hip-hop, "aesthetically engaging" to some, but not necessarily to others. We could easily expand this experiment to encompass all other gardens of the isles, the meditation gardens of the Apennine School of Living Philosophy, and in all probability—many other locations, whose inhabitants, or frequenters, have developed an intimate, caring, respectful, awe-filled relationship with the places, to which, in one way or another, they choose to tend. And yet, such intimacy—which, in David and Ethan Fenner's view, is what makes gardens "special"—is immune to the language adopted by the authors; intuiting its existence and importance, they nonetheless must fail to approximate it.

To explain it simply—anyone who has ever experienced intimacy understands that it is not reducible to the context: preparing for an intimate date, once could decorate the house, set the table, light candles, kindle the embers in the fireplace, burn incense, purchase expensive wine, serve it in elegant glasses, cook divine food and serve it spectacularly, make sure that the selection of music complements the whole, and even draw a scrumptious bubble bath for two, taking care that the scent of the salts is pleasing. All of these intentional efforts *may*, but do not *have to*, lead to the materialization of intimacy: if the partner, when he or she arrives, is preoccupied with a major problem at work, suffers from a headache, or experiences a spell of depression—the *atmosphere*, which extends beyond each of the individuals *and* beyond space, will not be conducive to intimacy. Likewise, a visit to the beautiful gardens of Peterhoff, Versailles, or Schönbrunn (which meet all of the ten criteria proposed by the Fenners)—ticketed, and limited by opening hours, and oftentimes guided—*may*, but does not have to, evoke a sense of an intimate relation to nature in a visitor. Furthermore, one can imagine that in the high season, crowds of tourists and their ceaseless, joyful, multilingual chatter, might make such a visit burdensome: it may even affect the *gardeners*, who, tired with oppressive crowd, will more likely count the time till the end of the shift than relish in the intimate relationship with nature, to which they tend. *Atmosphere* is key to intimacy—and such an atmosphere, conducive to the formation of a profound, intimate relationship with nature and conditioning the *attunement* to its rhythms, pervades the gardens of the isles. Gardens, that—unlike Peterhoff, Versailles, or Schönbrunn—fail to meet the "requirements."

David Fenner's and Ethan Fenner's book, undoubtedly, is a most informative, inspiring, and much-needed contribution to the debate concerning the connectedness of the human and the world. Irrespective of its obvious value, however, it is an example of a "heroic attempt" to quantify the phenomenon of the garden, which differs from other spaces because of the "special" intimacy that it affords to humans "intersecting" with nature. Attempting to quantify the garden, the authors attempt to quantify *intimacy*—an indispensable element of eudaimonic well-being, and of personal, social, and environmental health. Intimacy, however, as one of the central factors determining the quality of life, is not quantifiable with reference to criteria of inclusion and exclusion: it is an atmospheric phenomenon, the profound understanding of which requires a solid metaphysical fundament. Experientially speaking, spaces meeting the criteria of the "true garden" are often less conducive to the flourishing of intimacy than those "untrue gardens," disqualified by the scholars, even though local communities, intimately connected to them, "know better," and their toponymy still sanctions them as "gardens." The transgressiveness of the metaphysics of the garden, which—in the model we propose—is an atmospheric phenomenon in itself, puts to test the arbitrariness of the quantitative criteria or exclusive definitions, simultaneously offering one an experiential formula for a "reality check." Such metaphysics, as first philosophy, reveals its potential as a touchstone of criticism, and as an antidogmatic shield.

Transgressions and Intimacies

The discussion with David and Ethan Fenner (upon whose expertise we will soon call again) opens space for further investigation of the ramifications of the metaphysics of the garden. At the onset, it is worth to scrutinize the transgressiveness of the garden in relation to selected oppositions—contrasts, or pairs of categories. As we already know, the proposed metaphysics acknowledges (at least partial) blurring of binary relationships, and recognizes the experiential reality of interfusion, interpenetration, and mutual permeation of entities and phenomena, which, unlike it is the case in dualist paradigms, no longer occupy rigid, unchangeable positions. In the idea of the garden, some of the rudimentary, traditionally established binaries collapse, the effect of which is not so much the elimination of all boundaries, but the revelation of new possibilities for intellectual discernments. As a result of the application of a variety of "cognitive cuts" at the fundamental level, the world reveals itself in alternative perspectives—within the framework of alternative, sometimes diverse, categorizations. The garden is often described as embodying the fusion, or even the identity, of culture and nature—where art becomes nature and nature becomes art. Culture, as a product of human activity, is thus a fusion of man and nature. This synthesis is not an attempt on the part of culture—or art—to subordinate nature, or to liberate it from the manacles of determinism. Nor is it about freeing nature from consumerist utility, which, as Rosario Assunto suggested, might be interpreted as a form of retaliation against utilitarianism. In his treatise *Filosofia del giardino e filosofia nel giardino*,[16] Assunto argues that the art of gardens constitutes nature as independent, freeing it from being merely a result of other factors, or a means to other ends. However, as scholars such as John Dixon Hunt indicate, the matter is significantly more nuanced:

> Gardens now take their place as a third nature in a scale or hierarchy of human intervention into the physical world: gardens become more sophisticated, more deliberate, and more complex in their mixture of culture and nature than agricultural land, which is a large part of Cicero's "second nature." By implication, the first nature becomes [...] the territory of unmediated nature, what today we might (provisionally and awkwardly) call wilderness.[17]

The key issue seems to be the softening of the "traditionally" clear opposition between nature and culture, nature and art, or nature and humans. The once contrastive relationship between them transitions into that of cooperation, entanglement, and even intimate interaction, in which, intertwined, they no longer collide. The garden thus becomes a transgressive world where nature and culture converse without mutually seeking to convert the interlocutor to one's own principles and rules. It is not about subordination or superiority, nor about liberation from determinism (which is as impossible as it is unnecessary), but about creative cooperation. The nature-culture distinction is traditionally viewed as the theoretical foundation of contemporary anthropology, yet by revisiting its assumptions and questioning its "first principles," one can demonstrate that the softening—or even transcendence—of this foundational opposition, not only possible, but actually happening in the garden, does not *question* the identity of the discipline, but, conversely, enriches it by opening new vistas. Adopting such an attitude, one discovers that there is no need to cultivate the language of contradictions and clear-cut distinctions: the metaphysics of the garden tends to blur such oppositions, recognizing that the garden can speak the language of both art and nature. This, in turn, substantiates the belief that unambiguity of expression, even unambiguity of language (vocabulary), is not an unquestionable dogma. Following Odo Marquard's suggestions, one can assert that philosophy need not always collaborate with the monomyth;[18] the conversation need not lead to the end of the discussion, marked by the formulation of a closing answer. The voice of nature and the voice of art are diverse voices, both of which the garden can offer. Therefore, the transgressive dissolution of opposition also applies to the pair of "creation" and "influence" (determination); in this light, the intertexture of former opposites emerges as liberated: creation (particularly artistic creation) becomes free from the *label* of an activity tantamount to "defying influence" and, at the same time, influence is no longer perceived as a force independent of creative formation.

As we remember, the term "garden" is historically associated with the concept of delineation and separation; a garden was marked off from the larger environment in which it was situated and, therefore, the earliest words designating "garden" explicitly referred to "enclosure." Originally, gardens served as "enclosures" intended to protect crops, provide food, and offer pleasant space, in which people could enjoy their rest. Thus, their initial functions were utilitarian or aesthetic. The above notwithstanding, the applicability of etymologically based analysis does have certain limitations:

> A garden being "bounded" is not, however, the same as a garden being walled or enclosed in the way expressed above. In antiquity, "enclosed" meant "walled" or something comparable, such as "fenced." Today this seems no longer the case. There are many examples of gardens that do not have walls or fences; there are many examples of gardens that are so large as to make this sort of enclosure an expensive if not impractical suggestion. [...] So the etymological route to defining The Garden, while interesting and inspiring of ideas we will discuss later, is a non-starter in terms of providing us with a usable and inclusive definition on which to base any further discussion.[19]

The etymological approach may indeed be insufficient to provide us with a comprehensive and inclusive definition of a garden, but it can be useful for emphasizing once again the intimate interweavings inherent to gardens. In some respects the internal and external aspects of the garden (liminal by nature) also intertwine: the garden's internal structure—a blend of culture and nature, action and influence—becomes a reflection of the world, a particular way of apprehending it; internality becomes (an image of) externality. Conversely, the world as such is internalized within the garden. What is most interesting is not that the world "separates itself from itself" in the process, but how it reveals itself within a finite space, in the lens of a demarcated place. Garden spaces may indeed differ from the spaces of everyday life, yet it is within the gardens that one can clearly perceive the dynamic, all-encompassing, non-divisible tissue of the living world-as-a-whole.

In the garden, fusing nature and art, nature's rhythm intertwines with creative action, allowing determination and creativity to converge. Furthermore, the garden *demands* our activity: passivity is a form of resignation, of slow abandonment of the garden. Yet, although our agency must be actualized within the garden, its actualization cannot merely be an arbitrary exercise: it must be an exercise of agency-within-the-existing-rhythms-of-nature. Moreover, our knowledge of nature and the degree to which we are able to attune to its pulse affect our actions: the greater the knowledge, the closer the attunement—the more value (rather than merely effect) our actions generate. Our respect for nature and our communion with it result in the emergence of space conducive to our creativity—but simultaneously, the garden, as a work of art, is an object of our admiration and contemplation. It is evident, then, that action (agency) intertwines with the recognition of nature's determinations, which consequently leads to creative freedom, that, in turn, gives rise to aesthetic values. Wondrously, the garden integrates agency, determination, freedom, and the contemplation of beauty *in attention* and *in amazement*—which is yet another manifestation of its transboundary character. Possibly, it is this trait, rather than any quantitative measures, that accounts for the garden's ontological, and aesthetic, uniqueness.

These "ontoaesthetic" considerations are valid inasmuch as we agree that the garden is "nature in intimate relation with the human," not simply "nature touched by the human hand." Of course, the scale and intensity of the human presence varies from location to location, but it seems undebatable that nature "unaffected" by human actions does not exist: from the Arctic cap to Mariana Trench, the effects of our interventions are more than palpable. Furthermore, even actions oriented toward the conservation of particular biotopes—like the establishment of strict nature reserves—irrespective of their ethical premises, remain interventionist. Restricting access to a particular area to warrant the undisruptedness of the natural processes within it is—in itself—an exercise in control, which renders "wilderness" somewhat subjectivized, humanized, subjected to human rules of intrusion. Yet, while overexploitation, overharvesting, and other manifestations of human agency in line with the Dominion Mandate transform wilderness into wastelands, conservationism (even if economically driven) is a form of stewardship. When stewardship goes hand in hand with profound, compassionate attunement—it transmogrifies into gardening, in which concept two agencies—human and atmospheric—coalesce, blurring differences. This is key to intimacy. A garden, therefore, *emerges*, rather than *is*. And when it does—so does amazement, so does creativity, so does purpose, so does *eudaimonia*.

Our transgressive metaphysics, acknowledging atmospheric fluctuations, allows us to profoundly understand that permanent attachment to concepts is not only a cognitive obstacle, but also a source of frustration: as we have already demonstrated, some spaces, universally recognized as gardens, may manifest themselves as closer in character to industrial processing plants, while others, universally recognized as "wild nature"—when they become objects of our stewardship and compassionate care, emerge as intimate gardens. At the same time, in light of our metaphysics, it seems necessary to attune to the sensitivities of others, attempting to define gardens, not to engage them in pointless polemics, but to attentively listen for the elements of their argumentation that uncover the metaphysical foundations of their methodology and debar their attachments. Rosario Assunto, for instance, in his treatise "Per una ontologia del Giardino"[20] proposes that we understand the garden as nature shaped by humans to express their spirit, in which sense nature becomes a mirror of human *purpose*: the expression of the inner self. Interestingly, at the heyday of American Renaissance, Ralph Waldo Emerson, sensitive to the synchronicity of human and natural rhythms, expressed an *almost* homophonous idea. But "almost" makes a world of difference, because in the American philosopher's view the need to *shape* nature to express our spirit is, in fact, superfluous:

> [N]ature is not always tricked in holiday attire, but the same scene which yesterday breathed perfume and glittered as for the frolic of the nymphs, is overspread with melancholy today. Nature always wears the colors of the spirit. To a man laboring under calamity, the heat of his own fire hath sadness in it. Then, there is a kind of contempt of the landscape felt by him who has just lost by death a dear friend. The sky is less grand as it shuts down over less worth in the population.[21]

Nature *always* "wears the color of the spirit." If that is the case, one could conclude that, based on an erroneous premise, Assunto is "attached" to an anthropocentric vision of the garden, assuming that nature, purposefully shaped, is a medium of expression of the human spirit. He errs on two accounts: first, it is possible to impose shape on plants and terrain, but it is impossible to shape nature; second, disregarding the atmospheric character of the spirit, Assunto overlooks the impermanence, fleetingness, and changibility of human spiritual states. If a garden is "nature shaped to express human spirit," then the "shape" given to "nature" by a particular agent at a particular time may reflect only the-shape-of-the-spirit-in-the-past. It may, at best, reflect the "there-and-then" spiritual state of the person, whose agency, at the time, was decisive. And although the particular "shape" imposed on "nature" might not correspond to this person's state of spirit tomorrow, or a year from now, or a decade later, nature will *always* wear the color of their spirit. In Assunto's vision, the dynamic entanglement of humans and nature is disrupted, tilting toward human agency. Importantly, however, Assunto's garden is a testimony to the self-limitation of the spirit, but also, when confronted with the premises of the transgressive metaphysics, it offers a useful insight into the limitations of attitudes tending toward the subjectivization of nature. Freedom and determinism, let us reiterate, need not be polar opposites: just as the garden may be perceived as an effect of the "subjectivization" of nature, it should also be understood as an effect of the "objectification" of the human in the atmospheric sense of the phrase—as beings recognizing themselves in nature's objectivity. Humans are not merely elements of nature, not merely audience of the grand spectacle, cast into the world's whirl—but neither are they demigods or masters of creation, entitled to exploit the world in which they exist. *Being in the garden* is *being the garden*, and thus both phrases signify belonging-to-the-whole, without claims to being entirely subjected or entirely dominating.

Our deeply ingrained thinking about the world in a substantial perspective compels us to view our cosmos as composed of things, beings, and substances. In this view, "classifying," "categorizing," "distinguishing," and "separating" objects, while delineating ontic areas, are all a part of a practice that we deem "natural." True, our subtler ideas—including those affecting our choices to seek knowledge in conceptually non-mediated experience—necessarily rely on basic distinctions and oppositions. Alternatively, however, one can choose to think of the world in terms of *functions* rather than *substances*. In such a perspective, we view reality processually, not objectually, acknowledging relations rather than simply "plotting points" on the world map. As our discussions demonstrate, to zetetically avoid errors of judgment, it is worth thinking of the garden-as-a-process, involving its emergence and disappearance, its liminality responsible for the dynamics of transformations, phenomenal interpenetration, non-violent conversation, natural rhythms and cycles, atmospheric fluctuation of intimacy and amazement—or, in other words, as a process energized by the activity of forces operating within the boundaries of self-awareness.

Let us sum up our findings: the key ideas concerning the metaphysics of the garden, outlined in this section, may be regarded as a set of theses appurtenant to first philosophy—a philosophy foundational to all other relations and references. Most importantly, however, such a metaphysical framework justifies the necessity of the shift in our thinking about the relationship between humans and nature, and encourages an eudaimonic change of our attitudes toward the world in which we live. Reconsidering the entirety of existence as a garden, we adopt a metaphysical perspective, in which we can recognize ourselves-as-gardens, where ontological oppositions—once sharp—blur, simultaneously exposing entanglements and fluctuating dependencies that, transgressing the limits of the inherited knowledge systems, could not be apprehended before.

Thus, the perspective of the self-and-the-external-world-as-a-space-of-influence is not a perspective, in which two radically different ontological entities—of which one controls the other, or "speaks on its behalf"—collide.

Such a perspective steers clear of the sharply defined subject-object opposition, perceived in the spirit of Cartesian dualism. Conceiving of the world as a garden does relieve us of the duty of placing ontological emphasis on distinctions, as these dissolve within the context of the garden. The metaphysics of the garden transforms stark dualism into the intimacy of correlation, the intertwining of influences, the atmospheric cognitions. The self does not end at the boundaries of the body—nor even at the limits of subjective appropriation; rather, the *self* is constituted by the *non-self*, by *the whole*. This perspective fosters reconciliation with nature, encouraging a return to the garden. One of the most memorable lyrics by Jonasz Kofta, contemporary Polish poet, emphasizes the importance of this viewpoint across multiple dimensions:

Pamiętajcie o ogrodach Przecież stamtąd przyszliście W żar epoki użyczą wam chłodu Tylko drzewa, tylko liście.	Do remember of the gardens Of your home, free from swelter In the heat of your century's hardships Only tree-leaves lend you shelter[22]

To awaken to our fundamental rootedness in the rhythms of nature, we must recognize that the garden, as we conceive it, possesses its own inner dynamics and inherent value, which cannot be reduced merely to utility or even aesthetics. The essence of the garden—and the essence of the extended-human—lies in its liminality. As many of the failed attempts to "grasp" this phenomenon demonstrate, the garden resists being defined by the gardener's domination over nature. Gardens transcend the dichotomy between the individual and the universal, reveal an ethical relationship between humans and nature, a relationship based on intimacy, compassion, and awe. As extended humans, or human-gardens, we are the space, where global processes, pulsing in sync with the rhythms of nature, become actualized. Through this, we, extended humans, discover our own uniqueness in the face of mortality.

Garden and Modernity

To further clarify the particularity of the proposed metaphysics of the garden, it seems necessary to contrast it with starkly different perspectives. One of the most vivid, polarly different, conceptions of the garden—elevated to the status of a symbol of modernity—has been formulated by Zygmunt Bauman in his *Modernity and the Holocaust*, the book awarded the European Amalfi Prize for Sociology and Social Sciences in 1989. The key passage concerning the problems of our interest reads as follows:

> Modern culture is a garden culture. It defines itself as the design for an ideal life and a perfect arrangement of human conditions. It constructs its own identity out of distrust of nature. In fact, it defines itself and

> nature, and the distinction between them, through its endemic distrust of spontaneity and its longing for a better, and necessarily artificial, order. Apart from the overall plan, the artificial order of the garden needs tools and raw materials. It also needs defense—against the unrelenting danger of what is, obviously, a disorder. The order, first conceived of as a design, determines what is a tool, what is a raw material, what is useless, what is irrelevant, what is harmful, what is a weed or a pest. It classifies all elements of the universe by their relation to itself. This relation is the only meaning it grants them and tolerates—and the only justification of the gardener's actions, as differentiated as the relations themselves. From the point of view of the design all actions are instrumental, while all the objects of action are either facilities or hindrances. Modern genocide, like modern culture in general, is a gardener's job. It is just one of the many chores that people who treat society as a garden need to undertake. If garden design defines its weeds, there are weeds wherever there is a garden. And weeds are to be exterminated. Weeding out is a creative, not a destructive activity. It does not differ in kind from other activities which combine in the construction and sustenance of the perfect garden. All visions of society-as-garden define parts of the social habitat as human weeds. Like all other weeds, they must be segregated, contained, prevented from spreading, removed and kept outside the society boundaries; if all these means prove insufficient, they must be killed.[23]

Thus, asserting that the modern culture is a garden culture, the disappointed philosopher seems to argue that modernity only "fashions itself" as a project of "ideal life" and "perfect arrangement of human conditions," yet, in fact, it rests upon a metaphysical design founded upon "distrust of nature." Consequently, the order of the garden is an artificial order—one continuously threatened by external forces—whose integrity must be defended by the gardener, the guardian of the order. Hence, gardening emerges as forceful, often violent, imposition of distinctions and boundaries—most evident in the practical applications of the binary of "inclusion" and "exclusion," based on the criterion of utility. In Bauman's perspective, modern society was treated a "garden," in which the gardener, equipped with instrumental reason, epitomized the hegemonic order, which "determines what is a tool, what is a raw material, what is useless, what is irrelevant, what is harmful, what is a weed or a pest," an egocentric order which "classifies all elements of the universe by their relation to itself." In this context, Bauman deploys his metaphor to demonstrate how the concept of "cultivation" is tantamount to arbitrary selection:

> I suggest, further, that the bureaucratic culture which prompts us to view society as an object of administration, as a collection of so many "problems" to be solved, as "nature" to be "controlled," "mastered"

> and "improved" or "remade," as a legitimate target for "social engineering," and in general a garden to be designed and kept in the planned shape by force (the gardening posture divides vegetation into "cultured plants" to be taken care of, and weeds to be exterminated), was the very atmosphere in which the idea of the Holocaust could be conceived, slowly yet consistently developed, and brought to its conclusion.[24]

Bauman's poignant metaphor remind us of the difference between Kurt Vonnegut's *Slaughterhouse Five* and Hreinn Friðfinnsson's *Husprojekt*, which we have discussed in Chapter 3 of this book. Bauman's use of "the garden" in the context of the modern culture clearly counters our point: flourishing and artificial orders—separated from equally imaginary and equally artificial *disorders*—are mutually exclusive. "Society-as-garden," if the garden is to be a "healing garden," cannot operate on the principle of exclusion, of fear, constant sense of threat from "disorderly" forces—wildness, indeterminacy, or unboundedness. On the contrary, the garden catering to atmospheric health is non-exclusive; it flourishes in nature's indeterminacy, affirming mystery, accepting unknowability, at ease with ineffability. Thus, the metaphysics of a healing garden must be, in a sense, apophatic.

> The great world-garden has split into innumerable little plots with their own little orders. In a world densely populated with knowledgeable and intensely mobile gardeners, no room seems to be left for the Gardener Supreme, the gardener of gardeners.[25]

More importantly, the central problem of contemporary life does not lie in the fact that the modern "gardener of gardeners" has been marginalized, ousted, and ultimately replaced in his position, by postmodern, numerous, non-intersecting, intensely mobile gardeners, but in the fundamentally flawed thinking about the garden and the gardener's role. Yet the multiplication of independent gardens, in alignment with the postmodernist model—that, while it deconstructs and collapses oppositions, paradoxically perpetuates, and simultaneously relativizes, the "centrally governed" garden model, which it opposes—is not tantamount to wholeness. Inheriting the fundamentals of its metaphysics from modernity, to which it is indebted, it is thus incapable of embracing, or becoming one, with the transgressiveness in the idea of the garden.

Unconsumability and Wildness

Intuitively, we all realize that, unless we think of the craters of erupting volcanos, it would be naïve to claim that "untouched" nature exists on Earth, no matter how seductive such a projection could be. Simultaneously, we know that the project of ultimate subjectivization of nature is equally absurd, the idea of which could be encapsulated as "unconsumability of nature."

From the point of view of healthy balance, however, the essence and scale of our intrusion, of which we wrote earlier, should be carefully considered. This need is testified to by the work of numerous scholars engaged in thus oriented discussions. Among the many arguments in that debate, two, addressing the fundamentals, seem to coincide with our reflections on the metaphysics of the garden.

The first of the two is Arne Naess's thesis, who proposes that *non-human life has intrinsic value*. This argument is absolutist in character. The second thesis pertains to epistemology: human self-restraint in their intrusions into nature—epitomized by the garden—is a function of man's recognition of his or her own limitations, including the thresholds of cognition. The complexity and diversity of nature are not fully transparent to humans; many processes within it remain hidden, which precludes our ability to infallibly predict the consequences of our interventions. Of course, *a posteriori*, we have been able to create certain predictive models, and ecologists constantly warn about the effects of our actions that pose a threat in the future, but this is not the central issue of the present argument. The key premise of it is that our thinking must account for all that we do *not know* about nature, which, as we know, "loves to hide." This suggests that any radical interventions we might possibly plan (even if our motivations are conservationist) should be considered with utmost caution. This argument is, essentially, skeptical, embodying a certain form of trust in nature as a whole. Both these arguments—the thesis proposed by Naess and the argument concerning the recognition of the limitations of human cognition—unequivocally lead to the conclusion that any conceptions (including sozological projects) transferable into actions, if they are to be conducive to healing, must be free from forms of subjugation and domination: in light of the transgressive metaphysics we propose, the recognition of the acceptable limits of one's intervention is a function of finding one's freedom and autonomy within mutual interdependence. Tending to the garden is also a caring for the re-creation of nature, in the same sense as tending to health is caring for the restoration of balance. To rephrase these thoughts succinctly, catering to *eudaimonia*, the transgressive metaphysics of the garden must be based on the profound acceptance of nature's indeterminacy. Celebrating wildness, such a perspective acknowledges human ignorance and, at the same time, the intrinsic value of the non-human life. To truly understand the world and oneself in it, one must expose himself or herself to unpredictable engagements with wildness and encounter the untamed elements, becoming attuned to nature irrespective of the fact that one can never fully comprehend it. Henry David Thoreau understood it well:

> As I came home through the woods with my string of fish, trailing my pole, it being now quite dark, I caught a glimpse of a woodchuck stealing across my path, and felt a strange thrill of savage delight, and was strongly tempted to seize and devour him raw; not that I was hungry then, except for that wildness which he represented. Once or twice,

> however, while I lived at the pond, I found myself ranging the woods, like a half-starved hound, with a strange abandonment, seeking some kind of venison which I might devour, and no morsel could have been too savage for me. The wildest scenes had become unaccountably familiar. I found in myself, and still find, an instinct toward a higher, or, as it is named, spiritual life, as do most men, and another toward a primitive rank and savage one, and I reverence them both. I love the wild not less than the good. The wildness and adventure that are in fishing still recommended it to me. I like sometimes to take rank hold on life and spend my day more as the animals do.[26]

When coupled with our contemplative, attentive, intellectual reflection on nature-as-a-whole, our atavisms—teaching us a lesson in our constant participation in the processes of life and enhancing our awareness of our own cognitive limitations—result in reverence: profound respect for human and non-human life alike. Personal, but also communal, engagement with wildness[27]—is thus a *sine-qua-non* condition of flourishing:

> Our village life would stagnate if it were not for the unexplored forests and meadows which surround it. We need the tonic of wildness,—to wade sometimes in marshes where the bittern and the meadowhen lurk, and hear the booming of the snipe; to smell the whispering sedge where only some wilder and more solitary fowl builds her nest, and the mink crawls with its belly close to the ground.[28]

Garden as the Poetry of Nature

The current renaissance of interest in gardening partly arises from our recent disconnection from nature, yet, irrespective of its origins, it testifies to our inherent, if sometimes subcutaneous, persisting connection. This resurgence in interest in nature provides a rare opportunity to help us better understand who we truly are: the more insistently we experience its absence, the more palpably we sense our unity with its fundamental rhythms. Only too often do we take what is "natural" for granted: the "obvious," paradoxically, is frequently ignored—it, quite literally "goes without saying." Yet, it is enough to tangibly experience institutionally imposed severance from nature, to develop a profound appreciation of her value. Hopefully, this realization will awaken genuine, compassionate care.

In the chapters above, we conceptualized the garden as an epitome of the intimacy of connections and relationships. This intimacy typically eludes our perception, which is often skewed by our tendency to view the world through the lenses of the modality of practice, science, or history (to use Michael Oakeshott's vocabulary). This is the reason why, throughout this "philosophical" book, we have consistently transgressed disciplinary borders: to communicate the essence of the intimacy, we found poetry to be the most appropriate language. For some, the garden is poetry of nature. One can concur with this statement, if poetry is to be understood broadly, as a refined, sublimated, perception of the world's complexity. The crudely drawn scientific categories and broadly traced trajectories of semantic and epistemic justifications may overlook the subtlety of the fabric of interwoven dependencies. And only a sensitivity of a poetic pedigree can reverberate with this pulsating reality. Poetry precedes—and complements—philosophy and scientific discourse alike.

The fundamental bond with the world, care for nature, and sensitivity to its subtleties, is the substance of poetry. Poetry is uniquely capable of insight into the world's structure through the organicity of perception, energizing the recognition of the profundity of fundamental distinctions to the point, when the blurring of the boundaries becomes not only obvious but also welcome. Thus, poetry's emotive powers allow us to penetrate the subtle entanglements of being-in-the-world, and to attune to the intimacy informing our complex relationships with it. The transgressive metaphysics, whose essentials we have only managed to sketch out, acknowledges the subtle structure of the blurred world of the garden, which, at its most profound—and most obvious level—defies schematic distinctions. The garden is not an enclosure: it is an opening of the self to the fascinating diversity of the world.

It may indeed be fantasy when I
Essay to draw from all created things
Deep, heartfelt, inward joy that closely clings;
And trace in leaves and flowers that round me lie
Lessons of love and earnest piety.
So let it be; and if the wide world rings
In mock of this belief, it brings
Nor fear, nor grief, nor vain perplexity.
So will I build my altar in the fields,
And the blue sky my fretted dome shall be,
And the sweet fragrance that the wild flower yields
Shall be the incense I will yield to Thee,
Thee only God! and thou shalt not despise
Even me, the priest of this poor sacrifice.[29]

Notes

1 See *The Selected Works of Arne Naess*, Vol. 1–10, edited by Harold Glasser and Alan Drengson, in cooperation with Arne Naess, Bill Devall, and George Sessions (Dordrecht: Springer, 2005).

2 Bill Devall and Alan Drengson, "Preface," in *The Selected Works of Arne Naess*, Vol. 10, edited by Harold Glasser and Alan Drengson, in cooperation with Arne Naess, Bill Devall, and George Sessions (Dordrecht: Springer, 2005), XVII.

3 Arne Naess, "The Deep Ecology Movement: Some Philosophical Aspects," in *The Selected Works of Arne Naess*, Vol. 10, edited by Harold Glasser and Alan Drengson, in cooperation with Arne Naess, Bill Devall, and George Sessions (Dordrecht: Springer, 2005), 33–55.

4 Arne Naess, "The Deep Ecology Movement: Some Philosophical Aspects"..., 37–38.

5 "Generic attributes are understood to be interwoven aptitudes and abilities such as academic inquiry and intellectual curiosity, the ability to accommodate diversity and alternative perspectives, the ability to create and defend ideas, and the ability to use communication as a vehicle for learning." Simon C. Barrie, "A Conceptual Framework for the Teaching and Learning of Generic Graduate Attributes," *Studies in Higher Education* 32, no. 4 (2007): 456.

6 Arne Naess, "The Deep Ecology Movement: Some Philosophical Aspects"..., 38. In his commentary, on page 41, Naess adds: "Some economists criticize the term quality of life because, they say, it is vague. On closer inspection, however, what they consider to be vagueness is actually the nonquantitative nature of the term. One cannot quantify adequately what is important for quality of life as discussed here, and there is no need to do so."

7 Henry David Thoreau, *Life Without Principle* (London: Global Grey, 2018), ebook.

8 Henry David Thoreau, *Journal*, November 4, 1852, in *Thoreau's Seasons*, https://stpeter.im/writings/thoreau/seasons-9.html (accessed: November 4, 2023).

9 "I should like to interpret the relationship at issue by recourse to the concept of equilibrium and the way in which we experience it. This concept already plays a major role in the Hippocratic writings. And in fact it is not merely the phenomenon of human health which almost cries out to be understood in terms of the natural condition of equilibrium. The concept of equilibrium also readily offers itself for our understanding of nature in general [...]. Nature is therefore something which as it were holds to its own course, and does so in and of itself." Hans-Georg Gadamer, *The Enigma of Health: The Art of Healing in a Scientific Age*..., 35–36.

10 David Fenner and Ethan Fenner, *The Art and Philosophy of the Garden* (New York: Oxford University Press, 2024), 1.

11 David Fenner and Ethan Fenner, *The Art and Philosophy of the Garden*...,1.

12 David Fenner and Ethan Fenner, *The Art and Philosophy of the Garden*..., 1.

13 "We will use the term 'garden' in two ways: 'garden' means any particular garden or any token of the type; 'The Garden' means the type, abstract kind, or ontological category that includes only and all gardens. Used in this second way, 'The Garden' is the name of a set of objects." David Fenner and Ethan Fenner, *The Art and Philosophy of the Garden*..., 2.

14 David Fenner and Ethan Fenner, *The Art and Philosophy of the Garden*..., 4.

15 David Fenner and Ethan Fenner, *The Art and Philosophy of the Garden*..., 53.

16 Rosario Assunto, *Filosofia del giardino e filosofia nel giardino: Saggi di teoria e storia dell'estetica* (Rome: Bulzoni, 1981), 14–32.

17 John Dixon Hunt, *Greater Perfections: The Practice of Garden Theory* (London: Thames and Hudson, 2000), 34.

18 Odo Marquard, *Abschied vom Prinzipiellen: Philosophische Studien* (Stuttgart: Reclam, 2000).

19 David Fenner and Ethan Fenner, *The Art and Philosophy of the Garden*..., 11–12.

20 Rosario Assunto, "Per una ontologia del giardino," in *Ontologia e teleologia del giardino* (Milan: Guerini e Associati, 1994), 19–38.

21 Ralph Waldo Emerson, *Nature* (Boston: James Munroe and Company, 1836), 14.

22 Jonasz Kofta, "Pamiętajcie o ogrodach" in *Pamiętajcie o ogrodach* (Warszawa: Czytelnik, 1991). Translated by Paweł Jędrzejko.

23 Zygmunt Bauman, *Modernity and the Holocaust* (Cambridge: Polity Press, 1989), 92. The bitter disappointment with the unfulfilled promise of the project of modernity, which became the share of those who experienced the atrocities of the Great War, culminates in the reflection of the generation, who witnessed, or were directly affected by, the systematic deployment of the principles of social Darwinism during World War II. Justifing social exclusion and violence, social Darwinism, framing such practices as outcomes of "natural" selection, driven by the principle of the survical of the fittest, has historically influenced various political ideologies and policies, such as *laissez-faire* capitalism, imperialism, colonialism, or eugenics. Bauman demonstrates that the social-Darwinist ideology transforms the garden into a murderous industry. Distrustful of nature and distrustful of spontaneity, such a "garden" is a bitter parody of "cultivation": a horrid concept fostering devastation, suffering, and eugenic inbreeding—polar opposites of flourishing, eudaimonic happiness, or health. Bauman's example powerfully shows how far away the modernist project of "garden-as-society" is from the metaphysics of the garden based on trust, attunement, and affirmation of nature's spontaneous dynamics. Objecting to the social-Darwinist fallacy, fundamental to the belief that "hegemonic gardening" is tantamount to the cultivation of a "natural order," Bauman, implicitly, points us toward the conclusion that in the context of modern, oppressive, yet still dominant systems, the relation of a "weed-free-garden" to the world is dangerously close to that of the world to a concentration camp, whose inmates survive or die, while the alleged *lebensraum* for the select few to thrive—is, in fact, on fire: fire, which will bring demise to all.

24 Zygmunt Bauman, *Modernity and the Holocaust*..., 18.

25 Zygmunt Bauman, *Modernity and the Holocaust*..., 219.

26 Henry D. Thoreau, *The Writings of Henry D. Thoreau. Walden: 150th Anniversary Edition*, edited by Lyndon Shanley, with an introduction by John Updike (Princeton: Princeton University Press, 2004), 210.

27 For a broader spectrum of possible readings of this term, consult, for instance: Wojciech Kalaga and Tadeusz Rachwał, eds., *The Wild and the Tame: Essays in Cultural Practice* (Katowice: Wydawnictwo Uniwersytetu Śląskiego, 1997), or Tadeusz Rachwał and Tadeusz Sławek, eds., *Word-Subject-Nature: Studies in Seventeenth- and Eighteenth-Century Culture* (Katowice: Wydawnictwo Uniwersytetu Śląskiego, 1996).

28 Henry D. Thoreau, *The Writings of Henry D. Thoreau. Walden: 150th Anniversary Edition*..., 317.

29 Samuel Taylor Coleridge, "To Nature," https://www.poemhunter.com/poem/to-nature/ (accessed: March 2, 2024).

Conclusions

On Health and Its (Not So) Obvious Connections

In the light of the multidimensionality of the argumentation leading to our formulation of the proposition of the transgressive metaphysics of the garden—whose essential tenets we have attempted to make tangible against the complex background of existing theories, philosophical stances, cultural practices, and personal experiences—it becomes clear that what we discuss is not "ideology," but a first philosophy, rooted in non-discursive, lived, commonplace experience—a philosophy whose fundamentals are already manifest in the life practice of many communities world-wide, but largely marginalized in the Western world. As numerous examples we have invoked demonstrate, coming to terms with the *atmospheric* reality of our existence coupled with the profound realization of our subcutaneous connection with the pulses of nature not only allow us to "live deliberately," but also to discover that our daily professional goals, our social duties, or family-related involvements, do not have to stand in our way toward eudaimonic well-being. Furthermore, a deep, attentive reflection on our own, unmediated experiences allows us to profoundly comprehend that the compassionate inclusion of both our community and of the whole of our environment in the horizon of our non-egotistic, eudaimonic purposes is the inevitable consequence of our comprehension of the ultimate indivisibility, or wholeness, of the universe. In fact, as we have demonstrated, in daily practice, such a welcoming attitude comes naturally, if only we allow ourselves to be attentive to how our bodies—elemental—resonate with the pulse of nature, of which everything and everyone is a part. To be healthy is to be in balance. To be in balance, one cannot dismiss that, which one needs to be in balance with. In simple terms, our well-being is thus not unrelated to the well-being of others, nor is it independent of the well-being of the forests, oceans, or air, upon the health of which our life unquestionably hinges. In other words: choosing to tend to the garden of the universe, we choose to flourish, and allow the world around us to bloom. These "simple terms" are one of the reasons why this book was so difficult to write: we knew that we could not make anyone take our words for granted. Running the risk of mockery, we needed to strip to the bone, expose our own frailties, to afford others a chance to feel the

DOI: 10.4324/9781003328469-14

connection, and then—we needed to offer those willing to attune a language, in which they could name what they themselves sense. Even if fleetingly.

This reflection, personal as it is, leads us to more practical conclusions. In our perspective, the concept of flourishing, shared by Western philosophers of Antiquity, the Buddha, and the lores of many other cultures, emerges as central. On the one hand, related, but not reducible, to purpose, it is the fundament of the understanding of health as well-being. On the other hand, the concept is essential as a prerequisite of the profound reassessment of the very idea of healing. If flourishing is a function of the attentive attunement to the rhythms of nature that (whether we acknowledge it or not) reverberate in us, the process of healing necessitates a context conducive to the patients' self-discoveries: to heal, the patients need to recognize their originary entanglement in the fabric of life, reconnecting with the "primitive present"—and, to help them attain this goal, their doctors/healers must grant them the possibility of such a recognition. Grounding the healing procedures in transgressive metaphysics opens a space for the patient to develop an intimate understanding, based on insights provoked by their personal relationship with nature, and thus allows them to tap into the metaphysical fundamentals of their own connectedness to the whole. Indeed, nature loves to hide, yet it reveals itself to each of us through direct, individual experience. Such an unmediated experience, whether spontaneous or guided—allows one to step into a liminal space of the intimacy at the intersection of oneself and the all-encompassing nature. Such an experience affords one a possibility to become attuned to nature's rhythms, which *always-already* pulse in our bodies. In this sense, for instance, our respiration may present itself to us as an indispensable element of the whole: if we do not limit our perception of it to the contraction and expansion of muscles responsible for the intake and exhaust of air in our lungs, we will notice that the process rather heavily depends on the fact that the air our muscles pump in an out contains oxygen replenished in the atmosphere as a result of photosynthesis. With such an understanding in mind, each deep breath of fresh air in the morning is an invitation to bring certain emotions into our field of experience, which thus becomes an extended body, immersed in the fullness of life. In such moments, we do not merely experience atmospheres: we *become* atmospheres, our body-walls no longer consequential.

***

In this book, we examined the manners in which observable behavioral patterns actualize certain fundamental ideas related to the concept of health. Addressing such an important—and complex—issue necessitates the adoption of a specific methodological approach. In our case, we began by identifying and analyzing the essential traits of the existing theoretical framework, within which the problems of our central interest have been located. This allowed us to outline a protocol highlighting the discrepancies between the existing state of knowledge and the findings resulting from our critical assessment of the dominant paradigms of thinking about health, well-being, and the human-in-the-universe. This strategy allowed us to clearly define the objective of our explorations, which led to the second phase of our study—one based on the direct experience of dissimilar, yet very much alike, gardens, where each of us discovered our ideas as embodied. The autoethnographic, experiential material that we were able to collect provided us with a fundament for revisions, and therefore substantiated our recapitulation of theoretically grounded findings, which should be seen as a prelude to the thorough rethinking of the relationship between humans and nature in the context of the conceptualizations of health.

To view the problems afresh, we found it necessary to return to the very sources of reflection, to what is most fundamental—the metaphysical assumptions underlying the human place in the world. Our considerations culminate in the final, third, part of our work; yet, although it brings us to the point when we were able to propose a new—transgressive—metaphysical model, we do not presume that our proposition definitively resolves all the issues, in which conceptualizations of health are entangled. Rather, the findings from our research lead us to the formulation of a new theoretical *perspective* on health—one in sync with the "grassroots" intuitions, informing the daily practice of the cultivation of atmospheric health across different climate zones, across different countries. Whether in Greece, Scandinavia, or Thailand, the fundamental assumption underlying these intuitions is that well-being requires the departure from the monadic concept of humanity, but also the adoption of a perspective, in which the human becomes "extended human," whose organs—such as lungs from our previous example—are not independent of human environment: the understanding of the function of lungs is only possible if their sense is *extended* to include the air (carrying pollens and countless other micro-objects), solar energy, carbon dioxide we breath out, water, minerals, and green plants that produce oxygen and render air breathable as a result of photosynthesis. Since breathing is a function of all of these factors (and more), the *sense* of our lungs *transcends the limits of their purpose*. If this is the case, then one may safely assume that human health, even in the purely physical dimension of existence, cannot be reduced to the limits of the human body.

Based on the observations derived from the analyses of case studies representing positions formed in different cultural traditions, the findings collected lead to our philosophical reassessment of health as a *process* (not: *a state*) and of the ramifications of such a change of perspective. Indicating parallels between the perceptions of processes responsible for human well-being as emphasized by the practitioners of traditional Greek lifestyle, *friluftsliv*, and *bhāvanā*, we seek to support the thesis that the pandemic-induced need to re-create one's bonds with nature may prove to be conducive to the awakening of profound awareness of the fact that cultivating both the environment and compassion-based communal relations one cultivates with the commonwealth of atmospheric health. Its conclusion paves the path for our proposition of a new model of garden-inspired transgressive metaphysics, offering the reader a smooth transition to our proposal of a new model of health. We posit a revised conceptual system, grounded in transgressive metaphysics, in which health manifests itself as a concept transcending the ego-logical notion of human-as-separate from the world. At the same time, as a function of the communion of the human and the environment, health is a property of the common/wealth.

We have demonstrated that thinking about a good life as of the cultivation of a garden does not reduce the gardener to a demiurge wielding power—a hegemon who "includes" or "excludes," "marginalizes" or "centralizes," arbitrarily distinguishing between "valuable" plants and "useless weeds." The concept of the garden may be—and has been—employed in a variety of ways (for instance, as in Bauman, as a metaphor of modern society). For us, however, the garden holds a different significance: rather than a metaphor, it is an operational, albeit atmospheric, concept, epitomizing the "extended human," whose rhythms are attuned to the natural rhythms and cycles, and whose compassionate care affects the whole of existence. A thus conceived garden distinctly affirms the world, stressing the interconnectedness of self and other, and doing away with the binarities of body and soul, culture and nature, etc. The metaphysics of the garden is a reflection on transgressiveness, wholeness, and mutual relations; it sensitizes us to the functional dimension of existence without disregarding its substantial aspects. The garden is a space of interdependence, a new agora, and a form of communality. Depending on the efficacy of the global immune system—which, like the garden, may be consciously tended to, but requires our attunement to its biological rhythms—health demands that we adopt a metaphysical stance in which "every atom belonging to me as good belongs to you." Summarized by Walt Whitman's adage, such garden-inspired metaphysics triggers a revision of ethics, which, inexorably, transmogrifies into "atmospheric ethics." Consequently, the communion of man and the universe inevitably involves his or her development of a profound sense of the agora—a union with others, which is the *sine-qua-non* condition of the common/wealth. Caring for the health of the common/wealth requires attention to the rhythms of animate and inanimate nature alike, coupled with mindfulness in actions taken with respect to the community; as we stressed it, neglect in any of these areas will inevitably adversely affect one's own well-being.

Our insights into the Scandinavian and Greek lifestyles, in the context of the organization of the lives of local communities, give us reasons to believe that continued advocacy for such an understanding of health is effective. Inspired by the success of such advocacy in Sweden, Norway, and Finland, we are inclined to assume that systematic promotion of the atmospheric ethics we propose may give rise to communities immune not only to pathogens of social or economic provenance (like institutional racism or greed-driven overexploitation of natural resources), but also—owing to immersion in nature and to constant exposure to limited doses of natural pathogens—communities consisting of individuals increasingly immune to biological contagions as well. Aware of the fact that direct "transplantation" of historically and culturally rooted, and geographically conditioned, social norms onto foreign soils is far more difficult than the exportation of laws, we believe that, instead of promoting "the Scandinavian model," "the Greek model," or the "Thai model," we should be the advocates of the transgressive metaphysics of the garden, which the above models embody. Sublimating the essentials of the first philosophy underlying the social practices in the areas we studied, we arrive at a metaphysical model, with which one can identify irrespective of their cultural background: after all, gravity affects everyone, everyone needs oxygen to live, and everyone experiences aging.

And thus our insight into the role of philosophy as a means to promote atmospheric health automatically renders it a means to promote the atmospheric ethics of compassion. At the brink of environmental catastrophe, we can—and should—envision new types of public institutions, whose philosophies would embrace co-existential, processual, and practice-related metaphysics. It is worth observing that the concept of flourishing we promote encompasses both *caring for vitality* and *cultivating wisdom*, which was precisely how care for individuals was managed in the garden of Epicurus. In today's world, however, the hospital is responsible for the former, and the school for the latter. In the future, these responsibilities might once again be unified within a single institution, thus contributing to the decompartmentalization of the human, who, in order to flourish, cannot be reduced to the limits of their body wall, or the stiflingly inoperational body/mind binary.

Consequently, the perspective that we have proposed in this book emphasizes the potential for a revision of the understanding of philosophy, which—centered around particular biotopes (forests, mountains, rivers, seas, etc.) and not around the consistency with the discourses developed by particular philosophical schools (whether Stoic, Epicurean, Emersonian, or Buddhist)—could transgress the limits of the university campuses. Such a philosophy could transmogrify into a form of communal reflection embracing a co-existential, processual, and practice-related metaphysics. Thereby, as such, the metaphysics of the garden may prove to be a valid theoretical fundament for garden-based education, garden-based economics, or, like in Scandinavia, garden-based legislature.

We do realize that the proposal presented in this book may seem radical. However, if there has ever been an opportune moment to reconsider the foundational principles of our own attitudes to health and well-being, it is now: the crisis of the pandemic, despite its dramatic consequences, carries the potential to awaken healthy criticism not only among philosophers, but also—and perhaps more importantly—within communities. And because we believe that true change requires reflection on what is fundamental, we do hope that our project of the transgressive metaphysics of the garden, energizing the revision of dominant concepts of health, will ultimately serve as a catalyst in the process of shifting the ossified perspective "just a little."

Bibliography

"5 Tips for Getting the Most Out of *Lagom* Lifestyle." Technogym Newsroom. https://www.technogym.com/us/newsroom/5-tips-lagom-lifestyle/#:~:text=What%20is%20Lagom%3F,and%20juncture%20of%20everyday%20life (accessed: August 5, 2024).

Adams, Tony E., and Stacy Holman Jones. "Telling Stories: Reflexivity, Queer Theory, and Autoethnography." *Cultural Studies ↔ Critical Methodologies* 11, no. 2 (2011): 108–116. DOI: 10.1177/1532708611401329

Aelianus, Claudius. *Varia historia*. IV.13 (Usener, fr. 473). Quoted in https://www.attalus.org/translate/epicurus3.html#us4f

Allen-Collinson, Jacquelyn. "Breathing in Life: Phenomenological Perspectives on Sport and Exercise." In *Routledge Handbook of Qualitative Research in Sport and Exercise*, edited by Brett Smith and Andrew Charles Sparkes, 315–330. London: Routledge, 2016.

Allen-Collinson, Jacquelyn. "Intention and Epoché-in Tension: Autophenomenography, Bracketing and a Novel Approach to Researching Sporting Embodiment." *Qualitative Research in Sport, Exercise and Health* 3, no. 1 (2011): 48–62. DOI: 10.1080/19398441.2010.541484

André, Karin, Katarina Axelsson, Elena Dawkins, and Åsa Gerger Swartling. *Drivkrafter för hållbar konsumtion på lokal nivå: Svenska kommuners roll och möjligheter*. Rapport 6960. Stockholm: Naturvårdsverket, February 2021.

Aristotle. *De Anima*, Book One, Chapter Three. n.d. https://psychclassics.yorku.ca/Aristotle/De-anima/de-anima1.htm (accessed: May 6, 2024).

Aristotle, *Physica* [*Physikē Akroasis*], in *Aristotelis: Physica*, ed. W. D. Ross, Oxford Classical Texts (Oxford: Oxford University Press, 1950). DOI: 10.1093/oseo/instance.00259270

Assunto, Rosario. *Filosofia del giardino e filosofia nel giardino: Saggi di teoria e storia dell'estetica*. Rome: Bulzoni, 1981a.

Assunto, Rosario. "Filosofia del giardino e filosofia nel giardino." In *Filosofia del giardino e filosofia nel giardino: Saggi di teoria e storia dell'estetica*. Rome: Bulzoni, 1981b.

Assunto, Rosario. "Il giardinaggio come operazione filosofica." In *Paradisi ritrovati. Esperienze e proposte per il governo del paesaggio e del giardino. Con un'introduzione di Rosario Assunto*, edited by Mariapia Cunico and Domenico Luciani, 7–12. Milan: Guerini e Associati, 1991.

Assunto, Rosario. *Ontologia e teleologia del giardino*. Milan: Guerini e Associati, 1994a.

Assunto, Rosario. "Per una ontologia del giardino." In *Ontologia e teleologia del giardino*. Milan: Guerini e Associati, 1994b.

Atleo, E. Richard (Umeek). *Tsawalk: A Nuu-chah-nulth Worldview*. Vancouver, BC: UBC Press, 2005.

Augé, Marc. *Non-Places: Introduction to an Anthropology of Supermodernity*. Translated by John Howe. London; New York: Verso, 1995.

Bańka, Józef. *Ja teraz. U źródeł filozofii człowieka współczesnego*. Katowice: Wydawnictwo "Śląsk", 1983.

Baratay, Eric. *Le point de vue animal. Une autre version de l'histoire*. Paris: Le Seuil Edition, 2012.

Barrie, Simon C. "A Conceptual Framework for the Teaching and Learning of Generic Graduate Attributes." *Studies in Higher Education* 32, no. 4 (2007): 439–458.

Bauman, Zygmunt. *Modernity and the Holocaust*. Cambridge: Polity Press, 1989.

Bengtsson, Bertil. *Allemansrätten—Vad säger lagen?* Stockholm: Naturvårdsverket, 2023.

Bergström, Lena, Marcus C. Öhman, Charlotte Berkström, Martin Isæus, Lena Kautsky, Birgit Koehler, Antonia Nyström Sandman, Hans Ohlsson, Richard Ottvall, Henriette Schack, and Magnus Wahlberg. *Effekter av havsbaserad vindkraft på marint liv: En syntesrapport om kunskapsläget 2021*. Rapport 7049. Stockholm: Naturvårdsverket, 2022.

Bergström, Lena, Ulf Bergström, Scott Cole, Linus Hasselström, Patrik Kraufvelin, Per-Olav Moksnes, Göran Sundblad, Tore Söderqvist, and Sofia A. Wikström. *Ekologisk kompensation i kustmiljön: Hur kan man uppväga förluster av biologisk mångfald och ekosystemtjänster i samband med mänsklig verksamhet i kustområdet?* Rapport 6994. Stockholm: Naturvårdsverket, 2021.

Berkel, Tazuko Angela van. *The Economics of Friendship: Conceptions of Reciprocity in Classical Greece*. Mnemosyne Supplements. Leiden; Boston: Brill, 2020.

Bille, Mikkel, and Siri Schwabe. *The Atmospheric City*. London: Routledge, 2023.

Boowa, Ñanadhammo. *The Path to Arahantship: A Compilation of Venerable Ācariya Mahā Boowa's Dhamma Talks About His Path of Practice*. Wat Pa Baan Taad: Ampher Meuang, 2005.

Boyd, Drick. "Autoethnography as a Tool for Transformative Learning About White Privilege." *Journal of Transformative Education* 6 (2011): 212–225. DOI: 10.1177/1541344608326899

Braswell, William. "Melville as a Critic of Emerson." *American Literature* 9, no. 3 (November 1937): 317–334. DOI: 10.2307/2919662

Buettner, Dan. *The Blue Zones, Second Edition: 9 Lessons for Living Longer from the People Who've Lived the Longest*. Washington, D.C.: National Geographic Books, 2012.

Buettner, Dan. *The Blue Zones: Lessons for Living Longer from the People Who've Lived the Longest*. Washington, D.C.: National Geographic, 2008.

Buettner, Dan. *The Blue Zones Solution: Eating and Living Like the World's Healthiest People*. Washington, D.C.: National Geographic, 2015.

Buettner, Dan. *The Blue Zones of Happiness: Lessons from the World's Happiest People*. Washington, D.C.: National Geographic, 2017.

Buettner, Dan. *Thrive: Finding Happiness the Blue Zones Way*. Washington, D.C.: National Geographic, 2010.

Buettner, Dan. *The Blue Zones Kitchen: 100 Recipes to Live to 100*. Washington, D.C.: National Geographic, 2019.

Buettner, Dan. *The Blue Zones Challenge: A 4-Week Plan for a Longer, Better Life*. Washington, D.C.: National Geographic, 2021.

Butler, Robert N. *Why Survive? Being Old in America*. New York: Harper and Row, 1975.

Butler, Robert N. *The Longevity Prescription: The 8 Proven Keys to a Long, Healthy Life*. New York: Avery/Penguin Group, 2010.
Cavafy, Constantine P. "Ithaka." Translated by Edmund Keeley. In *C. P. Cavafy, Collected Poems*. Translated by Edmund Keeley and Philip Sherrard. Princeton, NJ: Princeton University Press, 1975.
Chah, Ajahn. *A Tree in a Forest: A Collection of Ajahn Chah Similes*. Taoyuan: Yuan Kuang Publishing House, 1994.
Chamberlain, Lesley. *Nietzsche in Turin: The End of the Future*. London: Pushkin Press, 1997.
Chapman, Robert, and Havi Carel. "Neurodiversity, Epistemic Injustice, and the Good Human Life." *Journal of Social Philosophy* 53, no. 4, Special Issue: Analyzing Social Wrongs (Winter 2022): 614. DOI: 10.1111/josp.12456
Chojecki, Andrzej. *IO*. Kraków: Wydawnictwo Literackie, 1995.
Chojecki, Andrzej. *W stronę niewiedzy. Antyontologia*. Warszawa: Graf, 1991.
Cicero, Marcus Tullius. *De finibus bonorum et malorum*. II, 28, 91 (Usener, fr. 468). Quoted in https://www.attalus.org/translate/epicurus3.html#us4f (accessed: July 5, 2023).
Clement of Alexandria. *Stromata*. VI.2 (Usener, fr. 476).
Coleridge, Samuel Taylor. "To Nature." https://www.poemhunter.com/poem/to-nature/ (accessed: March 2, 2024).
Cooper, David Edward. *A Philosophy of Gardens*. Oxford: Clarendon Press, 2010.
Cotton, Anne. "Gardener of Souls: Philosophical Education in Plato's *Phaedrus*." In *Gardening: Philosophy for Everyone*, edited by Dan O'Brien, 233–234. Sussex: Wiley Blackwell, 2010.
Crosthwaite, Paul, ed. *Criticism, Crisis, and Contemporary Narrative: Textual Horizons in an Age of Global Risk*. New York: Routledge, 2011.
Danzig, Gabriel. "Why Socrates was not a Farmer: Xenophon's *Oeconomicus* as a Philosophical Dialogue." *Greece and Rome* 50, no. 1 (2003): 57–76.
Davidson, Arnold Ira. "Spiritual Exercises and Ancient Philosophy: An Introduction to Pierre Hadot." *Critical Inquiry* 16, no. 3 (1990): 475–482. DOI: 10.1086/448542
de Man, Paul. "Criticism and Crisis." In *Blindness and Insight: Essays in the Rhetoric of Contemporary Criticism*, rev. ed., 9–34. Minneapolis: University of Minnesota Press, 1983 [1967].
De Silva, Lily. *Essays on Buddhism Culture and Ecology*. Dehiwala: Buddhist Cultural Center, 2001.
Devall, Bill, and Alan Drengson. "Preface." In *The Selected Works of Arne Naess*, Vol. 10, edited by Harold Glasser and Alan Drengson, in cooperation with Arne Naess, Bill Devall, and George Sessions. Dordrecht: Springer, 2005.
Dezallier d'Argenville, and Antoine Joseph. *La théorie et la pratique du jardinage*. Paris: Jean Mariette, 1709.
Dezallier d'Argenville, Antoine Joseph. *The Theory and Practice of Gardening: Wherein is Fully Handled All that Relates to Fine Gardens, Commonly Called Pleasure-Gardens, as Parterres, Groves, Bowling-Greens, &c*. Translated by John James. London: Geo. James, 1712.
Dhammādharo, Lee. *The Words of Ajaan Lee*. Kindle ed. Publisher not provided, 2011.
Dharma Seed. https://dharmaseed.org/talks/68169/ (accessed: April 10, 2024).
Diogenes Laertius. *Lives of Eminent Philosophers*. X.11.; X.120a.
Dorter, Kenneth. "Imagery in Plato's Phaedrus." *Journal of the History of Philosophy* 9, no. 3 (1971): 279–288.
Dos Reis, Róbson Ramos, João Carlos Onofre Pinto, Bruno Nobre, Andreas Gonçalves Lind, and Ricardo Barroso Batista. "Thinking the Pandemic: Philosophical Perspectives." *Revista Portuguesa de Filosofia* 77, no. 2/3 (2021): 478–479. DOI: 10.17990/rpf/2021_77_2_0478

Durrell, Lawrence. *Prospero's Cell: A Guide to the Landscape and Manners of the Island of Corfu*. London: Faber and Faber, 2000.
Early Greek Philosophy, vol. 3, edited and translated by André Laks and Glenn Warren Most. Cambridge: Harvard University Press, 2016.
EAB. "Digital Student Activism: COVID-19." EAB. n.d. https://eab.com/resources/whitepaper/digital-student-activism-covid-19/ (accessed: May 3, 2024).
Emerson, Ralph Waldo. *Nature*. Boston: James Munroe and Company, 1836.
Enigma. *Return to Innocence*. Virgin Records, 1994. https://www.youtube.com/watch?v=Rk_sAHh9s08 (accessed: May 3, 2024).
Erler, Michael. "Autodidact and Student: On the Relationship of Authority and Autonomy in Epicurus and the Epicurean Tradition." In *Epicurus and the Epicurean Tradition*, edited by Jeffrey Fish and Kirk R. Sanders, 26. Cambridge: Cambridge University Press, 2011.
Fabjański, Marcin. *Embodied Nature and Health: How to Attune to the Open-source Intelligence*. London: Routledge, 2022.
Fabjański, Marcin. Interview by Przemysław Poznański. "Szukajcie dostrojenia do wszechświata." *Zupełnie inna opowieść*, March 17, 2020. https://zupelnieinnaopowiesc.com/2020/03/17/marcin-fabjanski-szukajcie-dostrojenia-do-wszechswiata-rozmawia-przemyslaw-poznanski/ (accessed: April 25, 2024).
Fabjański, Marcin. "Ścieżka w Lesie Mokrzańskim." *Ścieżki Nowej Obecności*. Fundacja Nowa Obecność. http://sciezki.org/sciezka-w-lesie-mokrzanskim/ (accessed: May 27, 2024).
Fabjański, Marcin. "Ścieżka w Sowim Jarze." *Ścieżki Nowej Obecności*. Fundacja Nowa Obecność. http://sciezki.org/sciezka-w-sowim-jarze (accessed: May 27, 2024).
Fabjański, Marcin. "Walking-Derived Metaphysics in Nietzsche's *Thus Spoke Zarathustra*." *Forum Philosophicum* 29, no. 1 (2024): 29–41. DOI: 10.35765/forphil.2024.2901.02
Fabjański, Marcin, and Lorenza Tiberio. "Presence-Centered Flourishing: A Proposal of Alternative Strategies to Promote Sustainable Living." *Journal of Educational, Cultural and Psychological Studies (ECPS)*, no. 27 (2023): 129. ISBN 978-88-5513-096-7. DOI: 10.7358/ecps-2023-027-f
Favara-Kurkowski, Marianna. "Reclaiming Time Aesthetically: Hadot, Spiritual Exercises and Gardening." *Eidos. A Journal for Philosophy of Culture* 5, no. 2 (2021): 7–21. DOI: 10.14394/eidos.jpc.2021.0002
Fenner, David, and Ethan Fenner. *The Art and Philosophy of the Garden*. New York: Oxford University Press, 2024.
Ferraiolo, Angela. "Fictional Characters as Adaptive Systems: The Malaspina and de Graaf Gardens." In *Proceedings of EVA London 2020*, 313–321. London: BCS Learning & Development Ltd., July 2020. DOI: 10.14236/ewic/EVA2020.57
Fisher, Charles S. *Meditation in the Wild: Buddhism's Origin in the Heart of Nature*. Winchester, UK; Washington, USA: Changemakers Books, 2013a.
Fisher, Christopher Scott. *Meditation in the Wild: Buddhism's Origin in the Heart of Nature*. Winchester, UK: Changemakers Books, 2013b.
Foucault, Michel. *Discipline and Punish: The Birth of the Prison*. New York: Vintage Books, 1995.
Foucault, Michel. *Histoire de la sexualité: 1: La volonté de savoir*. Paris: Gallimard, 1976.
Foucault, Michel. *Histoire de la sexualité: 2: L'usage des plaisirs*. Paris: Gallimard, 1984a.
Foucault, Michel. *Histoire de la sexualité: 3: Le souci de soi*. Paris: Gallimard, 1984b.
Freeland, Cynthia. "Imagery in the *Phaedrus*: Seeing, Growing, Nourishing." *Symbolae Osloenses* 84, no. 1 (2010): 62–72.
Gadamer, Hans-Georg. *Über die Verborgenheit der Gesundheit*. Berlin: Suhrkamp Verlag, 2010.

Gadamer, Hans-Georg. *The Enigma of Health: The Art of Healing in a Scientific Age*. Translated by Jason Gaiger and Nicholas Walker. Cambridge: Polity Press, 1996. ISBN 9780804726924.

Gadamer, Hans-Georg. *The Enigma of Health: The Art of Healing in a Scientific Age*. Translated by Jason Gaiger and Nicholas Walker. Stanford, CA: Stanford University Press, 1996.

Gardening Philosophy for Everyone: Cultivating Wisdom, edited by Dan O'Brien. Chichester, West Sussex: John Wiley & Sons Ltd., 2010.

Garner, Leslie. *The NHS: Your Money or Your Life*. New York: Penguin Books, 1979.

Gelter, Hans. "*Friluftsliv*: The Scandinavian Philosophy of Outdoor Life." *Canadian Journal of Environmental Education* 5 (Summer 2000): 77–92. Lulea University of Technology, Sweden.

Gilpin, William. *Observations on the River Wye and Several Parts of South Wales & Relative Chiefly to Picturesque Beauty; Made in the Summer of the Year 1770*. Richmond, Surrey: Richmond, 1973.

Gilpin, William. *Three Essays: On Picturesque Beauty; On Picturesque Travel; and on Sketching Landscape: To Which Is Added a Poem on Landscape Painting*. London: R. Blamire, 1792.

Global Youth Mobilization. *Global Youth Mobilization Interim Impact Report: February 2022*. https://globalyouthmobilization.org/wp-content/uploads/2022/03/Global-Youth-Mobilization-Interim-Impact-Report-February-2022.pdf

von Goethe, Johann Wolfgang. *Faust*. Translated and with an introduction by Walter Kaufmann. New York: Anchor Books, 1962. ISBN 978-0385031141.

von Goethe, Johann Wolfgang. *Faust*. Translated by A. S. Kline. *Poetry in Translation*, n.d. https://www.poetryintranslation.com/PITBR/German/FaustIProl.php (accessed: October 10, 2023).

Gooley, Tristan. *How to Read Water: Clues, Signs and Patterns from Puddles to the Sea*. London: Sceptre, an imprint of Hodder and Stoughton, an Hachette UK Company, 2016.

Gordo, Ángel, and Jan De Vos. "Psychologism, Psychologising and De-psychologisation." *Annual Review of Critical Psychology* 8 (2010): 3–7.

Hadot, Pierre. *Exercices spirituels et philosophie antique*. Paris: Études Augustiniennes, 1981.

Hadot, Pierre. *The Veil of Isis: An Essay on the History of the Idea of Nature*. Cambridge, MA: Harvard University Press, 2006.

Hadot, Pierre, and Arnold Ira Davidson. *Philosophy as a Way of Life: Spiritual Exercises from Socrates to Foucault*. Malden, MA: Blackwell, 1995.

Hamman, Gwyneth, and Itai Ivtzan. "30 Minutes in Nature a Day Can Increase Mood, Well-Being, Meaning in Life and Mindfulness: Effects of a Pilot Programme." *Social Inquiry into Well-Being* 2 (2016): 34–46. DOI: 10.13165/SIIW-16-2-2-04

Han, Byung-Chul. *The Palliative Society: Pain Today*. Medford, MA: Polity Press, 2021.

Hanson, Rick, and Richard Mendius. *Buddha's Brain: The Practical Neuroscience of Happiness, Love & Wisdom*. Oakland, CA: New Harbinger Publications, 2009.

Hawthorne, Nathaniel. *The English Notebooks*. Based upon the original manuscripts in the Pierpont Morgan Library and edited by Randall Stewart. New York: Modern Language Association of America; London: Oxford University Press, 1941.

Hennlock, Magnus, Mia Romare, Yuqing Zhang, Steve Harris, Bengt Steen, and Tomas Rydberg. *Styrmedel för livscykler: En integrerad modellansats mellan cirkulärekonomiska modeller och livscykelanalys – Slutrapport*. Rapport 6961. Stockholm: Naturvårdsverket, 2021.

Herbert, Zbigniew. "A Journey." Translated by John Carpenter and Bogdana Carpenter. *Salmagundi*, no. 185/186 (Winter – Spring) 2015: 373–375.

Hershock, Peter D. *Buddhism in the Public Sphere: Reorienting Global Interdependence*. London: Routledge, 2017.
Hesiod. *Theogony. Works and Days. Testimonia*. Edited and translated by Glenn W. Most. Loeb Classical Library. Cambridge, MA: Harvard University Press, 2006.
Holy Bible, King James Version. New York: American Bible Society, 1999.
Howell, Andrew J., Holli-Anne Passmore, and Kimberly Buro. "Meaning in Nature: Meaning in Life as a Mediator of the Relationship Between Nature Connectedness and Wellbeing." *Journal of Happiness Studies* 14 (2013): 1681–1696. DOI: 10.1007/s10902-012-9403-x
Huber, Machteld et al. "Health: How Should We Define It?" *British Medical Journal* 343, no. 7817 (2011): 235–237. DOI: 10.1136/bmj.d4163
Hunt, John Dixon. *Greater Perfections: The Practice of Garden Theory*. London: Thames and Hudson, 2000.
Hurt, Avery. "Gen Z: How Young People Are Changing Activism." *BBC Worklife*, August 3, 2022. https://www.bbc.com/worklife/article/20220803-gen-z-how-young-people-are-changing-activism
Izba Morska przy Sądzie Okręgowym w Szczecinie. Orzeczenie z dnia 9 maja 2003, Sygn. WMS 46/2002 w sprawie zatonięcia jachtu Oskard w dniu 16 września 2002. In Zbierajewski, Janusz. "Wypadki jachtów." https://wypadkijachtow.pl/oskard-2002-zatoniecie-zatoka-biskajska/ (accessed: May 5, 2024).
James, Susan. "The World Health Organization's Definition of Health: Social versus Spiritual Health." *Social Indicators Research* 38, no. 2 (1996): 181–192. DOI: 10.1007/BF00300458
Jarrett, David, Tadeusz Rachwał, and Tadeusz Sławek. *Geometry, Winding Paths and the Mansions of Spirit*. Katowice: Wydawnictwo Uniwersytetu Śląskiego, 1997.
Jędrzejko, Paweł. "'Belle Lettre' o (I)mperium (O)ntycznym Sauny—Andrzejowi Chojeckiemu." *Er(r)go. Teoria—Literatura—Kultura*, no. 4 (2002): 113–123.
Jędrzejko, Paweł. *Melville w kontekstach: Prolegomena do studiów melvillistycznych*. Katowice–Sosnowiec–Zabrze: BananaArt/MStudio/Ex-Machina, 2007.
Jędrzejko, Paweł. *Płynność i egzystencja: Doświadczenie lądu i morza a myśl Hermana Melville'a*. Katowice–Sosnowiec–Zabrze: BananaArt/MStudio/Ex-Machina, 2008a.
Jędrzejko, Paweł. "Poza murami Sandomierza—czyli refleksja nad językiem, kulturą i wzrokiem turysty. (Miniwykład z ćwiczeniami)." In *Archeoturystyka: nowy produkt turystyczny*, edited by Tomisław Giergiel, 37–45. Sandomierz: LOT Partnerstwo Ziemi Sandomierskiej, 2008b.
Jönsson, Ingemar, Linus Hasselström, Fredrik Bengtsson, Helena Björn, Scott Cole, Hannah Doherty, Frida Franzén, Elsie Kjeller, Sofia Lidfalk, Erik Lindblom, Anna Mellin, Ida Pettersson, and Tore Söderqvist. *Ekologisk kompensation som styrmedel i kommunal planering*. Rapport 6992. Stockholm: Naturvårdsverket, 2021.
Kalaga, Wojciech, and Tadeusz Rachwał, eds. *The Wild and the Tame: Essays in Cultural Practice*. Katowice: Wydawnictwo Uniwersytetu Śląskiego, 1997.
Karlsson, Mikael, Henrik Josefsson, Tim Schnoor, and Charlotta Zetterberg. *Utbytbar natur? Etiska och rättsliga aspekter på ekologisk kompensation: Slutrapport från ett forskningsprojekt*. Rapport 6997. Stockholm: Naturvårdsverket, 2021.
Kępiński, Antoni. *Psychopatie*. Kraków: Wydawnictwo Literackie, 2007.
Khema, Ayya. *When the Iron Eagle Flies: Buddhism for the West*. Somerville, MA: Wisdom Publications, 2000.
Kidd, Ian James, and Havi Carel. "Pathocentric Epistemic Injustice and Conceptions of Health." In *Overcoming Epistemic Injustice: Social and Psychological Perspectives*, edited by Benjamin R. Sherman and Stacey Goguen, 153–168. London: Rowman and Littlefield, 2019.
Kofta, Jonasz. "Pamiętajcie o ogrodach." In *Pamiętajcie o ogrodach*. Warszawa: Czytelnik, 1991.

Kołakowski, Leszek. "Wielkie i małe kłopoty humanistów." In *Kultura i fetysze*. Warszawa: Państwowe Wydawnictwo Naukowe, 1989.
Krishnamurti, Jiddu, and David Bohm. *The Ending of Time: Where Philosophy and Physics Meet*. San Francisco: Harper One, 2014.
Kronenberg, Leah. *Allegories of Farming from Greece and Rome: Philosophical Satire in Xenophon, Varro, and Virgil*. Cambridge: Cambridge University Press, 2015.
Kubok, Dariusz. "Comments on the Sources of Greek Philosophical Criticism." In *Forms of Criticism in Philosophy and Science*, edited by Dariusz Kubok, 9–31. Katowice: Wydawnictwo Uniwersytetu Śląskiego, 2015.
Kubok, Dariusz. *Krytycyzm, sceptycyzm i zetetycyzm we wczesnej filozofii greckiej*. Katowice: Wydawnictwo Uniwersytetu Śląskiego, 2021.
Kubok, Dariusz. "Philosophy of / as a Journey." *Forum Philosophicum* 29, no. 1 (2024): 9–27. DOI: 10.35765/forphil.2024.2901.01
Larson, James S. "The World Health Organization's Definition of Health: Social versus Spiritual Health." *Social Indicators Research* 38, no. 2 (June 1996): 181–192.
Larsen, Rasmus Kløcker, Anna Skarin, Marita Stinnerbom, Jonas Vannar, Moudud Alam, Mikael Kuhmunen, Rebecca Lawrence, Jakob Nygård, Kaisa Raitio, Per Sandström, Stefan Sandström, Jonas Stinnerbom, Jenny Wik-Karlsson, and Carl Österlin. *Omtvistade landskap: Navigering mellan konkurrerande markanvändning och kumulativa effekter*. Rapport 6908. Stockholm: Naturvårdsverket, January 2020.
Leirhaug, Petter Erik. "The Role of Friluftsliv in Henrik Ibsen's Works." Paper presented at the Henrik Ibsen: The Birth of "Friluftsliv," A 150 Year International Dialogue Conference Jubilee Celebration, North Troendelag University College, Levanger, Norway, September 14–19, 2009. https://norwegianjournaloffriluftsliv.com/doc/172010.pdf (accessed: June 5, 2024).
Live to 100: Secrets of the Blue Zones. Directed by Angus Wall. Featuring Dan Buettner. Netflix, 2023. Based on the books by Dan Buettner.
Louv, Richard. *Last Child in the Woods: Saving Our Children from Nature-Deficit Disorder*. Chapel Hill, NC: Algonquin Books of Chapel Hill, 2005.
Marquard, Odo. *Abschied vom Prinzipiellen: Philosophische Studien*. Stuttgart: Reclam, 2000.
Massachusetts Institute of Technology. "Sensors Woven into a Shirt Can Monitor Vital Signs: Comfortable, Form-Fitting Garments Could Be Used to Remotely Track Patients' Health." April 23, 2020. Story by Anne Trafton. https://news.mit.edu/2020/sensors-shirt-monitor-vital-signs-0423 (accessed: May 22, 2024).
Meister Eckhart. "Sermon Eighty-Seven: Beati pauperes spiritu, quia ipsorum est regnum coelorum." In *The Complete Mystical Works of Meister Eckhart*, translated and edited by Maurice O'C. Walshe. Revised with a Foreword by Bernard McGinn. New York: The Crossroad Publishing Company, 1979.
Melville, Herman. *Correspondence*. Edited by Lynn North. Evanston and Chicago: Northwestern University Press and the Newberry Library, 1991.
Melville, Herman. *Moby-Dick, or The Whale*. Edited by Harrison Hayford, G. Thomas Tanselle, and Hershel Parker. Evanston and Chicago: Northwestern University Press and the Newberry Library, 1988.
Melville, Herman. *Redburn. White-Jacket. Moby-Dick*. New York: Library of America, 1983.
Mikosz, Ryszard. *Wprowadzenie do prawa środowiska*. Katowice: Wydawnictwo Uniwersytetu Śląskiego, 2021.
Miller, Mara. *The Garden as an Art*. Albany, NY: State University of New York Press, 1993.
Miłosz, Czesław. "Gift." In *New & Collected Poems 1931–2001*. Allen Lane, The Penguin Press, 2001. https://www.scottishpoetrylibrary.org.uk/poem/gift-0/ (accessed: May 5, 2024).

Miłosz, Czesław, and Ireneusz Kania. "Ciągle poszukujemy klucza. O Swedenborgu rozmawiają Czesław Miłosz i Ireneusz Kania." In *Emanuel Swedenborg: uczony i mistyk*, by Signe Toksvig. Translated into Polish by Ireneusz Kania. Kraków: Universitas, 2002. Pages 1–9, 14–15, 20–21.
de Montaigne, Michel. *The Essays of Montaigne, Complete*. Translated by Charles Cotton. Edited by William Carew Hazlitt. Project Gutenberg, May 28, 2001. https://www.gutenberg.org/ebooks/3600 (accessed: June 5, 2024).
Naess, Arne. "The Deep Ecology Movement: Some Philosophical Aspects." In *The Selected Works of Arne Naess*, Vol. 10, edited by Harold Glasser and Alan Drengson, in cooperation with Arne Naess, Bill Devall, and George Sessions, 33–55. Dordrecht: Springer, 2005a.
Naess, Arne. *The Selected Works of Arne Naess*. Vol. 1–10. Edited by Harold Glasser and Alan Drengson in cooperation with Arne Naess, Bill Devall, and George Sessions. Dordrecht: Springer, 2005b.
Naturvårdsverket. *Uppföljning av målen för friluftslivspolitiken 2023*. Rapport 7123. Stockholm: Naturvårdsverket, December 2023.
Nehamas, Alexander. *The Art of Living: Socratic Reflections from Plato to Foucault*. 1st ed. Berkeley: University of California Press, 2000.
New Presence Foundation. http://www.sciezki.org (accessed: May 27, 2024).
Nordenfelt, Lennart. *Action, Ability and Health: Essays in the Philosophy of Action and Welfare*. Dordrecht: Springer, 2014.
Nordström, Jonas, Johanna Alkan Olsson, Helena Hanson, Yann Clough, Mark Brady, Celina Alentun, Eva Constance Hedenfelt, Love Frykman, Joel Gunnarsson, Cecilia Hammarlund, Edwin Klint Bywater, Lisa Lorentzi Wall, Linda Lundmark, and Fredrik Wilhelmsson. *Ekologisk kompensation: Upptag och integrering bland svenska aktörer och kvantifiering av de samhällsekonomiska effekterna*. Rapport 7008. Stockholm: Naturvårdsverket, September 2021.
Norsk Friluftsliv. "Friluftsliv i Skolen." https://friluftsliviskolen.no/om-prosjektet/ (accessed: August 5, 2023).
Nyanaponika Thera. *Abhidhamma Studies: Buddhist Exploration of Consciousness and Time*. Boston, MA: Wisdom Publications, 1997a.
Nyanasobhano Bhikkhu. *Longing for Certainty: Reflections on the Buddhist Life*. Somerville, MA: Wisdom Publications, 2002.
Nyanatiloka Thera. *Buddhist Dictionary: Manual of Buddhist Terms and Doctrines*. Kandy: Buddhist Publication Society, 1997b.
Nyström Sandman, Antonia, Anna Christiernsson, Frida Gidhagen Fyhr, Mats Lindegarth, Patrik Kraufvelin, Per Bergström, Per Nilsson, Ronny Fredriksson, Ulf Bergström, and Hedvig Hogfors. *Grön infrastruktur i havet: Landskapsperspektiv i förvaltningen av Sveriges marina områden*. Rapport 6930. Stockholm: Naturvårdsverket, June 2020
Ólafsson, Stefán. "Well-Being in the Nordic Countries: An International Comparison." *Icelandic Review of Politics & Administration* 9, no. 2 (December 2013): 345–372. DOI: 10.13177/irpa.a.2013.9.2.5
Olfert, Michelle D., et al. "The Role of Social Media in Youth Activism During COVID-19." *Journal of Adolescent Health* 70, no. 2 (2022): 213–220. DOI: 10.1016/j.jadohealth.2021.11.021
Oisin, Anthoni. "Friluftsliv: The Norwegian Prescription for Health and Happiness." *News-Medical*. https://www.news-medical.net/health/Friluftsliv-The-Norwegian-Prescription-for-Health-and-Happiness.aspx (accessed: June 7, 2024).
Pamojjo, Pramote. *To See the Truth: The Teachings of Venerable Pramote Pamojjo*. Chonburi: Suan Santidham, 2013.
Pärt, Tomas. *Våtmarker i odlingslandskapet: Restaurering och anläggning för ökad mångfald av våtmarksfåglar*. Rapport 6933. Stockholm: Naturvårdsverket, November 2020.

Perrone, Jane. "How Coronavirus Changed Gardening Forever." *Financial Times*, September 11, 2020. https://www.ft.com/content/c3abc0bb-7ade-4ae1-8d2b-d3fa6be53416 (accessed: July 5, 2023).
Pineda Duque, Juan Carlos, and Rodrigo Sandoval-Almazán. "Digital Activism in Times of COVID-19: Lessons from Latin America." *Social Sciences* 3, no. 1 (2023): 9. https://www.mdpi.com/2673-995X/3/1/9 (accessed: July 5, 2023).
Plato. *Phaedrus*. Translated with an Introduction and Notes by Robin Waterfield. Oxford: Oxford University Press, 2003.
Prince, Cali (née Vandyk-Dunlevy). "The Unexpected Effects of Autoethnographic Embodied Texts and Performances." In *Proceedings of the 2022 International Symposium on Autoethnography and Narrative*, edited by Silvia M. Benard Calva, Lisa P. Z. Spinazola, Rishi Raj, Christina L. Ivey, David F. Purnell, Cody M. Clemens, Marlen Harrison et al., 107–116. International Association of Autoethnography and Narrative Inquiry, 2022.
Pritchard, Annabel, Miles Richardson, David Sheffield, and Kelly McEwan. "The Relationship Between Nature Connectedness and Eudaimonic Well-Being: A Meta-analysis." *Journal of Happiness Studies* 21 (2020): 1145–1167. DOI: 10.1007/s10902-019-00118-6
Pulina, Giovanni. *Guida filosofica dell'Italia*. Bologna: Diogene Multimedia, 2018.
The Qur'an. Translated by M. A. S. Abdel Haleem. Oxford: Oxford University Press, 2004.
Rachwał, Tadeusz, and Tadeusz Sławek, eds. *Word-Subject-Nature: Studies in Seventeenth- and Eighteenth-Century Culture*. Katowice: Wydawnictwo Uniwersytetu Śląskiego, 1996.
Raff, Marilyn. *The Intuitive Gardener: Finding Creative Freedom in the Garden*. Golden, CO: Fulcrum Publishing, 2002.
Ray, Meghan T. "Cultivating the Soul: The Ethics of Gardening in Ancient Greece and Rome." In *Gardening Philosophy for Everyone: Cultivating Wisdom*, edited by Dan O'Brien, 30–31. Chichester, West Sussex: John Wiley & Sons Ltd., 2010.
Röös, Elin, Sarah Säll, and Emma Moberg. *Effekter av en klimatskatt på livsmedel*. Rapport 6965. Stockholm: Naturvårdsverket, 2021.
Rubin, Rick. *The Creative Act: A Way of Being*. New York: Penguin, 2023.
Rufener, Abbie Elizabeth. *Blurring the Lines: The Intermingling of Garden and Theater in Seventeenth Century France*. Theses and Dissertations, no. 1615. Brigham Young University—Provo, 2008. https://scholarsarchive.byu.edu/etd/1615 (accessed: june 4, 2024).
Rura, Przemysław. "Doświadczenia zmysłowe w życiu codziennym: Autoetnografia wielozmysłowa." *Kultura i Społeczeństwo* 61, no. 3 (2017): 149–160. DOI: 10.35757/KiS.2017.61.3.9
Schiller, Johann Christoph Friedrich. *Gedichte. Aus: Das Lied von der Glocke* (1799). https://www.friedrich-schiller-archiv.de/inhaltsangaben/das-lied-von-der-glocke-zusammenfassung-friedrich-schiller (accessed: July 31, 2023).
Schmidt, Dennis J. "The Garden of Letters. Reading Plato's *Phaedrus* on Reading." In *International Yearbook for Hermeneutics* vol. 12, edited by Günter Figal, 61–76. Tübingen, Germany: Mohr Siebeck, 2013.
Schmitz, Hermann. "Atmospheric Spaces." *Ambiances: Environnement sensible, architecture et espace urbain*, in *Redécouvertes*, 2016. Translation by Margret Vince. DOI: 10.4000/ambiances.711 (accessed: April 19, 2019).
Schmitz, Hermann. *Der Leib*. Berlin: De Gruyter, 2011.
Schmitz, Hermann. *New Phenomenology: A Brief Introduction*. Sesto San Giovanni: Mimesis International, 2019.
Schmitz, Hermann, Rudolf Owen Müllan, and Jan Slaby. "Emotions Outside the Box—The New Phenomenology of Feeling and Corporeality." *Phenomenology and the Cognitive Sciences* 10 (June 2011): 241–259. DOI: 10.1007/s11097-011-9195-1

Schreiber, Paul. "Queequeg's Coffin and the Vortex: The Spatiality of Ishmael's Narrative in *Moby-Dick*." In *Proceedings from the Sixth Nordic Conference for English Studies, Tromsø, May 25–28, 1995*, edited by Gerd Bjørboude and Gøril Rague. Tromsø: University of Tromsø, 1995.

Sellars, John. "Self or Cosmos: Foucault versus Hadot." In *The Late Foucault: Ethical and Political Questions*, edited by Marta Faustino and Gianfranco Ferraro, 1–18. London: Bloomsbury, 2020.

Seneca, Lucius Annaeus. *Ad Lucilium Epistulae Morales*. Volume I. Translated by Richard M. Gummere. London: William Heinemann; New York: G. P. Putnam's Sons, 1920.

Seneca, Lucius Annaeus. "Moral Letters to Lucilius, Letter 48: On Quibbling as Unworthy of the Philosopher," paragraph 12. n.d. https://en.wikisource.org/wiki/Moral_letters_to_Lucilius/Letter_48 (accessed: July 5, 2023).

Shilton, Trevor, Michael Sparks, David McQueen, Marie-Claude Lamarre, and Suzanne Jackson. "How Should We Define Health? Proposal for New Definition of Health." *British Medical Journal* 343/7821 (2011). DOI: 10.1136/bmj.d5359

Silko, Leslie Marmon. *Ceremony*. Introduction by Larry McMurtry. London and New York: Penguin Books, 2006.

Soga, Masashi, Kevin J. Gaston, and Yuichi Yamaura. "Gardening is Beneficial for Health: A Meta-analysis." *Preventive Medicine Reports* 5 (2017): 92–99. DOI: 10.1016/j.pmedr.2016.11.007

Soga, Masashi, Kevin J. Gaston, T. F. Koyanagi, K. Kurisu, and K. Hanaki. "Urban Residents' Perceptions of Neighbourhood Nature: Does the Extinction of Experience Matter?" *Biological Conservation* 203 (2016): 143–150. DOI: 10.1016/j.biocon.2016.09.020

Sörme, Louise, Åsa Nyblom, Joacim Rosenlund, Hanna Matschke Ekholm, Lars Viklund, and Åsa Stenmarck. *Matavfall i butik – Analys av samhällsförändringar*. Rapport 6901. Stockholm: Naturvårdsverket, November 2019.

Stasiuk, Andrzej. *Fado*. Wołowiec: Wydawnictwo Czarne, 2006.

Steenbergen, Clemens, and Wouter Reh. *Architecture and Landscape: The Design Experiment of the Great European Gardens and Landscapes*. Basel; Boston: Birkhäuser Publishers, 2003.

Stevens, James A. "Friendship and Profit in Xenophon's *Oeconomicus*." In *The Socratic Movement*, edited by Paul A. Vander Waerdt, 209–237. Ithaca, NY: Cornell University Press, 1994.

Stobaeus, Johannes. *Anthology*. III, 17, 22 (Usener, fr. 469). Quoted in https://www.attalus.org/translate/epicurus3.html#us4f (accessed: July 5, 2023).

Stobaeus, Johannes. *Anthology*. III, 17, 33 (Usener, fr. 181). Quoted in https://www.attalus.org/translate/epicurus.html (accessed: July 5, 2023).

Swapnil Mishra, James A. Scott, Daniel J. Laydon, Seth Flaxman, Axel Gandy, Thomas A. Mellan, H. Juliette T. Unwin, Michaela Vollmer, Helen Coupland, Oliver Ratmann, Melodie Monod, Harrison H. Zhu, Anne Cori, Katy A. M. Gaythorpe, Lilith K. Whittles, Charles Whittaker, Christl A. Donnelly, Neil M. Ferguson, and Samir Bhatt. "Comparing the Responses of the UK, Sweden and Denmark to COVID-19 Using Counterfactual Modelling." *Scientific Reports* 11 (2021): 16342. DOI: 10.1038/s41598-021-95699-9

Tesson, Sylvain. *Un été avec Homère*. Paris: Équateurs, 2018.

Theroux, Paul. *The Happy Isles of Oceania: Paddling the Pacific*. New York: Putnam, 1992.

Thomas Aquinas. *Commentary on Aristotle's De Anima*. Translated by O. P. Kenelm Foster and O. P. Sylvester Humphries. New Haven: Yale University Press, 1951b. Html edition by Joseph Kenny, O.P. https://isidore.co/aquinas/english/DeAnima.htm (accessed: May 6, 2024).

Thoreau, Henry David. *Life Without Principle*. Ebook. London: Global Grey, 2018.

Thoreau, Henry David. *Journal*, November 4, 1852. In *Thoreau's Seasons*. https://stpeter.im/writings/thoreau/seasons-9.html (accessed: November 4, 2023).

Thoreau, Henry David. *Walden: A Fully Annotated Edition*. Edited by Jeffrey S. Cramer. New Haven: Yale University Press, 2004a.

Thoreau, Henry David. *The Writings of Henry D. Thoreau. Walden: 150th Anniversary Edition*. Edited by Lyndon Shanley, with an introduction by John Updike. Princeton: Princeton University Press, 2004b.

Tiyavanich, Kamala. *In the Cool Shade of Compassion: The Enchanted World of the Buddha in the Jungle*. Boulder, CO: Shambhala, 2018.

Toksvig, Signe. *Emanuel Swedenborg, Scientist and Mystic*. New York: Swedenborg Foundation, 1948.

Tolkien, J. R. R. *The Fellowship of the Ring, Being the First Part of the Lord of the Rings*. Harper Collins e-books. https://www.harpercollins.com/products/the-fellowship-of-the-ring-j-r-r-tolkien (accessed: August 28, 2023).

Trigwell, John, Alexander J. Francis, and Kathryn L. Bagot. "Nature Connectedness and Eudaimonic Well-Being: Spirituality as a Potential Mediator." *Ecopsychology* 6 (2014): 241. DOI: 10.1089/eco.2014.0025

Tseng, Wen-shing, Q. Y. Lu, and P. Y. Yin. "Psychotherapy for the Chinese: Cultural Considerations." In *Chinese Society and Mental Health*, edited by T. Lin, W. Tseng, and E. Yeh, 292. Hong Kong: Oxford University Press, 1995.

Ulrich, Roger. "View Through a Window May Influence Recovery from Surgery." *Science* 224 (New York, N.Y.) (1984): 420–421. DOI: 10.1126/science.6143402 (accessed: July 12, 2024).

UN Women. "Persisting in the Pandemic: Youth Activism During COVID-19." December 2021. https://www.unwomen.org/en/news-stories/feature-story/2021/12/persisting-in-the-pandemic-youth-activism-during-covid-19 (accessed: July 12, 2024).

United Nations. "Young People Key to COVID Recovery and a Sustainable Future." United Nations Department of Economic and Social Affairs. https://www.un.org/en/desa/young-people-key-covid-recovery-and-sustainable-future (accessed: July 12, 2024).

Venturi Ferriolo Massimo. *Oltre in giardino: Filosofia di Paesaggio*. Torino: Giulio Einaudi Editore, 2019.

Vincent, Howard P. *The Trying Out of Moby-Dick*. Kent, OH: Kent State University Press, 1980.

Walker, Merle G. "The One and Many in Plato's *Parmenides*." *The Philosophical Review* 47, no. 5 (1938): 488–516. DOI: 10.2307/2181200

Wang, Daoyan, and Thomas MacMillan. "The Benefits of Gardening for Older Adults: A Systematic Review of the Literature." *Activities, Adaptation & Aging* 37 (2013): 153–181. DOI: 10.1080/01924788.2013.784942

Ware, John E., Robert H. Brook, Alison Davies-Avery, Kathleen N. Williams, Anita L. Stewart, William H. Rogers, and Stewart A. Johnson. "Choosing Measures of Health Status for Individuals in General Populations." *American Journal of Public Health* 71 (1981): 620–625. DOI: 10.2105/AJPH.71.6.620

von Weizsäcker, Viktor. "Die Schmerzen." In *Der Arzt und der Kranke: Stücke einer medizinischen Anthropologie*, Gesammelte Schriften, vol. 5. Frankfurt am Main: Suhrkamp, 1987.

Whitman, Walt. *Song of Myself*. https://www.poetryfoundation.org/poems/45477/song-of-myself-1892-version (accessed: May 4, 2024).

Widenfalk, Lina, Jonas Josefsson, Malgorzata Blicharska, Marcus Hedblom, Tomas Pärt, Thomas Ranius, Olof Widenfalk, and Erik Öckinger. *Ekologisk kunskap för ekologisk kompensation: Syntes av forskningsprojekten Systematisera ekologisk kunskap för att effektivisera ekologisk kompensation och När kan ekologisk kompensation bidra till att bevara biologisk mångfald och ekosystemtjänster?* Rapport 6995. Stockholm: Naturvårdsverket, September 2021.

Wikström, Sofia A., Andreas Bryhn, Matilda Valman, Gustaf Almqvist, Thorsten Blenckner, Örjan Bodin, Annika Nilsson, and Henrik Österblom. *Ekosystemansatsen – praktiska erfarenheter från svensk havs- och vattenförvaltning*. Rapport 6934. Stockholm: Naturvårdsverket, August 2020.

Wittgenstein, Ludwig. *Zettel*. Translated by G. E. M. Anscombe. Edited by G. E. M. Anscombe and G. H. von Wright. Oxford: Basil Blackwell, 1967.

Wood, Emma. "Not All Green Space Is Created Equal: Biodiversity Predicts Psychological Restorative Benefits from Urban Green Space." In *One Health: The Well-being Impacts of Human–Nature Relationships*, edited by Elizabeth Brymer et al. Lausanne: Frontiers Media, 2017. DOI: 10.3389/978-2-88945-384-3

Wood, Philip H. N. "Health and Disease and Its Importance for Models Relevant to Health Research." In *Scientific Approach to Health and Healthcare*, edited by B. Z. Nizetic et al. Copenhagen: World Health Organization, 1986.

World Health Organization. "The Preamble to the WHO Constitution." https://www.who.int/about/governance/constitution (accessed: June 29, 2023).

World Health Organization. "Health and Well-Being." Global Health Observatory. https://www.who.int/data/gho/data/major-themes/health-and-well-being (accessed: June 29, 2023).

Yarmol-Matusiak, Erica A., Lauren E. Cipriano, and Saverio Stranges. "A Comparison of COVID-19 Epidemiological Indicators in Sweden, Norway, Denmark, and Finland." *Scandinavian Journal of Public Health* 49, no. 1 (February 2021): 69–78. DOI: 10.1177/1403494820980264

Zhang, Yanhua. "Negotiating a Path to Efficacy at a Clinic of Traditional Chinese Medicine." *Culture, Medicine and Psychiatry* 31, no. 1 (2007): 73–100. DOI: 10.1007/s11013-006-9039-6

Zottola, Angela, and Claudio de Majo. "The Anthropocene: Genesis of a Term and Popularization in the Press." *Text & Talk* 42, no. 4 (2022): 453–473. DOI: 10.1515/text-2020-0080

Index

Pages followed by "n" refer to notes.

For Product Safety Concerns and Information please contact our EU representative GPSR@taylorandfrancis.com Taylor & Francis Verlag GmbH, Kaufingerstraße 24, 80331 München, Germany